Second Edition

CALCULUS

A PRACTICAL APPROACH

Kenneth Kalmanson

Patricia C. Kenschaft

Montclair State College

Worth Publishers, Inc.

CALCULUS: A PRACTICAL APPROACH, SECOND EDITION

DESIGN: MALCOLM GREAR DESIGNERS, INC.

PRINTED IN THE UNITED STATES OF AMERICA
LIBRARY OF CONGRESS CATALOG CARD NO. 77-81756
ISBN 0-87901-083-5
FIRST PRINTING, JANUARY, 1978

WORTH PUBLISHERS, INC.
444 PARK AVENUE SOUTH
NEW YORK, NEW YORK 10016

Preface to the Second Edition

It has been gratifying to learn about the successful use of our first edition at so many colleges and universities. Before we began work on the second edition of *Calculus: A Practical Approach*, we asked the advice of over 300 professors who had used the first edition in their courses. We have incorporated many of their suggestions for improvement in the second edition and have made additional changes based on our personal experience with the book.

Many users have confirmed what we ourselves have found to be true, that students do indeed relearn the necessary algebra while studying the first chapter. A number of people, however, requested a more detailed algebra review, so we have greatly enlarged the review chapter and have included many more exercises. The review still appears at the back of the book in order not to deter qualified students (those who know the equivalent of two years of algebra) from devoting a complete semester to calculus. However, we have added a diagnostic test preceding Chapter 1 and have keyed the answers to appropriate sections of the algebra review in order to help students identify and strengthen those areas where remedial work is needed.

The most extensive addition to the second edition is the inclusion of Chapter 7 on trigonometry, including calculus. The chapter does not presuppose any prior knowledge of the subject, and, again, only two years of algebra is assumed. Students of both the biological and social sciences can profit from the many models of periodic phenomena contained in this chapter, as well as a basic knowledge of triangle trigonometry. Our treatment is practical and intuitive, in keeping with the rest of the book.

We have added a section on Newton's method for finding the zeros of a function. This section can be used to complement the derivative tests for extreme values and curve sketching. It can also be used to introduce the student to iterative methods. Similarly, we have added an appendix on L'Hôpital's rule for finding limits. This appendix expands the students' practical grasp of limits, just as Appendix I augments the students' theoretical understanding of the subject.

The second edition devotes three sections to exponential and logarithmic functions to allow the students to become better acquainted with these important tools. We have also introduced the idea of semi-logarithmic curve sketching.

Answers to all the A and C exercises are still included in the text, but the answers to the B exercises are now found only in the Instructor's Manual. Since the B exercises are similar to the A exercises, this gives the individual instructor the flexibility to assign homework for which all, or none, or some of the answers have been provided. All answers to the sample tests and quizzes are still found in the back of the text.

Exercises for which a hand calculator is suitable have been added and are clearly marked with an "HC" in the left-hand margin. Thus the course can be enriched with the newly available hand-calculator approach to calculus, or these problems can be easily omitted.

We have also tried to attain greater clarity by making improvements of a more local nature throughout the text, as, for example, in our more simplified treatment of "asset growth." Dozens of new applications, especially in the biological sciences, have been added, serving to illustrate the relevance of calculus to the modern world.

Special thanks are due to Robert Bixby of Northwestern University, Joyce Longmans of Drexel University, and Joseph Rosenstein of Rutgers University, all of whom read the manuscript and made detailed and very helpful suggestions throughout. We also wish to thank Claire May, who read page proofs and checked all the mathematical calculations.

Since this revision draws upon the suggestions of hundreds of people who have used the first edition, it would be impossible to thank each contributor here. But the authors want to express their appreciation to all who cared enough to contribute to this endeavor.

May calculus continue to provide insight and excitement to a new generation of students!

January 1978   Kenneth Kalmanson
Upper Montclair, N.J.   Patricia C. Kenschaft

Preface to the First Edition

Calculus is rapidly becoming part of the language of business, economics, biology, and the social sciences. More and more colleges and graduate schools are requiring a practical knowledge of calculus as a prerequisite for study in these fields, and students are finding calculus increasingly useful in their intermediate-level courses.

This text is designed to meet the needs of such students. Applications from business and economics appear throughout the book; they are used to enliven the discussion and to furnish a pattern of illustrations for the mathematical techniques. Applications from biology and the social sciences are also included in each chapter.

Throughout the book, we have tried to keep the writing practical and straightforward. For example, the phrase "if it exists" has been omitted in several places where it would have been included in a more rigorous exposition. Anything that is discussed in this text is assumed to exist. The presentation of each concept is intuitive, but more rigorous discussions (including proofs) are in the appendix.

Each section of the book has been used as one lesson at Montclair State College. Nearly every section is accompanied by three exercise sets. Either exercise set A or set B constitutes a night's assignment; the exercises in both sets are patterned on worked-out examples in the section. Exercises are graded in difficulty (following the development of the text) and involve a minimum of algebraic manipulation and numerical computation. Set C contains supplementary exercises in biology and the social sciences in addition to a few exercises requiring more originality. Answers to exercises (and solutions to some) are included at the end of each section.

Self-testing for this course can be done either once or twice per chapter, using the sample tests included after each chapter and the quizzes after each half-chapter. These self-tests will enable students to pinpoint quickly any lack of understanding. Answers to the quizzes and tests are at the back of the book.

We assume that our readers have had two years of high school algebra or the equivalent. This does not mean that they must remember all of elementary algebra, but rather that they are sufficiently familiar with it so they can relearn it quickly with the help of a teacher. Although the first chapter does review the most important topics from algebra, and there is a more extensive review, no book can answer all questions as they arise; only a good teacher can do that.

Covering one section a day with occasional "breathers" for reviews and tests, we have been able to cover all but seven of the forty-two sections in a three-credit, one-semester course. There are many ways to eliminate a few sections to form a well-rounded three-credit course, but we would suggest that the following are indispensable in any basic calculus course: Chapter 1, and Sections 2.1, 2.2, 2.3, 2.4, 2.6, 2.7, 3.1, 3.3, 3.4, 3.5, 3.8 (briefly), 4.1, 4.2, 4.4, 4.5, 4.6, and 4.7. By taking a more leisurely pace and more thoroughly digesting each topic, it is possible to stretch out the material for a year.

We wish to express our special appreciation to Robert Bixby of the University of Kentucky (who read the entire manuscript twice), whose suggestions have been most valuable. We are also grateful to James Van Valen, who read the entire manuscript and checked the computations, and to the people at Worth Publishers for all of their help.

We thank the many professors and students, too numerous to name, who have helped us with their comments while using preliminary versions of the text here at Montclair State. We also want to thank the following, each of whom has contributed something special to the book: William J. Adams of Pace University; Charles Barnhill of Montgomery Community College; Chuan-Yu Chen of Montclair State College; Frederick Chichester of Bloomfield College; William S. Davis III of Montclair State College; Harold S. Engelsohn of Kingsborough Community College; Garret J. Etgen of the University of Houston; Larry Goldstein of the University of Maryland; David Henderson of Cornell University; Herbert Hethcote of the University of Iowa; Eugene F. Krause of the University of Michigan; Joseph Krebs of Boston College; Stanley Lukawecki of Clemson University; Bruce Meserve of the University of Vermont; Dale E. Walston of the University of Texas; Frank Wattenberg of the University of Massachusetts; and Donald Wright of the University of Cincinnati. And finally, we want to thank our volunteer proofreaders: Carl Christiansen, Mary Hepler, and Linda Pachucki.

January 1975   Kenneth Kalmanson
Upper Montclair, N.J.   Patricia C. Kenschaft

Note to the Student

Calculus is the study of change, both instantaneous change (differential calculus) and total change (integral calculus). Change is such a pervasive part of modern life that calculus has become useful, even essential, in the study of many diverse fields. This book includes not only a presentation of calculus itself, but also many examples of how it can be applied to business and economics, and suggestions of how it is used in biology, medicine, sociology, psychology, paleontology, and anthropology.

During the first two centuries of its existence, calculus was closely allied with physics. The underlying concepts were imperfectly understood in those days, but calculus was so practical that physicists used it anyway.

Today, many people besides physicists recognize how handy calculus can be. And they, like the earlier physicists, can understand it and use its power without exploring all of its underlying concepts or technical details. This text is written for such people.

Calculus has never had the reputation of being easy. To learn calculus you *must* do homework regularly; this is true even of very good students. It helps to have a friend in the class with whom you can discuss problems that puzzle you. But you do not have to be especially gifted mathematically to learn some of the practical uses of calculus; many students with only an average high school mathematics record have successfully learned calculus using this book.

You will probably need to spend about an hour reading each section and about an hour doing the accompanying problems (either set A or set B). Some sections (such as 2.1) will require more time to read; others (such as Section 2.2) will require more time on computation. But if you spend about two hours conscientiously preparing each lesson, you can look forward to learning, enjoying, and using one of the most beautiful subjects ever created by the human mind.

Contents

Applications

Exercises Using Hand Calculators

Chapter 1

Exercises
1.1.C *problem* 5
1.4.C *problem* 5
1.5.A *problems 2, 6*   B *problems 2, 6*   C *problem 7*
1.6.C *problem* 1

Chapter 2

Exercises
2.1.A *problem 2*   B *problem 2*
2.4.C *problem* 7
2.5.C *problem* 3
2.9.A *problems 3, 4*   B *problems 3, 4*

Chapter 3

Exercises
3.1.C *problem* 3
3.2.C *problem* 3
3.3.C *problems 3, 4*
3.8.A *problem 8*   B *problem 8*

Chapter 4

Exercises
4.7.A *problem 6*   B *problem 6*

Chapter 6

Exercise
6.4.C *problem* 5

Chapter 7

Exercises
7.1.C *problem* 2
7.2.A *problem 10*   B *problem 10*
7.4.C *problems 1, 2*

Diagnostic Test

This course is designed for people who have passed two years of high school algebra. The text does not assume that you will actually remember everything you learned there; many algebraic details and reminders are included in Chapter 1. But it is expected that you once knew algebra well enough so that you can relearn it quickly.

If you had trouble with algebra, or if you took it a long time before opening this book, you may want some extra review. It may be enough merely to do both the set A and set B problems in each section until you have caught up with your classmates. But if you want more practice, there is an extensive Review of Algebra included at the end of the book.

The following quiz may help you evaluate how much algebra you recall. The answers for this quiz are on page 415 at the back of the book. If seeing an answer does not refresh your memory as to how it was obtained, perhaps you should spend some time studying the appropriate part of the Review of Algebra section.

1. All the numbers that are on the "number line" are called the _____ numbers.

2. (a) $-3 + (-4) =$ (b) $-3 - 4 =$ (c) $-3 + 4 =$ (d) $3 - 4 =$

3. (a) $(-3)(-6) =$ (b) $(-3)(6) =$ (c) $3(-6) =$

4. Does $\dfrac{-1}{2} = -\dfrac{1}{2}$?

5. (a) $|3| =$ (b) $|-3| =$

6. Make a graph and plot the points $(2, 3)$ and $(2, -3)$.

7. What is the distance between $(-2, 4)$ and $(-5, -1)$?

8. Solve: (a) $3x - 4 = x + 6$ (b) $-5x + 7 = -2x + 13$

9. (a) $x^2 \cdot x^3 =$ (b) $x^7/x^4 =$ (c) $x^0 =$ (d) $x^1 =$ (e) $x^{-2} =$ (f) $2^{-1} =$
   (g) $9^{1/2} =$ (h) $8^{1/3} =$ (i) $8^{-2/3}$ (j) $27^{-4/3}$

10. (a) $3(4x^2 - 2x + 7) =$ (b) $(a + b)(c + d) =$ (c) $(a + b)(a - b) =$
    (d) $(a + b)^2 =$ (e) $3/x + x/2 =$ (f) $(2x + 3)(4x - 2) =$

11. Factor: (a) $x^2 - 9$ (b) $x^2 - 7x + 6$ (c) $2x^2 - 7x + 6$ (d) $x^3 - x$
    (e) $4x - x^3$ (f) $2x^3 + 3x^2 + x$

12. Solve: (a) $x^2 - 9 = 0$ (b) $x^2 - 7x + 6 = 0$ (c) $2x^2 - 7x + 6 = 0$
    (d) $x^3 - x = 0$ (e) $4x - x^3 = 0$ (f) $2x^3 + 3x^2 + x = 0$

13. Simplify: (a) $\dfrac{x^2 + 3x}{x} =$ (b) $\dfrac{x^2 - 9}{x + 3} =$

14. Solve: (a) $x - 2 < 0$ (b) $2x < 6$ (c) $-3x + 4 > 4x - 3$

15. Solve: (a) $x^2 - x < 0$ (b) $x^2 - x > 0$ (c) $x^2 - 4 < 0$

Chapter 1

FUNCTIONS

1.1 Functions as Models

Calculus, a word to conjure with! And yet, all that it means, according to one dictionary definition, is "a branch of mathematics involving calculations." But so are algebra and arithmetic, two subjects with which you are familiar. What, then, distinguishes calculus? The answer, in one word, is "functions." More precisely, calculus studies the change in functions. Just as equations and numbers are the tools of algebra and arithmetic, respectively, functions are the tools of calculus.

Functions, like equations and numbers, are tools that we use to analyze abstract patterns in our everyday life. But equations and numbers deal with relatively stable concepts, such as a certain quantity of apples. Real life, however, is active, dynamic. Real life involves change. In calculus there might be one quantity of apples if one condition prevails and another quantity if the conditions are altered.

In our restless world "the conditions" are constantly being altered. What, then, could be a more significant subject to study than change itself? Especially in the management and social sciences, where informed, responsible decisions require a knowledge of shifting patterns, it is important to be familiar with the mathematical tools that model change. Functions are these tools. Functions model changing financial conditions much in the way that a bookkeeper's ledger describes transactions, or a monetary table describes exchange rates.

A function is a certain kind of relationship. In this book we shall often refer to functions that describe financial relationships. But the concept has far more general applications, so here we give a somewhat general definition.

1.1.1 Definition

A <u>function</u> is a rule of correspondence between two sets (often sets of numbers), A and B, such that for each element in the set A the function assigns precisely one corresponding element in the set B.

1.1.2   Definition

The set A above is called the <u>domain</u> of the function. The set of values in B that corresponds to elements in A is called the <u>range</u>. If x assumes values in A, it is called the <u>independent variable</u>. If y represents values in B, it is called the <u>dependent variable</u>. (However, we sometimes use letters other than x and y for these variables.)

How the function is defined or expressed can vary greatly. We might think of a function as a mysterious machine that changes values of x into values of y.

Figure 1.1–1

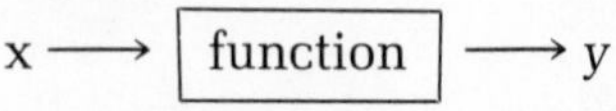

$$x \longrightarrow \boxed{\text{function}} \longrightarrow y$$

Or we might think of a function as being a set of arrows, exactly one for each member of the set A, connecting that member to some member of the set B.

Figure 1.1–2

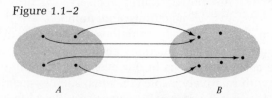

A   B

We can also think of a function as a chart or table that gives a value of y for each value of x. Such tables are common in defining monetary exchange rates.

1.1.3   Example

Here is a table that shows the relationship that existed between the American dollar and the Japanese yen in July, 1973:

Conversion Table

$	→ Yen
$ 0.10	26
0.20	52
0.25	65
0.50	130
0.75	195
1.00	260
1.50	390
10.00	2,600

For each value of the independent variable (the number of dollars you take to the bank), there is precisely one value of the dependent variable (the number of yen the bank will give to you). Therefore, the chart defines a function. The *domain* of this function is the set of numbers in the left column. The numbers in the right column constitute the *range*. (See also Example 1.1.12 in biology and sociology.)

Another way to express a function is as an equation that shows us how to calculate y for each value of x. Extending the domain of the function above, we can write the function

$$y = 260x$$

Or it can be written

$$f(x) = 260x \qquad (Read \text{ "f of x equals 260x."})$$

The latter method may be new to you. It looks similar to the previous equation, and you may wonder why we use the awkward notation "$f(x)$" instead of merely "y." The reason is that the f signifies the function itself, and later we shall need to refer explicitly to functions. Therefore, it is good to learn to handle the notation now.

To find the value of a function $f(x)$ for a particular x, *substitute* that particular number for x wherever x appears in the equation.

1.1.4   Example

If $f(x) = x^2$, find
(a) $f(0)$
(b) $f(3)$
(c) $f(-2)$

Solution:
(a) $f(0) = 0^2 = 0$
(b) $f(3) = 3^2 = 9$
(c) $f(-2) = (-2)^2 = 4$

If you want to find the value of a function $f(x)$ for some number that is expressed in terms of a letter, or in terms of a combination of numbers and letters, substitute that letter, or that combination, for x wherever x appears in the equation.

1.1.5   Example

If $f(x) = x^2$, find
(a) $f(a)$
(b) $f(a + 1)$
(c) $f(x + 1)$
(d) $f(2a + 1)$
(e) $f(2x + 1)$

Solution:

(a) $f(a) = a^2$

(b) $f(a + 1) = (a + 1)^2 = a^2 + 2a + 1$

(c) $f(x + 1) = (x + 1)^2 = x^2 + 2x + 1$

(d) $f(2a + 1) = (2a + 1)^2 = 4a^2 + 4a + 1$

(e) $f(2x + 1) = (2x + 1)^2 = 4x^2 + 4x + 1$

1.1.6   Example

If $f(x) = 3x - 2$, find

(a) $f(0)$

(b) $f(3)$

(c) $f(-2)$

(d) $f(a + 1)$

(e) $f(x + 1)$

Solution:

(a) $f(0) = 3 \cdot 0 - 2 = -2$

(b) $f(3) = 3 \cdot 3 - 2 = 9 - 2 = 7$

(c) $f(-2) = 3(-2) - 2 = -6 - 2 = -8$

(d) $f(a + 1) = 3(a + 1) - 2 = 3a + 3 - 2 = 3a + 1$

(e) $f(x + 1) = 3(x + 1) - 2 = 3x + 3 - 2 = 3x + 1$

1.1.7   Example

If $f(x) = \dfrac{x - 1}{2x + 1}$, find

(a) $f(0)$

(b) $f(3)$

(c) $f(-2)$

(d) $f(a + 1)$

(e) $f(x + 1)$

Solution:

(a) $f(0) = \dfrac{0 - 1}{2 \cdot 0 + 1} = -1$

(b) $f(3) = \dfrac{3 - 1}{2 \cdot 3 + 1} = \dfrac{2}{7}$

(c) $f(-2) = \dfrac{-2 - 1}{2(-2) + 1} = \dfrac{-3}{-3} = 1$

(d) $f(a + 1) = \dfrac{(a + 1) - 1}{2(a + 1) + 1} = \dfrac{a + 1 - 1}{2a + 2 + 1} = \dfrac{a}{2a + 3}$

(e) $f(x + 1) = \dfrac{(x + 1) - 1}{2(x + 1) + 1} = \dfrac{x + 1 - 1}{2x + 2 + 1} = \dfrac{x}{2x + 3}$

There are many standard types of functions in the world of business and economics. We list a few here. These functions will be used throughout this book in our applications.

1. The cost function (see Examples 1.1.8–1.1.11)
    x is the number of items produced.
    $f(x)$ is the cost of producing x items.
2. The revenue function
    x is the number of items sold.
    $f(x)$ is the revenue from selling x items.
3. The profit function
    x is the number of items sold.
    $f(x)$ is the profit from selling x items.
4. The demand function
    x is the quantity of a commodity on the open market.
    $f(x)$ is the price per unit that consumers will pay.
5. The supply function
    x is the quantity of a commodity on the open market.
    $f(x)$ is the price per unit that sellers will set.
6. The production function
    x is the quantity of some input (such as hours of labor) that a company invests in production of an item.
    $f(x)$ is the quantity of that item produced.

When a function, expressed as an equation or a graph, is used to describe a real-life relationship (such as one of the above), we say that the function is a model of the real-life situation.

The word "model" may call to mind different ideas to different people. Some will think of modeling clay; others, a human form (female?) displaying fashions. In both cases a specific entity is used to idealize a general situation. Functions are used to idealize the quantitative aspects of a situation, resembling the way that a fashion model idealizes how a certain outfit will "look"—even though it will not look the same on any two people. Similarly, functions as models are rarely entirely accurate; they are only approximately true. But by abstracting the basic pattern from a complicated situation, they are useful in discovering new insights; this is their value.

1.1.8   Example

Suppose that a manufacturer of watches discovers that he must spend $3,000 a month on such things as insurance, rent, and committed salaries whether or not he produces any watches. Furthermore, on each watch he does make, he must spend $2 for material and added labor costs. Thus making one watch costs $3,002, making two watches costs $3,004, making three costs $3,006, and so forth. We express the above in an equation by

$$C(x) = 2x + 3{,}000$$

where x is the number of watches manufactured and $C(x)$ is the cost of manufacturing x watches. We say that \$3,000 in the above equation is the "fixed cost." The \$2 is called the "marginal cost," the cost of making one extra item.

1.1.9   **Example:** *The Linear Cost-Output Model*

The previous example was a specific case of a common business model. The general linear cost-output equation is

$$C = mx + b \qquad \text{or} \qquad C(x) = mx + b$$

where

$$
\begin{aligned}
x &= \text{number of items produced} \qquad (\text{clearly, } x \geqslant 0) \\
m &= \text{marginal cost (cost of producing one extra item)} \\
b &= \text{fixed cost over a period of time} \\
C &= \text{total cost of producing } x \text{ items}
\end{aligned}
$$

We think of m and b as being constants (possibly unknown) and of x and C as being variables. mx is usually called the variable cost, so we have

$$\text{Total cost} = \text{variable cost} + \text{fixed cost}$$

1.1.10   **Example**

(a) Write the linear cost-output equation for producing pairs of shoes if the fixed monthly costs are \$1,500 and the marginal cost per pair is \$1.50.
(b) What is the cost of manufacturing 1,000 pairs of shoes?
(c) What is the cost of manufacturing 2,500 pairs of shoes?

Solution:

(a) $C = 1.5x + 1{,}500$
(b) $C(1{,}000) = 1.5(1{,}000) + 1{,}500 = \$3{,}000$
(c) $C(2{,}500) = 1.5(2{,}500) + 1{,}500 = \$5{,}250$

1.1.11   **Example**

Suppose that the fixed monthly cost of knitting hats in a home business is \$300 and the additional cost per hat is \$1.25.
(a) Write the cost-output equation.
(b) Find the cost of producing 150 hats in a month.

Solution:

(a) $C = 1.25x + 300$

(b) $C(150) = 1.25(150) + 300 = \$487.50$

1.1.12   Example in Biology, Pre-Med, or Sociology

Suppose that only dried soybeans and dried lentils are available to supply a person's minimum daily requirement of protein. For a given consumption of soybeans (in grams), the following table shows how many grams of lentils must be consumed:

Soybeans (grams)	Lentils (grams)
0	7,000/26
10	6,650/26
20	6,300/26
40	5,600/26
100	3,500/26

The function expressed in this chart may also be written

$$y = -\frac{35}{26}x + \frac{7,000}{26}$$

If x is the quantity of soybeans eaten, then y in this equation gives the amount of lentils that must be used as a supplement. Later in this chapter we show how to obtain the equation from the table.

SUMMARY

Calculus is the study of functions and the way that they change. In this section we defined the word "function." There are at least three ways that specific functions can be described — through charts, equations, and graphs. We mentioned several functions that are used to model real-life situations and closely examined the linear cost-output function. The $f(x)$ notation was introduced and used to evaluate functions at specific values of the independent variable, x. These ideas form the core of this book and its approach to a most practical subject — calculus.

(In each section of this book, either exercise set A or set B constitutes a lesson's homework in the opinion of the authors. Exercise sets C contain supplementary problems.)

EXERCISES   1.1. A

1. If $f(x) = 3x - 5$, find $f(0)$, $f(10)$, $f(-1)$, $f(x + 1)$, and $f(2x + 3)$.
2. Evaluate $f(1)$, $f(-3)$, $f(x + 1)$, and $f(2x + 3)$ if $f(x) = x^2 + 3x$.

3. Find $f(3x - 5)$ if $f(x) = x^2 + 3x$.
4. Compute $f(x^2 + 3x)$ when $f(x) = 3x - 5$.
5. Shown here is part of a conversion table from U.S. dollars to Taiwan dollars in July, 1973. It describes a function. Write an equation that describes the same function.

U.S.\$ → N.T.\$	
\$0.10	\$ 3.79
0.15	5.69
0.20	7.58
0.25	9.48
0.30	11.37

6. Suppose that the cost of producing x baseball mitts is given by $C = \$1.50x + \$300$.
   (a) What is the fixed cost?
   (b) What is the marginal cost?
   (c) What is the cost of producing 400 mitts?
   (d) What is the cost of producing 250 mitts?
7. Suppose that the cost of producing x baseball bats is given by $C(x) = \$0.85x + \$250$. Answer the same four questions as in problem 6.
8. Suppose that the fixed costs for a costume company are \$10,000 per month and that the marginal cost per costume is \$1.
   (a) Write the linear cost-output equation for monthly costume production.
   (b) What is the cost of producing 10,000 costumes?
   (c) What is the cost of producing 20,000 costumes?
9. Suppose that the fixed cost of manufacturing a certain pill is \$50,000 and the marginal cost per pill is \$0.005.
   (a) Write the linear cost-output equation for manufacturing these pills.
   (b) What is the cost of producing 10,000 pills?
   (c) What is the cost of producing 20,000 pills?

EXERCISES 1.1. B

1. If $f(x) = 3x^2$, find $f(0)$, $f(10)$, $f(-1)$, $f(x + 1)$, and $f(2x + 3)$.
2. Evaluate $f(1)$, $f(-3)$, $f(x + 1)$, and $f(2x + 3)$ if $f(x) = 4x - 2$.
3. Find $f(3x^2)$ if $f(x) = 4x - 2$.
4. Compute $f(4x - 2)$ when $f(x) = 3x^2$.
5. The table here, which shows the conversion from U.S. dollars to German marks in November, 1973, describes a function. Write an equation that approximates that function.

U.S.\$ → Marks	
\$ 5	12.20
10	24.40
15	36.59
20	48.79
25	60.98

6. Suppose that the cost of producing x calendars is given by $C(x) = 0.09x + 300$.
   (a) What is the marginal cost?
   (b) What is the fixed cost?
   (c) What is the cost of producing 2,000 calendars?
   (d) What is the cost of producing 3,500 calendars?
7. Suppose that the cost of producing x radios is given by $C(x) = 3,000 + 50x$. Answer the same four questions as for problem 6.
8. Suppose that the fixed costs for a typewriter company are $25,000 per month and the marginal cost per typewriter is $40.
   (a) Write the linear cost-output equation for monthly typewriter production.
   (b) What is the cost of producing 100 typewriters?
   (c) What is the cost of producing 1,000 typewriters?
9. Suppose that the fixed cost for manufacturing paper is $2,500 and the additional cost for each extra sheet is $0.001.
   (a) Write the linear cost-output equation for manufacturing this paper.
   (b) What is the cost of manufacturing 100,000 sheets?
   (c) What is the cost of manufacturing 200,000 sheets?

EXERCISES 1.1. C

1. If $f(x) = x^3$ and $g(x) = 3x + 2$, find $f(g(x))$ and $g(f(x))$.
2. If $f(x) = 3x^2 + 2$ and $g(x) = x - 1$, find $f(g(x))$ and $g(f(x))$.
3. Which of the following are functions such that $y = f(x)$?
   (a) y is the amount of income tax paid by x last year.
   (b) y is the person who paid x dollars of income tax last year.
4. Use the data of Example 1.1.12 to determine how many grams of lentils are needed to supplement a diet of 50 grams of soybeans.
HC 5. Use your hand calculator to complete the following table if $f(x) = (x - 1)/(2x + 1)$ (as in Example 1.1.7). Give answers that are accurate to four decimal places.

x	1	7	60	197	1,325
$f(x)$					

ANSWERS 1.1. A

1. $f(0) = -5$, $f(10) = 25$, $f(-1) = -8$, $f(x + 1) = 3x - 2$, $f(2x + 3) = 6x + 4$
2. $f(1) = 4$, $f(-3) = 0$, $f(x + 1) = x^2 + 5x + 4$, $f(2x + 3) = 4x^2 + 18x + 18$
3. $9x^2 - 21x + 10$
4. $3x^2 + 9x - 5$
5. $y = 37.9x$
6. (a) $300 (b) $1.50 (c) $900 (d) $675
7. (a) $250 (b) $0.85 (c) $590 (d) $462.50
8. (a) $C(x) = x + 10,000$ (b) $20,000 (c) $30,000
9. (a) $C(x) = 0.005x + 50,000$ (b) $50,050 (c) $50,100

ANSWERS 1.1. C

1. $f(g(x)) = 27x^3 + 54x^2 + 36x + 8$; $g(f(x)) = 3x^3 + 2$
2. $f(g(x)) = 3x^2 - 6x + 5$; $g(f(x)) = 3x^2 + 1$

3. (a) is a function; (b) is not.

4. 5,250/26 grams

5.

x	1	7	60	197	1,325
$f(x)$	0	0.4	0.4876	0.4962	0.4994

1.2   Graphs as Models

You remember that in high school a relationship between two variables x and y was often expressed by a graph. It is customary for a horizontal line to be the x-axis and a vertical line, the y-axis. The point where the axes cross is called the origin and is designated by (0, 0). All other points

Figure 1.2–1

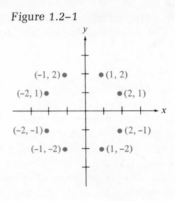

on the plane are uniquely named by an ordered pair of numbers; the first number is the x-coordinate and the second is the y-coordinate.

Ordered pairs of numbers are often the solution set of an equation. Thus each equation (an algebraic object) is made to correspond to its graph (a geometric object). This technique sheds new insight into both algebra and geometry.

Sometimes we say that an equation *is* a certain curve (its graph), intentionally confusing the algebra and the geometry to simplify the language.

1.2.1   Example

Graph $y = x$.

Solution:

The solution set to this equation includes the following pairs of numbers:
(1, 1) (2, 2) (0, 0) (−1, −1) (−3, −3) (10, 10) (−π, −π)

Figure 1.2–2

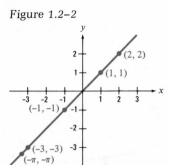

The domain of $f(x) = x$ consists of all real numbers, corresponding to the entire x-axis. When we plot these points and all others that satisfy the equation $y = x$, we get a straight line through the origin at 45° to the axes. Conversely, every point on this line is described by a pair of numbers that satisfy the equation $y = x$.

1.2.2   **Example**

The equation $y = x^2$ corresponds to the graph in Figure 1.2–3. Some-

Figure 1.2–3

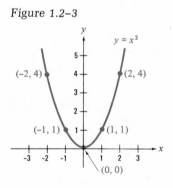

times we say that the curve in Figure 1.2–3 "is" $y = x^2$. The domain of $f(x) = x^2$ consists of all real numbers, corresponding to the entire x-axis.

1.2.3   **Example**

The equation $y^2 = x$ corresponds to the following graph. Sometimes we say that the graph "is" $y^2 = x$.

    The question now arises as to whether $y^2 = x$ defines a function. The equation $y^2 = x$ yields exactly one value of x for each value of y, so x is a function of y whose domain consists of all real numbers, corresponding to the entire y-axis.

Figure 1.2–4

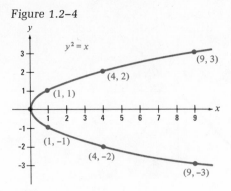

But, solving the equation for y in terms of x, we get two "answers" for each positive x:

$$y = \sqrt{x} \quad \text{and} \quad y = -\sqrt{x}$$

Thus the equation $y^2 = x$ implicitly gives us two *separate* functions of x, where x is nonnegative. (The square-root function is not defined for negative x.)

It is easy to tell by looking at a graph whether it defines a function of x. For a given value of x, we merely locate it on the x-axis and think of the vertical line passing through that point. If it cuts the graph at precisely one point, then the value of x corresponds to precisely one value of y; if it cuts the graph at more than one point, then there is more than one y-value corresponding to that value of x, so the graph does not define a function of x. Thus <u>if any vertical line cuts the graph at more than one point, the graph does not express y as a function of x</u>. So think of moving an imaginary vertical line to the right and left over the graph; if at any time it cuts the graph at more than one point, the graph does not describe a function of x.

By reversing the roles of the variables in the above paragraph, we see that <u>if any horizontal line cuts the graph at more than one point, then x is not a function of y</u>; we do not have $x = f(y)$.

1.2.4   **Example**

In which of the following graphs is y a function of x? In which of the following graphs is x a function of y?

Part (a) describes a function $y = f(x)$ because every vertical line intersects the graph in exactly one point. The domain of this function is assumed to be the set of all real numbers. This graph also describes a function $x = g(y)$ because every horizontal line also crosses the graph in precisely one point; for every y there is exactly one x.

Figure 1.2–5

(a)   (b)   (c)

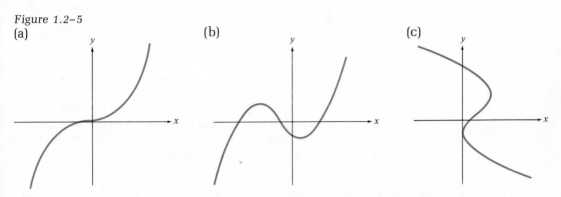

Part (b) describes a function $y = f(x)$ because every vertical line crosses the graph at exactly one point. But it is not a function $x = g(y)$ because there is at least one horizontal line (the x-axis, for instance) which cuts the graph at more than one point.

Part (c) describes a function $x = g(y)$ but not a function $y = f(x)$, by similar reasoning to that in part (b).

1.2.5   Example

In which of the following graphs is y a function of x? In which of the following graphs is x a function of y?

Figure 1.2–6

(a)   (b)   (c)

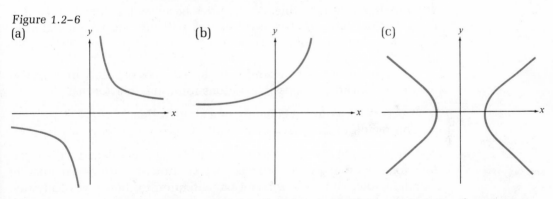

Part (a) describes a function such that $y = f(x)$; its domain is assumed to consist of all the real numbers except zero. This graph is also a function such that $x = g(y)$; its domain also consists of all real numbers except zero.

Part (b) describes a function such that $y = f(x)$ with a domain of all real numbers. It is also a function $x = g(y)$, but this time the domain can only be the set of positive numbers because the x-axis and all horizontal lines below it do not cross the graph at any point.

Part (c) describes a graph that does not describe a function $y = f(x)$ or $x = g(y)$; there exist both vertical and horizontal lines which cross the graph at two points.

Graphs, as models, are often useful in picturing a real-life situation. They enable us to see at a glance things that an algebraic equation may not reveal.

1.2.6   Example

Let $q = f(p)$ be the quantity of gasoline Ye Thoughtful Gas Station can sell if it sets its price per gallon at p. When the price is near zero, people will come from miles around to buy at Ye Thoughtful. When the price gets into the reasonable range (that is, competitive with surrounding sta-

Figure 1.2–7

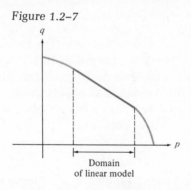

Domain
of linear model

tions), the function will decrease proportionally, which means that it will look like a straight line. As the price becomes much more than that of competing stations, even loyal buyers will turn elsewhere and sales will drop to zero.

We sometimes restrict attention to those values of x for which a given function is linear. Two possible reasons for doing this are that

1. It is easy to write the equation of a straight line.
2. The region of linearity is frequently the region of practical interest.

Since straight lines are the easiest functions to graph and write equations for, we often simply pretend that the function can be extended to a straight line, since we will not pay attention to the "edges" anyway. Thus, Figure 1.2–7 in Example 1.2.6 becomes Figure 1.2–8.

Figure 1.2–8

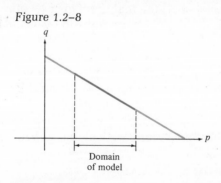

Domain
of model

Depending on the scale, which we have not indicated before, this graph might be

$$q = 10{,}000 - 0.5p$$

where q indicates the quantity demanded if the price per item is p.

Figure 1.2–9

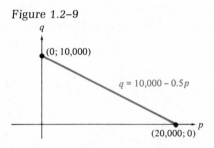

We notice that q is expressed as a function of p; it is clear that for every p there is precisely one q. From the graph it is also clear that for each q there is precisely one p. Solving in the equation for p in terms of q, we have

$$q + 0.5p = 10{,}000$$
$$0.5p = 10{,}000 - q$$
$$p = 20{,}000 - 2q$$

The functions

$$q = 10{,}000 - 0.5p \qquad \text{and} \qquad p = 20{,}000 - 2q$$

are called _inverse functions_ of each other. Figure 1.2–10 shows the inverse of the function graphed previously; it is obtained by interchanging the p and q axes (that is, reflecting the graph over the line at 45° to the axes through the origin from the lower left to the upper right).

Figure 1.2–10

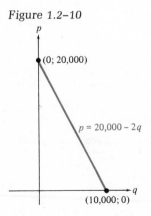

The fact that p decreases as q increases in the above graph (and, indeed, in any other demand function) reflects the fact that as the quantity supplied to the market increases, the price will decrease if all other factors (such as fashion, consumer income, and advertising) remain constant.

1.2.7   Example

Write the inverse function of $p = 2 - 0.5q$ and graph both functions by plotting the points where $p = 0$ and where $q = 0$.

Solution: Figure 1.2–11

$$p + 0.5q = 2$$
$$0.5q = 2 - p$$
$$q = 4 - 2p$$

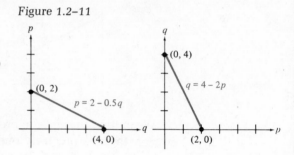

1.2.8   Example

Write the inverse function of $q = 6 - 3p$ and graph both functions by plotting the points where $p = 0$ and where $q = 0$.

Solution:

$$q + 3p = 6$$
$$3p = 6 - q$$
$$p = 2 - \tfrac{1}{3}q$$

Figure 1.2–12

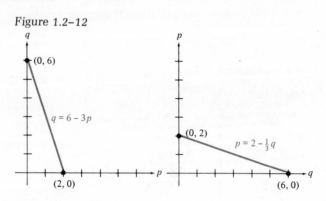

1.2.9   Example in Biology

Not all graphical models are linear (straight lines), of course. Another common graph is the <u>sigmoid</u>, or <u>S-curve</u>. The following graph shows how many fruit flies were living in a certain environment after a given number of days in one experiment. There is an equation that describes the sigmoid curve; we shall encounter it later. The experimenters observed that the population never rose above 340; in fact, this S-curve is eventually a close approximation to the horizontal line $y = 340$. Notice that the population increased gradually at first, then very rapidly around 24 days. Finally, the rate of change of the population becomes negligible.

Figure 1.2–13

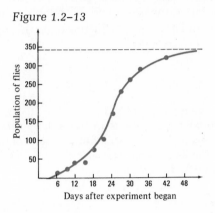

1.2.10   **Example in Biology**

Another nonlinear model is given by the *exponential function*, that is, one of the form $y = c \cdot a^t$, for appropriate constants, c and a. This kind of function is useful in modeling *unchecked population growth*. For example, a certain species of bacterium reproduces itself by dividing in half every minute. Unless the offspring are killed off by lack of nutrients, predators, or other natural checks on population, the population function will have the form $P = P_0 \cdot 2^t$, where t is the number of minutes that have elapsed since time $t = 0$ and P_0 is the initial population of the bacteria when $t = 0$. Now you can see why you get sick so quickly once you catch "the bug"! After only 6 minutes the population is multiplied by $2^6 = 64$ — the graph is off the page.

Figure 1.2–14

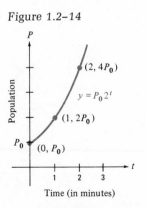

SUMMARY

After quickly reviewing what graphs are, we showed how the graph of an equation can be used to discover whether that equation describes a function of x (if no vertical line cuts the graph in more than one point) or of y (if no horizontal line cuts the graph in more than one point). Then we graphed several price–quantity demand relationships, showing how they can be thought of either as a function of price or a function of quantity. These two functions are inverses of each other, and their graphs are reflections over the line at 45° to the axes through the origin from the lower left to the upper right.

EXERCISES   1.2. A

1. Plot the following points: $(-2, 3)$, $(\frac{1}{2}, -3)$, $(-3, -2)$.
2. Graph the following equations in the order given. Try to do most of them without plotting more than two points; the goal of this exercise is to get a feel for the function as a whole, not just as a set of points. Try to visualize the graphs before you draw them.

   (a) $f(x) = x$ (e) $f(x) = -x$
   (b) $f(x) = 2x$ (f) $f(x) = -2x$
   (c) $f(x) = 3x$ (g) $f(x) = -3x$
   (d) $f(x) = 4x$

3. Which of the following graphs picture a function $y = f(x)$? Which show a function $x = f(y)$?

(a)

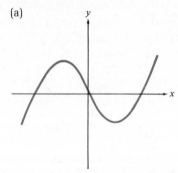

(b)

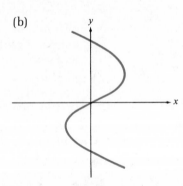

(c)

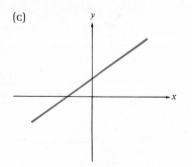

(d)

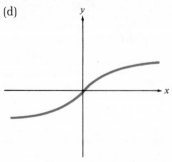

(e)

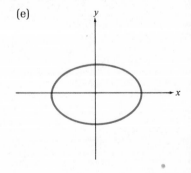

4.   Write the inverse of the function $p = 8 - 4q$ and graph both functions.
5.   Find the inverse of the function $q = 2 - 0.2p$ and graph both functions.

EXERCISES   1.2. B

1.   Plot the following points: $(2, -1)$, $(\frac{1}{3}, -2)$, $(-4, 2)$.
2.   Graph the following equations in the order given. Try to do most of them without plotting more than two points; the goal of this exercise is to get a feel for the function as a whole, not just as a set of points. Try to visualize the graphs before you draw them.

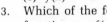

(a) $f(x) = \frac{1}{2}x$	(f) $f(x) = -\frac{1}{2}x$
(b) $f(x) = \frac{1}{3}x$	(g) $f(x) = -\frac{1}{3}x$
(c) $f(x) = \frac{1}{4}x$	(h) $f(x) = -\frac{1}{4}x$
(d) $f(x) = \frac{1}{5}x$	(i) $f(x) = -\frac{1}{5}x$
(e) $f(x) = -x$	

3.   Which of the following graphs show a function $y = f(x)$? Which picture a function $x = f(y)$?

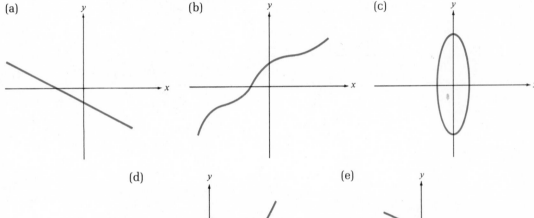

(a)   (b)   (c)

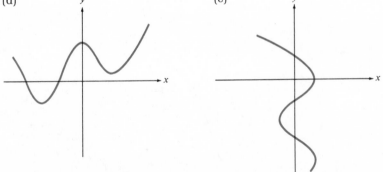

(d)   (e)

4.   Write the inverse of the function $p = 6 - 2q$ and graph both functions.
5.   Compute the inverse of the function $q = 2 - 0.4p$ and graph both functions.

EXERCISES   1.2. C

1.   Roughly sketch
     (a) $f(x) = x$ (b) $f(x) = x^2$ (c) $f(x) = x^3$

(d) $f(x) = x^4$ (e) $f(x) = x^5$ (f) $f(x) = x^6$

Which of these functions have an inverse?

2.   Write the equation for the inverse of each of the functions in problem 2 of Exercises 1.2.A. What relationship does the graph of the inverse have to the graph of the original function?

3.   Example 1.2.5(b) is the graph of an exponential function. Depending on the scale of values, it might be $f(x) = 10^x$, or $f(x) = c^x$, where c is any positive constant greater than one. In any case, its inverse is called the "logarithm." Make a rough sketch of the graph of the logarithm. What is its domain?

ANSWERS   1.2. A

1.

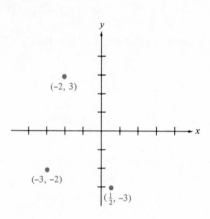

2.

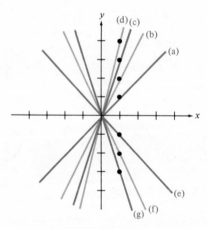

3.   (a), (c), and (d) are functions such that $y = f(x)$ because each vertical line crosses the graph at exactly one point. (b), (c), and (d) are functions such that $x = f(y)$ because each horizontal line crosses the graph at one point. [In (c) and (d) we assume that the graph extends "infinitely far" in the manner indicated.] (e) does not describe a function of either x or y.

4. $q = 2 - 0.25p$

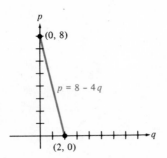

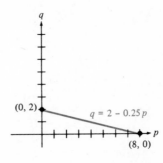

5. $p = 10 - 5q$

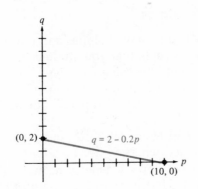

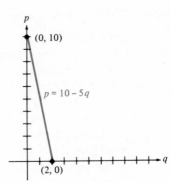

ANSWERS   **1.2. C**

1.   (a), (c), and (e) have an inverse. The others do not.

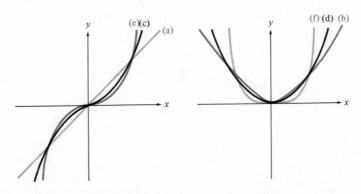

2.   (a) $g(y) = y$ (b) $g(y) = \frac{1}{2}y$ (c) $g(y) = \frac{1}{3}y$ (d) $g(y) = \frac{1}{4}y$ (e) $g(y) = -y$
     (f) $g(y) = -\frac{1}{2}y$ (g) $g(y) = -\frac{1}{3}y$
     The graph of an inverse of a function is its reflection in the line $y = x$ (the
     line at 45° to the axes through the origin going from the lower left to the
     upper right).

3.

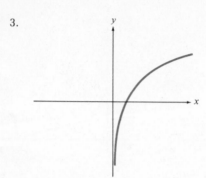

1.3   Slopes and Straight Lines

This section is primarily a review of slopes and the slope-intercept form for straight lines, two concepts that you studied in high school and are fundamental to understanding calculus. The derivative, which is the subject matter of Chapters 2 and 3, is the generalization of the slope of a straight line to a curved line. If the graph of a function is a straight line, its derivative is its slope.

In doing problem 2 of Exercises 1.2.A and 1.2.B you may have noticed that the lines $y = mx$, where m was positive, went up as you move to the right, and they went down whenever m was negative. It is easy to convince yourself that this is always so. Two points that always lie on the line $y = mx$ are $(0, 0)$ and $(1, m)$. This means that if $m > 0$, we go up from the origin, and if $m < 0$, we go down.

Figure 1.3–1

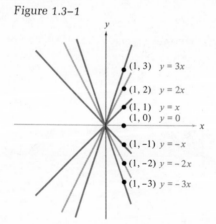

The equation $y = mx$ describes a line through the origin with slope m. The slope of a line measures how fast it is rising or falling.

1.3.1   Definition

If (x_1, y_1) and (x_2, y_2) are any two points on the Cartesian plane such that $x_1 \neq x_2$, the slope between them is $\dfrac{y_2 - y_1}{x_2 - x_1}$.

Figure 1.3-2

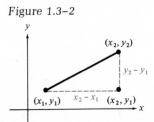

In Figure 1.3-2 we have taken the upward (or positive) direction for the change in y and the right (or positive) direction for the change in x only for convenience and definiteness. We could just as easily have taken both directions negative (downward and to the left). We would have gotten the same answer.

1.3.2   Example

Find the slope between the following points:
(a) (1, 1) and (3, 5)
(b) (3, −4) and (−5, −2)

Solution:

(a) $\dfrac{y_2 - y_1}{x_2 - x_1} = \dfrac{5 - 1}{3 - 1} = \dfrac{4}{2} = 2$

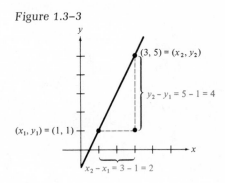

Figure 1.3-3

(b) $\dfrac{y_2 - y_1}{x_2 - x_1} = \dfrac{-2 - (-4)}{-5 - 3} = \dfrac{2}{-8} = -\dfrac{1}{4}$

Figure 1.3-4

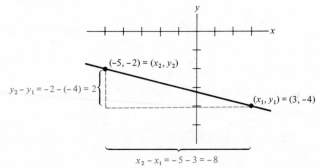

1.3.3   Rule

If (x_1, y_1) and (x_2, y_2) are any two points on the line $y = mx$, then the slope between them is m.

Figure 1.3–5

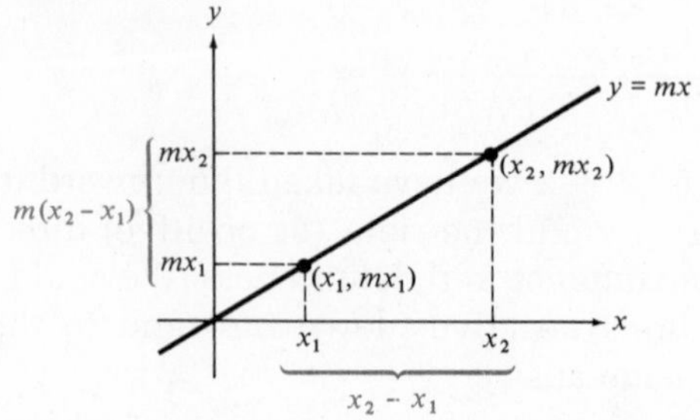

Since the points lie on the line, we have $y_1 = mx_1$ and $y_2 = mx_2$. Thus

$$\frac{y_2 - y_1}{x_2 - x_1} = \frac{mx_2 - mx_1}{x_2 - x_1} = \frac{m(x_2 - x_1)}{x_2 - x_1} = m$$

This rule says that as x changes from x_1 to x_2, y changes m times as much as it goes from $f(x_1)$ to $f(x_2)$. Or it says that $(y_2 - y_1)$ is m times $(x_2 - x_1)$. In other words, the change in y is always m times the change in x.

It is clear from Figure 1.3–1 that if the slope of the line is positive, the line goes up as x moves to the right. The function described by such a line is said to be increasing.

Similarly, if the slope of a line is negative, the line goes down as x moves to the right. We say that the function described by such a line is decreasing.

1.3.4   Rule

The y-axis is described by the equation $x = 0$. All other lines through the origin are described by the equation

$$y = mx$$

where m is the slope of the line.

We now consider straight lines that do not go through the origin.

1.3.5   Example

Graph $f(x) = 2x + 1$. (This is another way of writing $y = 2x + 1$.)

Solution:

Each value for $f(x)$ will be one more than the value of $g(x) = 2x$ at the same value of x. Thus the graph will be a straight line parallel to the graph of $g(x) = 2x$, but one unit higher at all points.

Figure 1.3–6

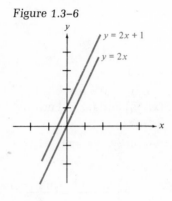

1.3.6   Example

Graph $f(x) = 2x + 2$.

1.3.7   Example

Graph $g(x) = 2x - 1$.

Figure 1.3–7

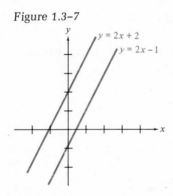

1.3.8   Rule

The graph of $f(x) = 2x + b$ is parallel to that of $g(x) = 2x$ and b units above it if b is positive. If b is negative, it is $|b|$ (the absolute value of b, that is, the magnitude of b) units below it.

Conversely, any line parallel to $g(x) = 2x$ and b units above it is described by the equation $f(x) = 2x + b$. A line parallel to $g(x)$ and b units below it is described by the equation $g(x) = 2x - b$, where $b \geq 0$.

1.3.9   Definition

The value of y where a line crosses the y-axis is called the y-intercept of the line.

1.3.10   Rule

The graph of $y = mx + b$ is a line with slope m and y-intercept b.

It is clear that if $x = 0$, $y = m \cdot 0 + b = b$, so the y-intercept is b.
Suppose now that (x_1, y_1) and (x_2, y_2) are any two points on the graph. Then

$$y_1 = mx_1 + b \qquad \text{and} \qquad y_2 = mx_2 + b$$

Hence

$$\frac{y_2 - y_1}{x_2 - x_1} = \frac{mx_2 + b - (mx_1 + b)}{x_2 - x_1} = \frac{mx_2 - mx_1}{x_2 - x_1} = \frac{m(x_2 - x_1)}{x_2 - x_1} = m$$

So the slope is m.

It is important for you to understand this rule. It means that the graph of $y = mx + b$ is parallel to the line $y = mx$ and a distance of $|b|$ above or below it depending on whether b is positive or negative. A little reflection should convince you that every straight line, except vertical lines, have a y-intercept and are parallel to some line of the form $y = mx$.

1.3.11   Example

What is the equation of each of the following graphs?

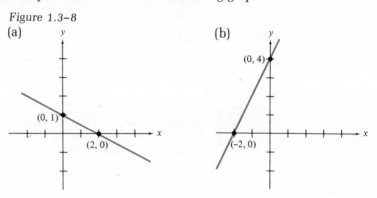

Figure 1.3–8
(a)
(b)

Solution:

(a) The line has y-intercept 1, so it is of the form $y = mx + 1$, where m is the slope. To find the slope, we observe that it goes through the points $(0, 1)$ and $(2, 0)$. Thus the slope is

$$\frac{0 - 1}{2 - 0} = \frac{-1}{2} = -\frac{1}{2}$$

Therefore, the equation is $y = -\frac{1}{2}x + 1$.

(b) Since the line goes through the points $(-2, 0)$ and $(0, 4)$, it has slope 2. Since the y-intercept is 4, the equation is $y = 2x + 4$.

1.3.12   Definition

An equation of a straight line in the form

$$y = mx + b$$

is said to have slope-intercept form. Notice that y must be alone on one side of the equation.

1.3.13   Example

Comparing the role of $y = mx + b$ in Rule 1.3.10 and in Example 1.1.9, we see that the linear cost-output model is in the slope-intercept form. The y-intercept, b, is the fixed cost. The slope, m, is the marginal cost.

A line is *horizontal* if and only if it has slope zero. Its equation is $y = b$, where b is some constant.

A *vertical* line does not describe a function of x and it does not have a slope. Its equation is $x = b$, where b is some constant.

One trick for remembering which is the numerator in the expression for the slope is to think of the phrase "rise over run." We list here some common expressions for the slope:

$$\text{slope} = \frac{y_2 - y_1}{x_2 - x_1} = \frac{y_1 - y_2}{x_1 - x_2} = \frac{\text{change in } y}{\text{change in } x} = \frac{\text{rise}}{\text{run}}$$

The following two facts were proved in high school algebra. Only the first is essential for this course, but we state them both here for your general information.

1. Two nonvertical lines are *parallel* if and only if they have the same slope.
2. Two lines, neither of which is vertical, are *perpendicular* if and only if their slopes are negative reciprocals of each other.

1.3.14   Example in Biology or Pre-Med

Hemoglobin is a substance in the blood which transfers oxygen from the lungs to other parts of the body. If the number x measures the concentration of oxygen in the air, and $C(x)$ denotes the rate of combination of hemoglobin and oxygen in the lungs, it has been observed that $C(x) = mx$ for some constant m depending upon the pressure of oxygen and other factors. Thus, if we draw a graph showing the concentration of oxygen in the air on the x-axis and the rate of combination of hemoglobin and oxygen $C(x)$, on the vertical axis, the graph will be a straight line going through the origin with slope m.

SUMMARY

In this section we reviewed the ideas of the slope between two points and the slope of a line. We noticed that a line, $y = mx + b$, is determined by its slope m and its y-intercept b. We used this fact to graph lines and to relate them to one another.

EXERCISES 1.3. A

1. Graph
   (a) $f(x) = 2x$ (c) $f(x) = \frac{1}{2}x$
   (b) $f(x) = -2x$ (d) $f(x) = -\frac{1}{2}x$
2. Graph the following equations in the order given:
   (a) $y = -x$ (e) $y = -x - 1$
   (b) $y = -x + 1$ (f) $y = -x - 2$
   (c) $y = -x + 2$ (g) $y = -x - 3$
   (d) $y = -x + 3$
3. Describe the line whose equation is $y = mx$.
4. Find the slope between
   (a) $(-1, 3)$ and $(4, -9)$ (c) $(4, 15)$ and $(-2, -3)$
   (b) $(-5, -9)$ and $(6, -3)$
5. Write the equations of the following lines:

(a)

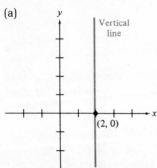

(b)

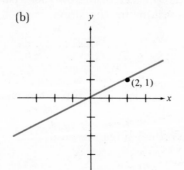

(c)

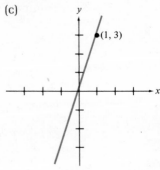

(d)

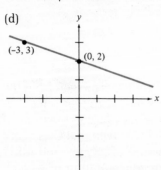

(e)

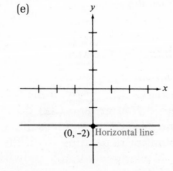

(f)

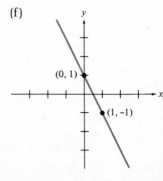

6. Which of the graphs in problem 5 describe functions of x? Tell which of the functions are increasing and which are decreasing.
7. Does the line $y = \frac{2}{3}x$ rise or fall as x moves right?
8. What is the equation of the line going through $(-4, 2)$ and the origin?
9. (a) What is the slope of the line going through $(-3, -6)$ and the origin?
   (b) Write the equation of the line.
10. (a) Prove that $y = 2$ has slope zero.
    (b) Prove that every horizontal line has slope zero. Use the fact that such lines are of the form $y = b$.
11. Write the equation of the line that is parallel to $y = 3x + 7$ and has y-intercept -4.
12. Write the equation of the line that is perpendicular to $y = -4x$ and goes through the point $(4, -1)$.

EXERCISES   1.3. B

1. Graph
   (a) $f(x) = 3x$ (c) $f(x) = \frac{1}{3}x$
   (b) $f(x) = -3x$ (d) $f(x) = -\frac{1}{3}x$
2. Graph the following equations in the order given:
   (a) $y = -2x$ (e) $y = -2x - 1$
   (b) $y = -2x + 1$ (f) $y = -2x - 2$
   (c) $y = -2x + 2$ (g) $y = -2x - 3$
   (d) $y = -2x + 3$
3. Describe the line whose equation is $y = mx$.
4. Find the slope between
   (a) $(-2, 4)$ and $(3, -6)$ (c) $(3, 12)$ and $(-1, 3)$
   (b) $(-2, -3)$ and $(1, -6)$
5. Write the equation of each of the following lines:

(a)

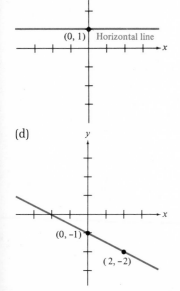

(b)

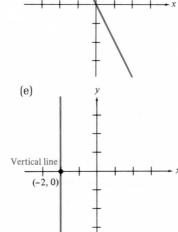

(c)

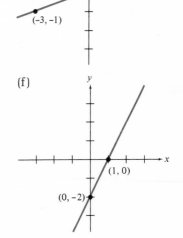

6. Which of the graphs in problem 5 describe functions such that $y = f(x)$? Tell which of the functions are increasing and which are decreasing.

7. Does the line $y = -\frac{1}{2}x$ rise or fall as x moves right?

8. What is the equation of the line going through $(3, -2)$ and the origin?

9. What is the equation of the line going through $(-7, -14)$ and the point $(0, 7)$?

10. (a) Prove that $y = 3$ has slope zero.
    (b) Prove that every horizontal line has slope zero. (Use the fact that such lines are of the form $y = b$.)

11. Write the equation of the line that is parallel to $y = -2x + 1$ and has y-intercept 4.

12. Write the equation of the line that is perpendicular to $y = \frac{2}{3}x + \pi$ and has y-intercept 12.

ANSWERS   1.3. A

1.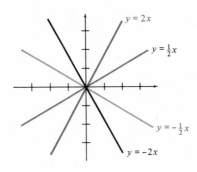

2.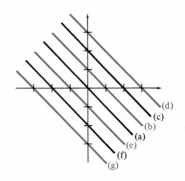

3. It is a line through the origin with slope m.

4. (a) $\dfrac{-9 - 3}{4 - (-1)} = \dfrac{-12}{5}$ (b) $\dfrac{-9 - (-3)}{-5 - 6} = \dfrac{-6}{-11} = \dfrac{6}{11}$ (c) 3

5. (a) $x = 2$ (b) $y = \frac{1}{2}x$ (c) $y = 3x$ (d) $y = -\frac{1}{3}x + 2$ (e) $y = -2$
   (f) $y = -2x + 1$

6. All except (a) describe functions. (b) and (c) are increasing. (d) and (f) are decreasing. (e) is a *constant function*; it is a function whose value is constantly equal to -2.

7. It has a positive slope, so it rises.

8. $y = -\frac{1}{2}x$

9. (a) 2   (b) $y = 2x$

10. (a) $\dfrac{y_2 - y_1}{x_2 - x_1} = \dfrac{2 - 2}{x_2 - x_1} = 0$
    (b) Since y is always b, $y_2 - y_1 = b - b = 0$, so the slope is zero since its numerator is.

11. $y = 3x - 4$

12. $y = \frac{1}{4}x - 2$

SAMPLE QUIZ   **Sections 1.1, 1.2, and 1.3**

Each problem counts 10 points.
1. Define: A function is _____ .
2. If $f(x) = 2x^2 + 5$, what is $f(-3)$?
3. Suppose that the total cost of producing x tables is given by $C = 50x + 2,000$. Then in this expression, 50 is the _____ cost and 2,000 is the _____ cost.
4. Which of the following graphs define a function such that y is a function of x?

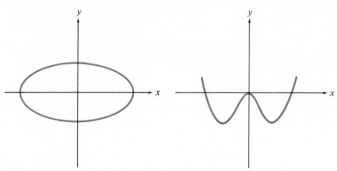

5, 6.   Suppose that the quantity of a given item that will be sold if the price is p is described by $q = 3 - 0.25p$.
5. Graph this function.
6. Write the equation of the inverse function (that is, for a given quantity show what the corresponding price will be).
7. If a straight-line graph goes up as x goes to the right, then the slope of the line is _____ and we say that the function described by the line is _____ .
8. Graph $y = -2x + 3$.
9. Find the slope of the line connecting the points $(-1, -2)$ and $(-3, -1)$.
10. Graph on the same coordinate system: $y = 2x$, $y = -2x$, $y = \frac{1}{2}x$, and $y = -\frac{1}{2}x$.

The answers are at the back of the book.

1.4   Determining Linear and Quadratic Models

We have shown that functions of x whose graphs are straight lines are of the form

$$f(x) = mx + b$$

These are called <u>linear functions</u>. In some sense they are the simplest functions and they, therefore, yield the simplest models. The next simplest functions are the <u>quadratic functions</u>. They are generally written in the form

$$f(x) = ax^2 + bx + c$$

where a, b, and c are constants and $a \neq 0$.

Linear functions are so common in real-world models that it is important to be able to manipulate them easily. Such functions are most often determined in one of five ways.

1. An equation
2. A graph
3. Slope and y-intercept
4. Slope and a point (not necessarily the y-intercept)
5. Two points

If you think about it, you should be able to convince yourself that each of the last three descriptions determines precisely one straight line. Our next project will be to examine how, given one of the above five descriptions, we can find the other four.

You have studied the interrelationships among (1), (2), and (3) exhaustively in high school algebra and in Section 1.3. We turn now to (4) and (5). There is another general expression for a line that can be used (although it is not necessary) when handling these.

If (x_1, y_1) is a point on a straight line with slope m, then for any other point (x, y) on this line $(x \neq x_1)$, $m = \dfrac{y - y_1}{x - x_1}$. Multiplying both sides of this equation by $(x - x_1)$ gives the following definition.

Figure 1.4–1

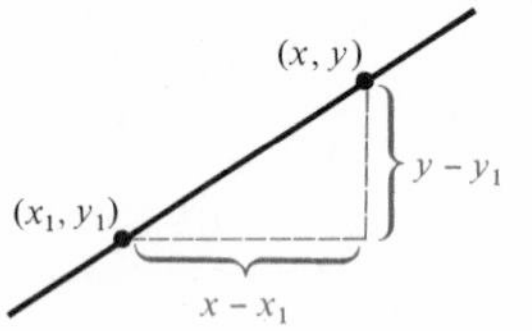

1.4.1   Definition

For all (x, y) on the line,

$$m(x - x_1) = y - y_1 \qquad \text{or} \qquad y - y_1 = m(x - x_1)$$

is called the <u>point-slope formula</u> for that line.

1.4.2   Example

Find an equation in slope-intercept form of the straight line with slope 4 passing through the point $(2, 6)$.

Solution 1:

We can substitute into the point-slope formula to solve this problem:

$$4(x - 2) = y - 6$$
$$4x - 8 = y - 6$$
$$4x - 2 = y$$

Figure 1.4–2

Solution 2:

We can also find this equation by directly using the slope-intercept form of a linear equation:

$$y = mx + b$$

Since the slope of the given line is 4, we know that the desired equation is of the form

$$y = 4x + b$$

To say that the line "passes through the point (2, 6)" is to say that $x = 2$, $y = 6$ satisfies the equation. (Note that we are intentionally confusing the line with its equation.) In other words, for the proper b, it is true that

$$6 = 4 \cdot 2 + b$$

But there is only one b for which this last equation is true. We use elementary algebra to find what it is:

$$6 = 8 + b$$
$$-2 = b$$

So the final form of the equation is again $y = 4x - 2$.

Both solutions yield the same answer and they take about the same number of steps. You can use whichever method you prefer to solve problems like this one, but you should become thoroughly familiar with at least one method, since problems using this skill will occur repeatedly in this book.

It is obvious that two points determine a straight line. It is not so obvious how to develop the equation for the line given the two points.

One technique is to use them to find the slope and then proceed as in Example 1.4.2 (either solution).

1.4.3   Example

Find an equation of the straight line through the points (1, 5) and (3, 9).

Solution:

First we find the slope:

$$\frac{y_2 - y_1}{x_2 - x_1} = \frac{9 - 5}{3 - 1} = \frac{4}{2} = 2$$

Then we proceed as before; it does not matter which point we use:

$$y = 2x + b$$
$$5 = 2 \cdot 1 + b, \text{ using the point } (1, 5)$$
$$5 - 2 = b$$
$$3 = b$$

So the answer is

$$y = 2x + 3$$

It is easy to check that (3, 9) also satisfies this equation.

Another technique is to use the two points (x_1, y_1) and (x_2, y_2) to write the equations in two unknowns, m and b. These equations, as below, are then solved simultaneously for m and b, the slope and intercept, respectively:

$$y_1 = mx_1 + b$$
$$y_2 = mx_2 + b$$

1.4.4   Example

Use the technique that we have described above to find an equation of a line through the points (1, 5) and (3, 9), the same two points as in Example 1.4.3.

Solution:

$$5 = m \cdot 1 + b$$
$$9 = m \cdot 3 + b$$

Subtracting the second equation from the first, we obtain

$$-4 = -2m, \text{ which implies that } m = 2$$

Substituting $m = 2$ into the first equation, we obtain

$5 = (2)(1) + b$, which implies that $b = +3$

Hence the equation of the line is $y = 2x + 3$, as in Example 1.4.3.

The ability to go from one to the other among these five methods for expressing linear functions is useful in practical applications. Suppose, for example, that a problem calls for determining the (linear) cost-output equation given the fixed cost and the cost of producing some specific quantity of the items. Then you must realize that the y-intercept and another point have been given and use this information to get the equation. Remember, two data points are sufficient to establish a linear model.

1.4.5   **Example**

If it costs \$2,000 to manufacture 1,000 hats this month and fixed costs are \$500 per month, about how much will it cost to produce 1,200 hats next month? Assume that the linear cost-output equation is valid.

Solution:

In mathematical language, $b = 500$ and (1,000; 2,000) satisfies the equation $C = mx + b$. There are at least two ways of finding m.

(a)
$$2,000 = m \cdot 1,000 + 500$$
$$2,000 - 500 = m \cdot 1,000$$
$$1,500 = 1,000m$$
$$1.5 = m$$

(b) Recognize that (1,000; 2,000) and (0, 500) are two points on the line and use them to calculate the slope:

$$\frac{y_2 - y_1}{x_2 - x_1} = \frac{2,000 - 500}{1,000 - 0} = \frac{1,500}{1,000} = 1.5$$

By either method we conclude that the equation is

$C = 1.5x + 500$

To find how much it costs to produce 1,200 hats we now substitute $x = 1,200$:

$C = 1.5(1,200) + 500 = 1,800 + 500 = 2,300$

1.4.6   **Example**

If it costs \$2,700 to produce 1,000 lawn chairs one month, and it costs \$3,100 to produce 1,200 lawn chairs the following month, how much will it cost to produce 1,300 lawn chairs the next month, assuming that the linear cost-output model is valid?

Solution:

Using the technique of Example 1.4.4, we obtain two equations in m and b:

$$2{,}700 = 1{,}000m + b$$
$$3{,}100 = 1{,}200m + b$$

Subtracting the second equation from the first, we obtain

$$-400 = -200m \qquad \text{which implies that } m = 2$$

Substituting $m = 2$ in the first equation, we obtain

$$2{,}700 = 1{,}000(2) + b \qquad \text{or} \qquad b = 700$$

Hence the cost equation is

$$C = 2x + 700$$

Letting $x = 1{,}300$, we get $C = 2(1{,}300) + 700$, so it will cost $3,300 to produce 1,300 chairs.

Quadratic Models

The graph of a quadratic function of x is a parabola, which has the appearance of one of the following graphs:

Figure 1.4–3

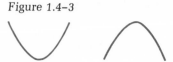

Unlike a line (which is determined by two points), a parabola is determined by three points, not on a straight line. Given any three such data points, we can make a quadratic model by writing three equations in three unknowns, using a technique similar to that in Example 1.4.4:

$$y_1 = ax_1^2 + bx_1 + c$$
$$y_2 = ax_2^2 + bx_2 + c$$
$$y_3 = ax_3^2 + bx_3 + c$$

When these are solved simultaneously for a, b, and c, we obtain a quadratic equation that is satisfied by the original three data points.

1.4.7   **Example**

Determine the quadratic function $y = ax^2 + bx + c$ which passes through the points (1, 2), (2, 6), and (3, 12).

Solution:

A quick sketch will convince you that these three points are not on a straight line. If we substitute them into the expression $y = ax^2 + bx + c$, we have

$$2 = a + b + c$$
$$6 = 4a + 2b + c$$
$$12 = 9a + 3b + c$$

Subtracting the first equation from the second and third, we obtain

$$4 = 3a + b$$
$$10 = 8a + 2b$$

Subtracting twice the first of these equations from the second, we now have

$2 = 2a$, which implies that $a = 1$

Substituting $a = 1$ into $4 = 3a + b$, we obtain

$4 = 3 + b$, which implies that $b = 1$

And substituting $a = 1$ and $b = 1$ into $2 = a + b + c$, we get

$$2 = 1 + 1 + c \qquad \text{or} \qquad c = 0$$

The required quadratic function, then, is

$$y = x^2 + x + 0 \qquad \text{or} \qquad y = x^2 + x$$

1.4.8   **Example**

If the overhead for producing no lawn chairs is $100, while 10 chairs cost $190 and 12 chairs cost $232, find a quadratic cost function which these data satisfy. Using this model, find the cost of 15 chairs.

Solution:

The three data points are (0, 100), (10, 190), and (12, 232). Substituting these into the equation $y = ax^2 + bx + c$, we get

$$100 = 0a + 0b + c$$
$$190 = 100a + 10b + c$$
$$232 = 144a + 12b + c$$

Solving these for a, b, and c, we get $a = 1$, $b = -1$, and $c = 100$, so

$$y = f(x) = x^2 - x + 100$$

is the quadratic model fitting these data.

$$f(15) = 225 - 15 + 100, \text{ so 15 chairs will cost } \$310$$

It is sometimes necessary to find the "roots" of a quadratic function, in other words, the values x_1 of x for which $ax_1^2 + bx_1 + c = 0$. This can often be done by factoring the expression $ax^2 + bx + c$; you practiced this in high school. But sometimes you must use the foolproof "quadratic formula." This formula can be derived by completing the square in a quadratic equation.

$$\text{First root} = \frac{-b + \sqrt{b^2 - 4ac}}{2a}$$

$$\text{Second root} = \frac{-b - \sqrt{b^2 - 4ac}}{2a}$$

When the quantity under the square root sign ($b^2 - 4ac$) is negative, then for our purposes there is no solution; this means that the parabola which is the graph of the function $f(x) = ax^2 + bx + c$ does not cross the x-axis.

Figure 1.4–4

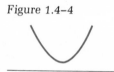

If the quantity under the square-root sign ($b^2 - 4ac$) is zero, then both the roots are the same; this means that the corresponding parabola is tangent to the x-axis.

Figure 1.4–5

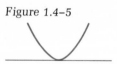

If the quantity under the square-root sign ($b^2 - 4ac$) is positive, then the quadratic formula yields two distinct roots; this means that the parabola crosses the x-axis at two points.

Figure 1.4–6

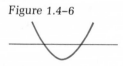

1.4.9   Example in Biology

Let us imagine an artery to be a cylindrical tube of constant diameter 2R. (See Figure 1.4–7.) Let r be the distance from the central axis of the artery to a certain point P within the artery. Then the velocity of the blood flowing past the point P is a quadratic function of the distance r, and it is given by

$$f(r) = k \cdot (R^2 - r^2) \qquad \text{units per second}$$

where k is a constant depending upon blood pressure, viscosity, and the length of the artery. This relationship was discovered by the French physician Jean Louis Marie Poiseuille in 1842.

Figure 1.4–7

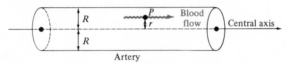

Artery

SUMMARY

Linear functions can be described in five ways: an equation, a graph, slope and y-intercept, point and slope, and two points. In this section we discussed how to understand the relationships among these five descriptions.

We also discussed how to develop mathematical models given real-world data. In particular, two data points determine a straight line and three data points not on a line determine a parabola.

EXERCISES   1.4. A

1. Find an equation of the line that goes through the point $(-12, 6)$ and has slope $\frac{1}{3}$.
2. Find an equation of the line passing through $(-1, -3)$ and $(-3, 9)$.
3. Suppose that the marginal cost for manufacturing a camera is $5 and the fixed cost per month is $2,500.
    (a) Write the cost-output equation.
    (b) Find the cost of producing 1,700 cameras in one month.
4. Suppose that it cost $1,000 to manufacture 200 belts and $1,100 to manufacture 400 belts.
    (a) Write the linear cost-output equation for these two pieces of information.
    (b) Using this model, how much will it cost to manufacture 450 belts?
5. (a) Write the linear cost-output equation if it costs $1,000 to produce 200 lamps and the marginal cost per lamp is $3.50.
    (b) How much does it cost to produce 400 lamps?
6. Find the solutions, if any, to the following equations.
    (a) $x^2 + 5x - 2 = 0$ (c) $6x - 9x^2 - 1 = 0$
    (b) $3x^2 - 2x + 5 = 0$ (d) $4x - 5x^2 + 1 = 0$

EXERCISES   1.4. B

1. Find an equation of the line that goes through the point $(2, -5)$ and has slope -2.

2. Find an equation of the line passing through $(-2, 4)$ and $(1, -2)$.
3. Suppose that the marginal cost for manufacturing a record is \$1 and the fixed cost per month is \$5,000.
    (a) Write the linear cost-output equation.
    (b) Find the cost of producing 1,400 records in one month.
4. Suppose that it cost \$100 to manufacture 50 pairs of mittens and \$150 to manufacture 150 pairs of mittens.
    (a) Assuming that the linear cost-output model is valid, write an equation for producing x pairs of mittens.
    (b) How much will it cost to make 300 pairs?
5. (a) Write the linear cost-output equation if it costs \$200 to produce 50 sweaters and \$300 to produce 150 sweaters.
    (b) How much does it cost to produce 500 sweaters?
6. Find the solutions, if any, to the following equations:
    (a) $8x - 9x^2 + 1 = 0$ (c) $4x - 4x^2 - 1 = 0$
    (b) $4x^2 - 2x + 3 = 0$ (d) $x^2 + 4x + 2 = 0$

EXERCISES 1.4. C

1. Find a quadratic function $y = ax^2 + bx + c$ which satisfies the data points $(0, 20)$, $(2, 18)$, and $(-2, 14)$.
2. Write a quadratic cost equation for the data in problem 4, Exercises 1.4.B, if there is no overhead [that is, if $f(0) = 0$].
3. Suppose that the fixed cost of manufacturing widgets is \$7, it costs \$16 to manufacture one widget, and it costs \$23 to manufacture two widgets.
    (a) Write the quadratic model that satisfies these data.
    (b) What is the cost of manufacturing three widgets?
4. Use the data of Example 1.4.9. Suppose that the radius of the artery is 2 millimeters and the distance of a point from the central axis of the artery is 1 millimeter. Find the velocity of the flow of blood past the point in mm/sec if the constant $k = 1$.
HC 5. Animal physiologists have determined that the surface area of a horse (in square meters) is given, in terms of its weight W (in kilograms), by the formula $f(W) = (0.1)W^{2/3}$. This is not, of course, a quadratic function, since we are taking the cube root of the square of W. However, if your hand calculator has a y^x key, you can use it to complete the following table:

Weight and Surface Area of a Horse

weight, W (in kg)	200	250	300	350	400
surface area, $= (0.1)W^{2/3}$ (in m²)					

ANSWERS 1.4. A

1. $y = \frac{1}{3}x + 10$
2. $y = -6x - 9$
3. (a) $C = 5x + 2,500$ (b) 11,000
4. (a) $C = \frac{1}{2}x + 900$ (b) \$1,125

5. (a) $C = 3.5x + 300$ (b) $1,700
6. (a) $x = -(5 \pm \sqrt{33})/2$ (b) no real solution (c) $x = \frac{1}{3}$ only
    (d) $x = +1$ and $x = \frac{-1}{5}$

ANSWERS 1.4. C

1. $y = -x^2 + x + 20$
2. $y = -0.01x^2 + 2.5x$
3. (a) $y = 10x - x^2 + 7$ (b) $28
4. 3 mm/sec
5. 3.42; 3.97; 4.48; 4.97, 5.43

1.5   Introduction to Limits

If functions are the flesh and blood of calculus, then "limits" are its life force, without which the subject is an empty bag of tricks. The aim of this section is to develop an intuitive feeling for this important concept as well as an elementary, but practical, working knowledge of how to calculate limits.

1.5.1   Definition

A function is said to approach a <u>limit</u> L at a certain value x_1 for x if $f(x)$ gets arbitrarily near to L as x gets near to, but not equal to, x_1. We write this

$$\lim_{x \to x_1} f(x) = L$$

Geometrically, this definition means that one can force the values of $f(x)$ to be within an arbitrarily specified a units of the number L if x is within b units of x_1. (See Figure 1.5–1.)

Figure 1.5–1

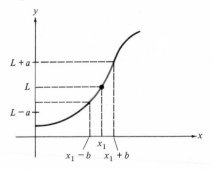

In formal mathematics courses a more careful definition is given (see Appendix I), but this will be enough for our needs.

If the graph of the function is an unbroken line or curve, we can find the limiting value merely by evaluating the function at the point $x = x_1$. Even without looking at the graph, it is often clear when this can be done.

1.5.2   Example

$$\lim_{x \to 4} (3x + 2) = 14$$

This says that as x gets near to 4, $3x + 2$ gets arbitrarily near to 14; this is obvious since the graph of $y = 3x + 2$ is a line and $3 \cdot 4 + 2 = 14$.

We read the expression in Example 1.5.2: "The limit as x approaches 4 of $3x + 2$ is 14."

1.5.3   Example

(a) $\lim\limits_{x \to 3} \dfrac{1}{x} = \dfrac{1}{3}$

(b) $\lim\limits_{x \to 2} (x^2 + 3x + 2) = 2^2 + 3(2) + 2 = 12$

Finding a limit sometimes involves more work than just plugging a number into a given expression. This is often true when substitution requires the breaking of an arithmetic law, such as the prohibition against dividing by zero.

1.5.4   Example

Find $\lim\limits_{x \to 2} \dfrac{x^2 - 4}{x - 2}$.

Solution:

We cannot merely substitute $x = 2$ into this expression because then we get "0/0," which is meaningless. But if x does not equal 2, the following manipulation is valid, since in that case we are not dividing by $2 - 2 = 0$.

$$* \qquad \frac{x^2 - 4}{x - 2} = \frac{(x - 2)(x + 2)}{x - 2} = x + 2 \qquad \text{if } x \neq 2$$

This says that $f(x) = (x^2 - 4)/(x - 2)$ is exactly the same as $g(x) = x + 2$ except where $x = 2$; that point is missing on the graph of $f(x)$, giving us a break in the line.

Figure 1.5–2

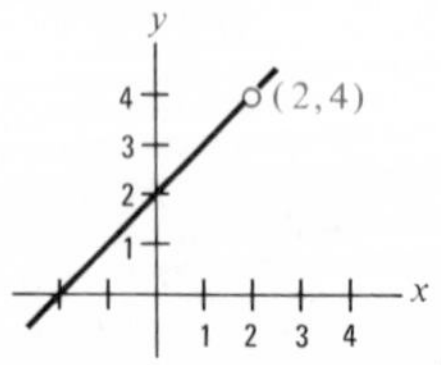

Since $\lim_{x \to 2} (x + 2) = 4$ (clearly), we also have $\lim_{x \to 2} (x^2 - 4)/(x - 2) = 4$. We can get the value 4 either from the graph or from the starred expression.

The trick in evaluating limits is to reduce the expression to another expression in which the troublesome number can be plugged in without difficulty. Then we evaluate the new expression at that number.

1.5.5   Example

Evaluate $\lim_{x \to 3} \dfrac{x^2 - 10x + 21}{x - 3}$.

Solution:

$$\frac{x^2 - 10x + 21}{x - 3} = \frac{(x - 7)(x - 3)}{x - 3} = x - 7 \qquad \text{if } x \neq 3$$

so

$$\lim_{x \to 3} \frac{x^2 - 10x + 21}{x - 3} = \lim_{x \to 3} (x - 7) = -4$$

Figure 1.5–3

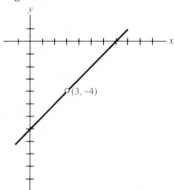

$(3, -4)$

In the previous example, if the function were defined to be anything but -4 where $x = 3$, we would say that the function was <u>discontinuous</u> at the point $x = 3$.

A function with no such breaks (discontinuities) in its graph is called a <u>continuous function</u>. Roughly speaking, a function is continuous if its graph can be drawn without taking your pencil from the paper. Although continuous functions are nice to work with, discontinuous functions play an important part in mathematical modeling.

1.5.6   Example: *The Postage Stamp Function*

Let $f(x)$ be the price of mailing a letter weighing x ounces. The postal rates in 1974 were such that

If $x = 1$ or x is less than 1, then $f(x) = 10$.
If $x = 2$ or x is between 1 and 2, then $f(x) = 20$.
If $x = 3$ or x is between 2 and 3, then $f(x) = 30$.
    etc.

Figure 1.5–4

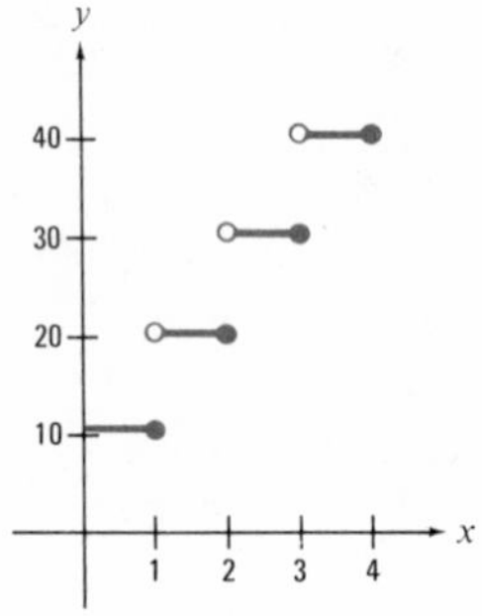

This can be written more precisely as

$$f(x) = 10 \qquad \text{if } 0 < x \leq 1$$
$$f(x) = 20 \qquad \text{if } 1 < x \leq 2$$
$$f(x) = 30 \qquad \text{if } 2 < x \leq 3 \text{ etc.}$$

The graph of $f(x)$ is Figure 1.5–4. (Notice that the scale is different on the two axes.)

A filled circle indicates that the point is included; an empty circle indicates that a point is missing even though every other point in the line is there up to that point. No matter by how little your letter exceeded 1 ounce, you would have been charged a full 20 cents for it if it weighed between 1 and 2 ounces!

Let us investigate the continuity, or lack of it, of the function we have described above. The postage stamp function is discontinuous at all points $x = 1, 2, 3, 4$, etc., since we must "lift the pencil" at these points when graphing this function.

Notice that this function *is* defined at the points $x = 1, 2, 3$, etc. It is not correct to say that lack of continuity means a gap or absence of a point (although such a situation does imply a discontinuity). For example, it is not true that either $\lim_{x \to 1} f(x) = 10$ or $\lim_{x \to 1} f(x) = 20$, even though $f(1) = 10$.

This is precisely the trouble! If we were to consider x approaching 1 only from the left or only from the right, then 10 and 20 would be acceptable limits. But in this course, we do not; hence we will say that there is no limit.

In order for $f(x)$ to be continuous at $x = 1$, three criteria have to be satisfied:

1. $f(1)$ must be defined.
2. $\lim\limits_{x \to 1} f(x)$ must exist and be unambiguous.
3. $\lim\limits_{x \to 1} f(x) = f(1)$.

Of these, only criterion 1 is satisfied by the postage stamp function.

Discontinuities tend to be trouble spots in mathematical analysis. It is, therefore, a good idea to be aware of their existence. Sometimes we approximate discontinuous functions by continuous functions because the latter are easier to work with.

1.5.7   Example in Psychology

Discontinuous functions appear in many fields. For example, a student encountering a new concept (especially in mathematics) may study hard and find that progress is steady but discouragingly slow. Then a teacher or a friend says something that "clicks," or a certain sentence in the text suddenly takes on meaning, leading the student to exclaim, "Aha, I see it now!" Psychologists call this the "Aha" or "Eureka" experience. After it occurs, learning progress proceeds at a much higher rate. We can graph this phenomenon as follows, where the discontinuity in the graph represents the "Aha" experience.

Figure 1.5–5

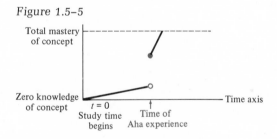

The expression $\lim\limits_{x \to \infty} f(x)$ [read "limit of $f(x)$ as x goes to infinity"] means the limit of $f(x)$ as x grows larger and larger without any finite bound. Even though infinity (∞) is not a number (at least not in this course), this expression often has a clear meaning.

1.5.8   Example

Find $\lim\limits_{x \to \infty} \dfrac{1}{x}$.

Solution:

Trying ever larger numbers ($x = 100$, $x = 1{,}000$, $x = 10{,}000$) should make it clear that

$$\frac{1}{100}, \frac{1}{1,000}, \frac{1}{10,000} \ldots$$

is getting nearer and nearer to zero. Thus we say that

$$\lim_{x \to \infty} \frac{1}{x} = 0$$

The key idea for finding limits as x approaches infinity is to look for the reciprocals of expressions that grow large as x → ∞. Such reciprocals approach zero (by reasoning similar to that in Example 1.5.8) and can be disregarded.

1.5.9   Example

Find

(a) $\displaystyle\lim_{x \to \infty} \frac{1}{x^2 + 1}$

(b) $\displaystyle\lim_{x \to \infty} \left(\frac{2}{x + x^2} + 7 \right)$

Solution:

(a) As x gets large, so does $x^2 + 1$. Hence its reciprocal, $1/(x^2 + 1)$, approaches zero.

(b) As x gets large, so does $x + x^2$, so its reciprocal approaches zero. Thus

$$\lim_{x \to \infty} \left(\frac{2}{x + x^2} + 7 \right) = 0 + 7 = 7$$

Sometimes it is helpful to remember some high school algebra while looking for reciprocals.

1.5.10   Example

Find

(a) $\displaystyle\lim_{x \to \infty} 2^{-x}$

(b) $\displaystyle\lim_{x \to \infty} \frac{x + 1}{x}$

Solution:

(a) $2^{-x} = 1/2^x$, and as x gets large, so does 2^x. Thus

$$\lim_{x \to \infty} 2^{-x} = 0$$

(b) $\displaystyle\lim_{x \to \infty} \left(\frac{x + 1}{x} \right) = \lim_{x \to \infty} \left(1 + \frac{1}{x} \right) = 1 + 0 = 1$

Sometimes the value of a function gets large without bound as x gets large. We then say $\lim\limits_{x \to \infty} f(x) = \infty$.

1.5.11   Example

Find

(a) $\lim\limits_{x \to \infty} 5x$

(b) $\lim\limits_{x \to \infty} \dfrac{x^2 - x}{x}$

Solution:

(a) $\lim\limits_{x \to \infty} 5x = \infty$ because as x gets large without bound, so does 5x.

(b) $\lim\limits_{x \to \infty} \dfrac{x^2 - x}{x} = \lim (x - 1) = \infty$, because as x gets large without bound, so does x − 1.

1.5.12   Example

We mention here an important limit that we will use in Chapter 3, although showing that it exists is far beyond the scope of this course. Consider

$$\lim_{x \to \infty} \left(1 + \frac{1}{x}\right)^x$$

It is not obvious that this limit exists, but the following calculations might help you believe that as x grows large, the expression $(1 + 1/x)^x$ approaches a number between 2 and 3. (The symbol $\approx$ below means "equals approximately.")

$$x = 1: \quad \left(1 + \frac{1}{1}\right)^1 \quad = 2$$

$$x = 2: \quad \left(1 + \frac{1}{2}\right)^2 \quad = \left(\frac{3}{2}\right)^2 = \frac{9}{4} = 2\frac{1}{4} = 2.25$$

$$x = 3: \quad \left(1 + \frac{1}{3}\right)^3 \quad = \left(\frac{4}{3}\right)^3 = \frac{64}{27} = 2\frac{10}{27} \approx 2.37$$

$$x = 4: \quad \left(1 + \frac{1}{4}\right)^4 \quad = \left(\frac{5}{4}\right)^4 = \frac{625}{256} = 2\frac{113}{256} \approx 2.44$$

$$x = 10: \quad \left(1 + \frac{1}{10}\right)^{10} \quad = \left(\frac{11}{10}\right)^{10} \approx 2.59$$

$$x = 100: \quad \left(1 + \frac{1}{100}\right)^{100} = \left(\frac{101}{100}\right)^{100} \approx 2.70$$

In fact, this limit does exist and has been calculated to many decimal places; to five places it is 2.71828. . . . Its usual name is "e." By definition

$$e = \lim_{x \to \infty} \left(1 + \frac{1}{x}\right)^x$$

The number "e" is another irrational number, like "π," that is useful in many fields and has a special name of its own. It does not look like a number, but it is!

In this book and many others using calculus the expression Δx (read "delta x" or "the change in x") is often used to indicate a small number. For example, Δx appears in limits where Δx goes to zero. It should be emphasized that Δx is one variable; we shall not use Δ by itself.

1.5.13   Example

(a) $\lim_{\Delta x \to 0} (3 + \Delta x) = 3 + 0 = 3$

(b) $\lim_{\Delta x \to 0} (3x + 2\ \Delta x) = 3x + 0 = 3x$

(c) $\lim_{\Delta x \to 0} [3x\ \Delta x - 4x^2 + 7(\Delta x)^2] = 0 - 4x^2 + 0 = -4x^2$

(d) $\lim_{\Delta x \to 0} \dfrac{3x\ \Delta x + (\Delta x)^2}{\Delta x} = \lim_{\Delta x \to 0} (3x + \Delta x) = 3x$

Notice that a function $f(x)$ is continuous at a point x_1 if and only if

1. $f(x_1)$ is defined.
2. $\lim_{\Delta x \to 0} f(x_1 + \Delta x)$ exists and is unambiguous.
3. $\lim_{\Delta x \to 0} f(x_1 + \Delta x) = f(x_1)$.

The deep concept of "limits" merits a study of several weeks in some courses for math majors. We have skimmed it briefly here since that is enough for the purposes of this course. If you want to read more about limits, turn to Appendix I.

SUMMARY

This section was devoted to a brief treatment of limits. Sometimes limits can be evaluated by substitution, but not always; often it is necessary to do some algebraic manipulation first, such as factoring and division. Sometimes the limit of a function does not exist at all. Occasionally we will speak of a variable "going to infinity," by which we merely mean that it gets large without bound — in this course ∞ is not a number.

Although we have only scratched the surface of the subject of limits, we have presented them in several different contexts to give you a "feel" for this concept. Such an intuitive grasp is needed for you to appreciate what calculus is about.

EXERCISES   1.5. A

1. (a) Find $\lim\limits_{x \to 5} \dfrac{x^2 - 25}{x - 5}$ using the method of this section.

HC   (b) Verify this limit by completing the following table; it is possible to give completely accurate answers.

x	5.1	5.01	5.001	5.0001
$\dfrac{x^2 - 25}{x - 5}$				

2. $\lim\limits_{x \to -3} \dfrac{x^2 - 9}{x + 3} =$

3. $\lim\limits_{x \to 1} \dfrac{x^2 - 1}{x - 1} =$

4. $\lim\limits_{x \to 3} \dfrac{x^2 + 4x - 21}{x - 3} =$

5. $\lim\limits_{x \to -1} \dfrac{x^2 + 5x + 4}{x + 1} =$

HC 6. (a) Complete the following table; make your answers accurate to five decimal places.

x	80	500	1,800	8,000
$\dfrac{x}{3x + 1}$				

(b) Use part (a) to guess $\lim\limits_{x \to \infty} \dfrac{x}{3x + 1}$.

7. $\lim\limits_{x \to \infty} \dfrac{1}{2x + 5} =$

8. $\lim\limits_{x \to \infty} \dfrac{2x - 7}{x} =$

9. $\lim\limits_{x \to \infty} 3^{-x} =$

10. $\lim\limits_{x \to \infty} \dfrac{1}{x^3 + 9} =$

11. $\lim\limits_{x \to \infty} \dfrac{x^2 + 3}{x} =$

12. $\lim\limits_{x \to \infty} \dfrac{4x^2 - 3x + 5}{x^2} =$

13. $\lim\limits_{x \to \infty} \dfrac{4x^3 - 3x + 5}{x^2} =$

14. $\lim\limits_{\Delta x \to 0} (4 + 3\,\Delta x) =$

15. $\lim\limits_{\Delta x \to 0} \dfrac{3x\,\Delta x + 2(\Delta x)^2}{\Delta x} =$

16. $\lim\limits_{\Delta x \to 0} \dfrac{4(\Delta x)^3 + 3x(\Delta x)^2 - 2x^2\,\Delta x}{\Delta x} =$

EXERCISES   1.5. B

1. (a) Find $\lim\limits_{x \to -2} \dfrac{x^2 - 4}{x + 2}$ using the method of this section.

HC   (b) Verify this limit by completing the following table:

x	−2.1	−2.01	−2.001	−2.0001
$\dfrac{x^2 - 4}{x + 2}$				

2. $\lim\limits_{x \to 2} \dfrac{x^2 - 4}{x - 2} =$

3. $\lim\limits_{x \to 4} \dfrac{x^2 - 16}{x - 4} =$

4. $\lim\limits_{x \to 3} \dfrac{5x^2 - 17x + 6}{x - 3} =$

5. $\lim\limits_{x \to -2} \dfrac{3x^2 + 9x + 6}{x + 2} =$

HC 6. (a) Complete the following table with answers accurate to five decimal places.

x	60	400	3,000	7,000
$\dfrac{x}{2x + 3}$				

(b) Use part (a) to guess $\lim\limits_{x \to \infty} \dfrac{x}{2x + 3}$.

7. $\lim\limits_{x \to \infty} \dfrac{1}{3x + 4} =$

12. $\lim\limits_{x \to \infty} \dfrac{2x^2 - 4x + 6}{x^2} =$

8. $\lim\limits_{x \to \infty} \dfrac{4x + 3}{x} =$

13. $\lim\limits_{x \to \infty} \dfrac{2x^3 - 4x + 6}{x^2} =$

9. $\lim\limits_{x \to \infty} 4^{-x} =$

14. $\lim\limits_{\Delta x \to 0} (3 + 4\,\Delta x) =$

10. $\lim\limits_{x \to \infty} \dfrac{1}{x^2 + 8} =$

15. $\lim\limits_{\Delta x \to 0} \dfrac{2x\,\Delta x + 4(\Delta x)^2}{\Delta x} =$

11. $\lim\limits_{x \to \infty} \dfrac{x^3 + 2}{x} =$

16. $\lim\limits_{\Delta x \to 0} \dfrac{5(\Delta x)^3 - x(\Delta x)^2 + 4x^2\,\Delta x}{\Delta x} =$

EXERCISES   1.5. C

1. $\lim\limits_{x \to \infty} e^{-x} =$

2. $\lim\limits_{x \to \infty} 2^{-x^2 + 3x} =$

3. $\lim\limits_{x \to \infty} \dfrac{x}{x + 1} =$

4. $\lim\limits_{x \to \infty} \dfrac{3x^2 + 3x - 4}{6x^2 - 2x + 5} =$

5. $\lim\limits_{x \to \infty} \dfrac{3x^2 + 3x - 4}{x^3 + 3x^2} =$

6. Suppose that we supply our Aha experience graph of Example 1.5.7 with percentage scores as in the figure here.

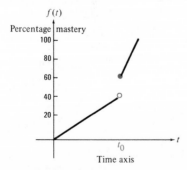

(a) If $f(t)$ denotes the percentage of the student's mastery at time t, what is $\lim_{t \to t_0^-} f(t)$, that is, the limit of $f(t)$ as t approaches t_0 from the left?

(b) What is $\lim_{t \to t_0^+} f(t)$, the limit of $f(t)$ as t approaches t_0 from the right?

(c) Why is $f(t)$ discontinuous at $t = t_0$? Base your discussion on that following Example 1.5.6.

HC 7. Find your own approximation of $e = \lim_{x \to \infty} (1 + 1/x)^x$ by filling in the following table (accurate to four decimal places):

x	17	96	527	1,200	2,000
$\left(1 + \dfrac{1}{x}\right)^x$					

ANSWERS 1.5. A

1. (a) 10 (b)

x	5.1	5.01	5.001	5.0001
$\dfrac{x^2 - 25}{x - 5}$	10.1	10.01	10.001	10.0001

2. -6 4. 10
3. 2   5. 3

6. (a)

x	80	500	1,800	8,000
$\dfrac{x}{3x + 1}$	0.33195	0.33311	0.33327	0.33332

(b) $\frac{1}{3}$

7. 0   12. 4
8. 2   13. ∞
9. 0   14. 4
10. 0   15. $3x$
11. ∞ 16. $-2x^2$

ANSWERS 1.5. C

1. 0

2. 0

3. 1

4. $\frac{1}{2}$

5. 0

6. (a) $\lim_{t \to t_0^-} f(t) = 40$ (b) $\lim_{t \to t_0^+} f(t) = 60$

    (c) $\lim_{t \to t_0^-} f(t) \neq \lim_{t \to t_0^+} f(t)$

7.

x	17	96	527	1,200	2,000
$\left(1 + \dfrac{1}{x}\right)^x$	2.6424	2.7043	2.7161	2.7172	2.7183

1.6 Marginal and Average Rates of Change

Thomas Malthus (1766–1834) observed long ago that the world's population is growing geometrically while our food supplies grow only arithmetically. This means that population grows much, much faster than food supply. The problem, he pointed out, was not the total change in the population or food supply, but their relative *rates of change*. Since functions are often used to describe the size of populations, food supply, and many other real-world quantities, we now turn to a central idea of differential calculus, their rates of change.

The expressions "change in x" and "change in y" appear so often in calculus that they are given a special name and notation. This section provides practice with this notation and shows one of its uses, representing and calculating the average rate of change.

1.6.1   Definition

$\Delta x = x_2 - x_1$. It is called the "change in x" or "delta x" or "the increment in x from x_1 to x_2."

If $y = f(x)$, then $\Delta y = y_2 - y_1 = f(x_2) - f(x_1) = f(x_1 + \Delta x) - f(x_1)$. It is called the "change in y" or "delta y" or "the increment in y."

In general, the expression for Δy depends on the particular function as well as the values of x_1 and x_2.

In this book Δ does not appear by itself; it is only used to show that some variable has changed.

1.6.2   Example

Evaluate Δx and Δy if $x_1 = 4$, $x_2 = 1$, and $f(x) =$
(a) x
(b) 3x
(c) 1 − x
(d) x^2
(e) $x^2 - 5x$

Solution:

In each case $\Delta x = x_2 - x_1 = 1 - 4 = -3$. (Yes, increments can be negative.) We note that $x_2 = x_1 + \Delta x$.
(a) $\Delta y = f(x_1 + \Delta x) - f(x_1) = f(1) - f(4) = 1 - 4 = -3$
(b) $\Delta y = f(x_1 + \Delta x) - f(x_1) = (3 \cdot 1) - (3 \cdot 4) = -9$
(c) $\Delta y = f(x_1 + \Delta x) - f(x_1) = (1 - 1) - (1 - 4) = 3$
(d) $\Delta y = f(x_1 + \Delta x) - f(x_1) = 1^2 - 4^2 = -15$
(e) $\Delta y = f(x_1 + \Delta x) - f(x_1) = [(1)^2 - 5 \cdot 1] - [4^2 - 5 \cdot 4] = 0$

There will be many uses for increments and the delta notation in the course of this text, but the first, as you may have guessed, will be for calculating the slope between two points, since

$$\frac{y_2 - y_1}{x_2 - x_1} = \frac{\Delta y}{\Delta x}$$

We shall never let Δx be zero in such fractions, since it is impossible to divide by zero. But Δy can be positive, negative, or zero.

Even if the function under consideration does not have a straight-line graph, it is often still useful to study the quantity $\frac{\Delta y}{\Delta x}$, because this tells the average change in y per unit change in x. We will call it a "difference quotient" because it is a quotient of differences.

1.6.3   Definition

The average rate of change (or marginal change) of $y = f(x)$ as x changes from x_1 to x_2 is the difference quotient

$$\frac{\Delta y}{\Delta x} = \frac{y_2 - y_1}{x_2 - x_1} = \frac{f(x_2) - f(x_1)}{x_2 - x_1} = \frac{f(x_1 + \Delta x) - f(x_1)}{\Delta x}$$

1.6.4   Example

Find the marginal change as x changes from 4 to 1 in each of the functions of Example 1.6.2.

Solution:
(a) $\frac{-3}{-3} = 1$
(b) $\frac{-9}{-3} = 3$
(c) $\frac{3}{-3} = -1$
(d) $\frac{-15}{-3} = 5$
(e) $\frac{0}{-3} = 0$

Our first application of the average rate of change is in a physical context since the ideas are most precise and understandable there. The average rate of change of distance with respect to time is just the average velocity. The following example was one of the first that motivated differential calculus.

1.6.5   Example

Suppose that we drop an object from a high place. The distance in feet that it will fall in t seconds is given by

$$f(t) = 16t^2$$

What is its average rate of change of distance (that is, its average velocity)
(a) During the first second?
(b) During the second second (that is, between times $t = 1$ and $t = 2$)?
(c) During the third second (that is, between times $t = 2$ and $t = 3$)?
(d) Over the first two seconds?
(e) Between times $t = 3$ and $t = 5$?
(f) Between times $t = t_1$ and $t = t_1 + \Delta t$?

Figure 1.6–1

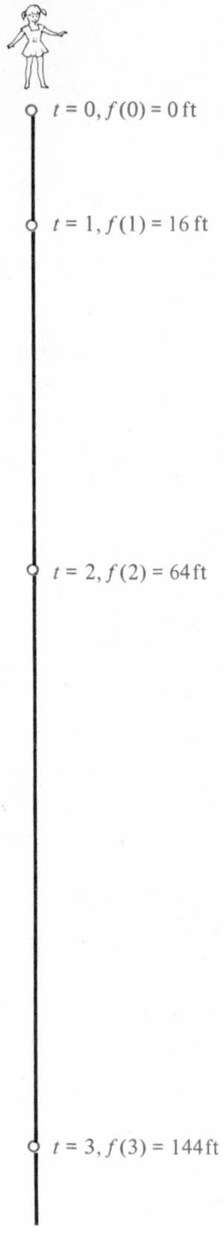

$t = 0, f(0) = 0\,\text{ft}$

$t = 1, f(1) = 16\,\text{ft}$

$t = 2, f(2) = 64\,\text{ft}$

$t = 3, f(3) = 144\,\text{ft}$

Solution:

(a) The object will fall 16 feet in the first second, so its average velocity during that second will be 16 feet per second. (This is often written 16 ft/sec.)

(b) At $t = 1, f(t) = 16$, and at $t = 2, f(t) = 64$. So the distance traveled in the second second is $64 - 16 = 48$ feet and the average rate of change in the distance (the average velocity) is

$$\frac{f(2) - f(1)}{2 - 1} = \frac{64 - 16}{1} = 48 \text{ ft/sec}$$

(c) $\dfrac{f(3) - f(2)}{3 - 2} = \dfrac{144 - 64}{1} = 80 \text{ ft/sec}$

We notice that the object's velocity increases as it falls, as we would expect.

(d) $\dfrac{f(2) - f(0)}{2 - 0} = \dfrac{64 - 0}{2 - 0} = 32 \text{ ft/sec}$

Notice that the average velocity over the first 2 seconds is greater than that over the first second, but less than that over the second second.

(e) $\dfrac{f(5) - f(3)}{5 - 3} = \dfrac{400 - 144}{2} = \dfrac{256}{2} = 128$

(f) $\dfrac{f(t_1 + \Delta t) - f(t_1)}{\Delta t} = \dfrac{16(t_1 + \Delta t)^2 - 16t_1^2}{\Delta t}$

[This is obtained by substituting first $(t_1 + \Delta t)$ and then t_1 wherever you see "t" in the expression "$f(t) = 16t^2$."]

$$= \frac{16t_1^2 + 32t_1\,\Delta t + 16(\Delta t)^2 - 16t_1^2}{\Delta t} \qquad \text{squaring and multiplying}$$

$$= \frac{32t_1\,\Delta t + 16(\Delta t)^2}{\Delta t} \qquad \text{subtracting}$$

$$= \frac{(\Delta t)(32t_1 + 16\,\Delta t)}{\Delta t} \qquad \text{factoring}$$

$$= 32t_1 + 16\,\Delta t \qquad \text{dividing}$$

The last part of Example 1.6.5 plays a crucial role in calculus. It says that in a "small" interval around time $t = t_1$, the average velocity will be "about $32t_1$." To be precise, if the interval is Δt long, the average velocity over the interval is $32t_1 + 16\,\Delta t$. We might say that the velocity "at" time $t = t_1$ "is" $32t_1$. (Notice that the velocity increases as t_1 increases, as we would expect.)

We shall return to similar ideas many times in Chapter 2 as we study the derivative. But now we shall do a similar problem in a financial situation. Our function will be the profit function; we shall consider the average rate of change or the marginal change in the profit.

1.6.6   Example

If the profit from selling x items is given by the function $P(x) = 100x - x^2 - 500$, find the <u>marginal change in profit</u> when sales change from
(a) $x_1 = 30$ to $x_2 = 60$
(b) $x_1 = 30$ to $x_2 = 35$

Solution:

(a) $\Delta x = 60 - 30 = 30$ and

$$\begin{aligned}
\Delta P = P(60) - P(30) &= [100(60) - 60^2 - 500] \\
&\quad - [100(30) - 30^2 - 500] \\
&= 6{,}000 - 3{,}600 - 500 - 3{,}000 + 900 + 500 \\
&= 300
\end{aligned}$$

so in this case

$$\frac{\Delta P}{\Delta x} = \frac{300}{30} = 10$$

(b) $\Delta x = 35 - 30 = 5$ and

$$\begin{aligned}
\Delta P = P(35) - P(30) &= [100(35) - 35^2 - 500] \\
&\quad - [100(30) - 30^2 - 500] \\
&= 3{,}500 - 1{,}225 - 500 - 3{,}000 + 900 + 500 \\
&= 175
\end{aligned}$$

so in this case

$$\frac{\Delta P}{\Delta x} = \frac{175}{5} = 35$$

The concept of marginal change is particularly useful when Δx is "small" and Example 1.6.6 illustrates why this is so. The relatively small increase in the number of items sold ($35 - 30 = 5$) in the last part contributed a more significant increase in the profit (175) than one might have suspected from the first part, where the increase in sales ($60 - 30 = 30$) was six times larger, but the increase in profits (300) was less than twice as large.

1.6.7   Example

If $P(x) = 100x - x^2 - 500$, find an expression for $\dfrac{\Delta P}{\Delta x}$ in terms of x_1 and Δx.
(See Example 1.6.6.)

Solution:

First we get an expression for ΔP.

$$\Delta P = P(x_1 + \Delta x) - P(x_1)$$

To get an expression for the first term, substitute "$x_1 + \Delta x$" wherever you see "x" in the expression "$P(x) = 100x - x^2 - 500$."

$$\Delta P = 100(x_1 + \Delta x) - (x_1 + \Delta x)^2 - 500 - (100x_1 - x_1^2 - 500)$$

Remembering that $(a + b)^2 = a^2 + 2ab + b^2$ and letting $x_1 = a$ and $\Delta x = b$, we get

$$\begin{aligned}\Delta P &= \cancel{100x_1} + 100\,\Delta x - [x_1^2 + 2x_1\,\Delta x + (\Delta x)^2] - \cancel{500} - \cancel{100x_1} \\ &\quad + x_1^2 + \cancel{500} \\ &= 100\,\Delta x - \cancel{x_1^2} - 2x_1\,\Delta x - (\Delta x)^2 + \cancel{x_1^2} \\ &= 100\,\Delta x - 2x_1\,\Delta x - (\Delta x)^2\end{aligned}$$

Now dividing by Δx, we get

$$\begin{aligned}\frac{\Delta P}{\Delta x} &= \frac{100\,\Delta x - 2x_1\,\Delta x - (\Delta x)^2}{\Delta x} = \frac{\Delta x(100 - 2x_1 - \Delta x)}{\Delta x} \\ &= 100 - 2x_1 - \Delta x\end{aligned}$$

Looking ahead, we notice that when Δx is "very small," $\dfrac{\Delta P}{\Delta x}$ is very near to $100 - 2x_1$. Economists call the expression $100 - 2x_1$ the <u>marginal profit</u> and mathematicians call it the <u>derivative of the profit</u>. We shall discuss both these concepts in more detail in Chapter 2.

1.6.8   Example in Biology or Sociology

Suppose that $P(t)$ is the size of a certain (plant, animal, or human) population at time t. As is common, we let t_1 and t_2 denote two particular times. Then

$$\frac{P(t_2) - P(t_1)}{t_2 - t_1}$$

is the <u>average rate of change</u> in the size of the population between time t_1 and time t_2.

SUMMARY

This section dealt with the fundamental concepts and notation which are the springboards to calculus—average rates of change and the increments of dependent and independent variables. The notation used to describe these ideas—the delta notation—is often confusing to beginners, but it is worth getting accustomed to because of its widespread use in books on many other subjects. You get some hint of these applications in our examples about velocity, marginal profit, and population change. In order to know calculus, you must first learn to speak the language of calculus. And now calculus is just around the bend—in the next section!

EXERCISES 1.6. A

1. Find Δx and Δy if $x_1 = 3$, $x_2 = 6$, and $f(x) =$
   (a) $x/2$
   (b) $x + 2$
   (c) $x^2 + x$
   (Hint: See Example 1.6.2.)
   (d) $x^2 + x + 2$
   (e) $4x^2 + 3x + 2$

2. Find Δx and Δy if $y = f(x) = x^2 + x + 2$ and
   (a) $x_1 = 1$ and $x_2 = 2$
   (b) $x_1 = 2$ and $x_2 = 1$
   (c) $x_1 = -1$ and $x_2 = -1$

3. Find the average rate of change, $\dfrac{\Delta y}{\Delta x}$, when x goes from $x_1 = 1$ to $x_2 = 3$ for each function $f(x)$ given in the left column of the table. Part (e) is completed as an example.

$f(x)$	Δx	$y_2 = f(x_2)$	$y_1 = f(x_1)$	$\Delta y = y_2 - y_1$	$\dfrac{\Delta y}{\Delta x}$
(a) $2x$					
(b) $x - 4$					
(c) x^2					
(d) $x^2 - 5x$					
(e) $2x^2 + 1$	2	19	3	16	16/2

4. Find an expression for Δy and $\dfrac{\Delta y}{\Delta x}$ in terms of Δx and x_1 for each $f(x)$ in the left column of the table. Part (f) is completed as an example.

$f(x)$	$f(x_1 + \Delta x)$	$\Delta y = f(x_1 + \Delta x) - f(x_1)$	$\dfrac{\Delta y}{\Delta x}$
(a) $x/2$			
(b) $x + 2$			
(c) $x^2 + x$			
(d) $x^2 + x + 2$			
(e) $4x^2 + 3x + 2$			
(f) $3x^2$	$3(x_1 + \Delta x)^2$	$6x_1 \Delta x + 3(\Delta x)^2$	$6x_1 + 3 \Delta x$

5. If $P(x) = 7x - x^2$ is the profit when x thousands of some item is sold, what is the average rate of change in the profit as x changes from $x_1 = 1$ to $x_2 = 3$?

EXERCISES 1.6. B

1. Find Δx and Δy if $x_1 = 1$, $x_2 = 5$, and $f(x) =$
   (a) $x/3$
   (b) $x + 3$

(c) $2x^2 - x$ (e) x^3

(d) $2x^2 - x - 3$

(Hint: See Example 1.6.2.)

2.   Find Δx and Δy if $y = f(x) = 2x^2 - x - 3$ and

(a) $x_1 = 2$ and $x_2 = 3$ (c) $x_1 = 3$ and $x_2 = 3$

(b) $x_1 = 3$ and $x_2 = 2$

3.   Find the average rate of change, $\dfrac{\Delta y}{\Delta x}$, when x goes from $x_1 = 0$ to $x_2 = 3$ for

each function $f(x)$ given in the left column of the table. Part (e) is completed as an example.

$f(x)$	Δx	$y_2 = f(x_2)$	$y_1 = f(x_1)$	$\Delta y = y_2 - y_1$	$\dfrac{\Delta y}{\Delta x}$
(a) $3x$					
(b) $x - 2$					
(c) x^3					
(d) $x^2 + 2x$					
(e) $3x^2 - 1$	2	26	2	24	12

4.   Find an expression for Δy and $\dfrac{\Delta y}{\Delta x}$ in terms of Δx and x_1 for each $f(x)$ in the

left column of the table. Part (f) is completed as an example.

$f(x)$	$f(x_1 + \Delta x)$	$\Delta y = f(x_1 + \Delta x) - f(x_1)$	$\dfrac{\Delta y}{\Delta x}$
(a) $x/3$			
(b) $x + 3$			
(c) $2x^2 - x$			
(d) $2x^2 - x - 3$			
(e) x^3			
(f) $2x^2$	$2(x_1 + \Delta x)^2$	$4x_1 \Delta x + 2(\Delta x)^2$	$4x_1 + 2 \Delta x$

5.   If $P(x) = 8x - x^2$ is the profit when x thousands of some item are sold, what is the average rate of change in the profit as x changes from $x_1 = 2$ to $x_2 = 4$?

EXERCISES 1.6. C

HC 1.   Approximate the velocity of a falling object "at" the end of the first second after it is dropped by computing the following (see Example 1.6.5):

t	2	1.5	1.1	1.01	1.001
$\dfrac{16t^2 - 16}{t - 1}$					

2. (a) If the profit from selling a certain commodity is given by $P(q) = -150 + 500q^2$, where q is the quantity sold, find an expression for the average rate of change of the profit, $\dfrac{\Delta P}{\Delta q}$, in terms of q and Δq.

(b) Explain what happens when $\Delta q \to 0$.

(c) Find the average change in profit $\dfrac{\Delta P}{\Delta q}$ when $\Delta q = 1$ and $q_1 = 1$.

(d) Find the average change in profit when $\Delta q = 1$ and $q_1 = 2$.

(e) Find the average change in profit when $\Delta q = 1$ and $q_1 = 3$.

(f) Find the average change in profit when $\Delta q = 4$ and $q_1 = 1$.

3. If $P(t) = (t + 2)^2$ gives the size, in hundreds, of a certain rabbit population at time t years, find the average rate of change in the size of this population between years:

(a) $t = 0$ and $t = 1$ (c) $t = 1$ and $t = 3$

(b) $t = 1$ and $t = 2$ (d) $t = 2$ and $t = 3$

4. A human being's rate of respiration (in breaths/minute) is a function of the partial pressure p of carbon dioxide in the lungs (in torr) given by $f(p) = -10.39 + 0.59p$. Find the average rate of change of the rate of respiration when the partial pressure of carbon dioxide is increased from 40 torr to $(40 + \Delta p)$ torr in breaths/min/torr. (One torr equals $\frac{1}{760}$ of an atmosphere.)

ANSWERS 1.6. A

1. $\Delta x = 3$ in all parts. Δy, for each part, is   (a) 1.5   (b) 3   (c) 30   (d) 30   (e) 117

2. (a) $\Delta x = 1$, $\Delta y = 4$ (b) $\Delta x = -1$, $\Delta y = -4$ (c) $\Delta x = 0$, $\Delta y = 0$

3. (a) 2   (b) 1   (c) 4   (d) -1

4. (a) $\Delta y = \dfrac{\Delta x}{2}$; $\dfrac{\Delta y}{\Delta x} = \dfrac{1}{2}$ (b) $\Delta y = \Delta x$; $\dfrac{\Delta y}{\Delta x} = 1$

(c) $\Delta y = 2x_1 \Delta x + (\Delta x)^2 + \Delta x$; $\dfrac{\Delta y}{\Delta x} = 2x_1 + \Delta x + 1$ (d) Same as (c)

(e) $\Delta y = 8x_1 \Delta x + 4(\Delta x)^2 + 3 \Delta x$; $\dfrac{\Delta y}{\Delta x} = 8x_1 + 4 \Delta x + 3$

5. 3

ANSWERS 1.6. C

1.

t	2	1.5	1.1	1.01	1.001
$\dfrac{16t^2 - 16}{t - 1}$	48	40	33.6	32.16	32.016

2. (a) $\dfrac{\Delta P}{\Delta q} = 1{,}000q + 500 \Delta q$ (b) $\lim\limits_{\Delta q \to 0} (1{,}000q + 500 \Delta q) = 1{,}000q$ (c) 1,500

(d) 2,500   (e) 3,500   (f) 3,000

3. (a) 500 per year   (b) 700 per year   (c) 800 per year   (d) 900 per year

4. 0.59 breath/min/torr.

SAMPLE QUIZ   **Sections 1.4, 1.5, and 1.6**

1. (10 pts) State five ways of determining a particular linear function.
2. (20 pts) If it costs $100 to produce 100 items and $150 to produce 300 items, write the linear cost-output equation for these data. How much does it cost to produce 600 items?
3. (30 pts) (a) $\lim\limits_{x \to 5} \dfrac{x^2 - 25}{x - 5} =$

    (b) $\lim\limits_{x \to \infty} \dfrac{1}{3^x} =$

    (c) $\lim\limits_{\Delta x \to 0} \dfrac{3x\,\Delta x + 5(\Delta x)^2}{\Delta x} =$
4. (20 pts) A freely falling object falls $16t^2$ feet in t seconds after it is dropped. What is its average velocity between time $t_1 = 1$ and $t_2 = 4$?
5. (20 pts) Suppose that the profit when x items are sold is given by $P(x) = 5x - 0.01x^2 - 20$. Find the average rate of change in the profit as x changes from $x_1 = 10$ to $x_2 = 20$.

The answers are at the back of the book.

SAMPLE TEST   **Chapter 1**

Each of the following is worth 10 points, except Question 7, which is worth 20 points.

1. Let $f(x) = 2x^2 - 1$
    (a) $f(0) =$ (c) $\lim\limits_{x \to 3} f(x) =$
    (b) $f(2) =$
2. Write "Yes" on each of the following graphs that describes a function such that $y = f(x)$ and "No" on each that does not.

(a)

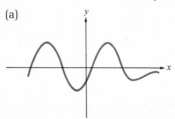

(b)

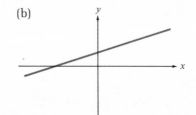

(c)

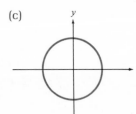

3. Suppose that $C = 5x + 200$ describes the cost of producing x objects.
    (a) How much does it cost to produce 50 items?
    (b) What is the marginal cost?
    (c) What is the fixed cost?
    (d) If this function were graphed, what would be the slope of the line? What would be the y-intercept?
4. Write the equation of the straight line parallel to $y = -2x$ and going through the point (0, 2).
5. Suppose that it costs $800 to make 6 items and $1,000 to make 10 items. Assuming that the linear cost-output model is valid, write an equation describing the cost of producing x items. How much will it cost to produce 15 items?

6. (a) $\lim\limits_{x \to \infty} \dfrac{1}{5x + 1} =$ (b) $\lim\limits_{\Delta x \to 0} \dfrac{4x\,\Delta x - \Delta x^2}{\Delta x} =$ (c) $\lim\limits_{x \to 1} \dfrac{x^2 + 4x - 5}{x - 1} =$

7. Let $f(x) = 2x^2 - 1$.
   (a and b) Find Δx and Δy as x goes from $x_1 = 2$ to $x_2 = 5$.
   (c) Find Δy in terms of x_1 and Δx if x changes from x_1 to $x_1 + \Delta x$.
   (d) Find the average rate of change as x changes from x_1 to $x_1 + \Delta x$ for arbitrary x_1 and Δx.

8. Describe the equations of lines going through the origin.

9. Suppose that the profit is described by $P(x) = 200 + 3x + 0.01x^2$. What is the average rate of change in the profit as x goes from 50 to 100?

The answers are at the back of the book.

Chapter 2

DERIVATIVES: PART I

2.1 Definition of Derivative

Differential calculus is the mathematical study of small, subtle change. The derivative of a function tells how the dependent variable tends to change "with respect to" the independent variable. In other words, the derivative describes how slight variations in the independent variable affect the dependent variable. For example, marginal cost (the derivative of the cost) describes how changing the number of items produced tends to change the total cost; marginal cost is the rate of change of total cost. Similarly, marginal revenue is the rate of change of total revenue, and marginal profit is the rate of change of total profit. (In this book we shall often use "cost" to mean "total cost"; similar comments apply to revenue and profit.)

If the function is linear (so that its graph is a straight line), then the derivative is exactly the slope of this line. For example, the marginal cost in the linear cost-output model was simply the slope of the line.

Figure 2.1–1

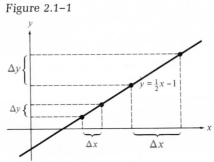

No matter which two points are chosen on this line,
it is always true that $\dfrac{\Delta y}{\Delta x} = \dfrac{1}{2}$

But many functions do not have a constant rate of change. Then "the slope" is not an adequate description of their rates of change. We are forced, instead, to compute the rate of change (called the derivative) at *each point on the curve*. To do this we find the slope of the tangent line, the straight line that skims the curve, at each point. (The dictionary says that a line is tangent to a curve if it touches the curve but does not go through it; we give a more mathematical definition of "tangent line" on page 69.)

Figure 2.1–2

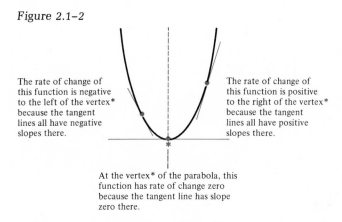

The rate of change of this function is negative to the left of the vertex* because the tangent lines all have negative slopes there.

The rate of change of this function is positive to the right of the vertex* because the tangent lines all have positive slopes there.

At the vertex* of the parabola, this function has rate of change zero because the tangent line has slope zero there.

Intuitively, we might say that the derivative at a point is approximately the rate of change in the function when the change in the independent variable is "very small." The average rate of change *between the points* where $x = x_1$ and $x = x_1 + \Delta x$ has been defined to be

$$\frac{f(x_1 + \Delta x) - f(x_1)}{\Delta x}$$

which, you remember, is the slope of the straight line through these points. We define the derivative to be the limit of this average rate of change as $\Delta x \to 0$.

Figure 2.1–3

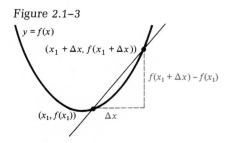

$y = f(x)$

$(x_1 + \Delta x, f(x_1 + \Delta x))$

$f(x_1 + \Delta x) - f(x_1)$

$(x_1, f(x_1))$ Δx

2.1.1   Definition

Let $y = f(x)$. If the following limit exists, it is called the derivative with respect to x of $f(x)$ at $x = x_1$:

$$\lim_{\Delta x \to 0} \frac{f(x_1 + \Delta x) - f(x_1)}{\Delta x}$$

This limit is often denoted $f'(x_1)$.

If the function is continuous, $\Delta x \to 0$ implies that $f(x_1 + \Delta x) \to f(x_1)$. Thus, as $\Delta x \to 0$, the point $(x_1 + \Delta x, f(x_1 + \Delta x))$ gets closer and closer to the point $(x_1, f(x_1))$. You should convince yourself that the derivative is a reasonable definition for the slope of the tangent line to the graph of a curved function.

Figure 2.1–4

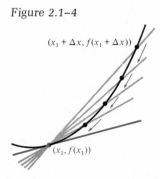

Think of the point $(x_1, f(x_1))$ as being a fixed pivot in Figure 2.1–4, and the point $(x_1 + \Delta x, f(x_1 + \Delta x))$ as being a second point sliding along the graph of the function. As it slides, the straight line that connects the two points also moves. As $\Delta x \to 0$, the point $(x_1 + \Delta x, f(x_1 + \Delta x))$ moves toward the point $(x_1, f(x_1))$, forcing the line determined by these two points to make a smaller and smaller angle with the tangent line. As this happens, the slope of the line through $(x_1, f(x_1))$ and $(x_1 + \Delta x, f(x_1 + \Delta x))$ approaches the slope of the tangent line to the curve at $(x_1, f(x_1))$.

Calculations

It is relatively easy to calculate the slope of the line between two points on a curve; if, as one point moves close to the other, the slope of the line connecting them approaches a limit, then Definition 2.1.1 says that this limit is the derivative of the function at the fixed point (that is, the slope of the line tangent to the curve at the fixed point). We shall now do some calculations. First we verify that the derivative of a straight line is indeed everywhere its slope; then we turn to examples that are more interesting.

2.1.2   **Example**

Find the derivative of $f(x) = mx + b$ at
(a) $x = 3$
(b) $x = x_1$

Solution:

(a) $f'(3) = \lim\limits_{\Delta x \to 0} \dfrac{f(3 + \Delta x) - f(3)}{\Delta x}$ by definition

$= \lim\limits_{\Delta x \to 0} \dfrac{m(3 + \Delta x) + b - (m3 + b)}{\Delta x}$ by substitution

$= \lim\limits_{\Delta x \to 0} \dfrac{\cancel{3m} + m\,\Delta x + \cancel{b} - \cancel{3m} - \cancel{b}}{\Delta x}$ by multiplication

$= \lim\limits_{\Delta x \to 0} \dfrac{m\,\cancel{\Delta x}}{\cancel{\Delta x}}$ by subtraction

$= \lim\limits_{\Delta x \to 0} m$ by dividing

$= m$ by taking the limit

Figure 2.1–5

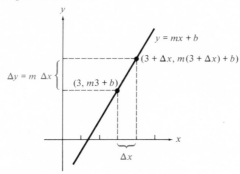

$\Delta y = m\,\Delta x$

$(3, m3 + b)$

$y = mx + b$

$(3 + \Delta x, m(3 + \Delta x) + b)$

Δx

(b) Part (b) is done in exactly the same way as part (a) except that x_1 replaces 3 in the computations; again the answer is m.

Thus the derivative of $f(x) = mx + b$ is indeed everywhere equal to m — as it should be! But the function $f(x) = x^2$ has a different derivative at different points. To find its values at various points, we can merely calculate its value for an arbitrary point x_1 and get a general expression for the derivative $f'(x_1)$, which can then be calculated for any particular x_1. In the next example we shall calculate the derivative first at the point $x_1 = 1$ because you may feel more comfortable with a number (rather than an x_1) while reading your first calculation of a derivative. This is done merely for your comfort; usually one would perform the computation of part (b) directly and then substitute in values, as is done in parts (c), (d), and (e).

2.1.3   **Example**

If $f(x) = x^2$, find
(a) $f'(1)$
(b) $f'(x_1)$

(c) $f'(-1)$
(d) $f'(2)$
(e) $f'(-2)$

Solution:

(a) $f'(1) = \lim\limits_{\Delta x \to 0} \dfrac{f(1 + \Delta x) - f(1)}{\Delta x}$

$\qquad = \lim\limits_{\Delta x \to 0} \dfrac{(1 + \Delta x)^2 - 1^2}{\Delta x}$ by substituting

$\qquad = \lim\limits_{\Delta x \to 0} \dfrac{1 + 2\,\Delta x + (\Delta x)^2 - 1}{\Delta x}$ by squaring

$\qquad = \lim\limits_{\Delta x \to 0} \dfrac{\Delta x(2 + \Delta x)}{\Delta x}$ by factoring

$\qquad = \lim\limits_{\Delta x \to 0} (2 + \Delta x) = 2$ by dividing

(b) $f'(x_1) = \lim\limits_{\Delta x \to 0} \dfrac{f(x_1 + \Delta x) - f(x_1)}{\Delta x} = \lim\limits_{\Delta x \to 0} \dfrac{(x_1 + \Delta x)^2 - x_1^2}{\Delta x}$

$\qquad = \lim\limits_{\Delta x \to 0} \dfrac{x_1^2 + 2x_1\,\Delta x + (\Delta x)^2 - x_1^2}{\Delta x} = \lim\limits_{\Delta x \to 0} \dfrac{\Delta x(2x_1 + \Delta x)}{\Delta x}$

$\qquad = \lim\limits_{\Delta x \to 0} (2x_1 + \Delta x) = 2x_1$

So we conclude that $f'(x_1) = 2x_1$; to find the value of the derivative of the function $f(x) = x^2$ at the point where $x = x_1$, we merely double the value of x_1. Thus the remaining three answers are immediate:

(c) $f'(-1) = 2(-1) = -2$
(d) $f'(2) = 2(2) = 4$
(e) $f'(-2) = 2(-2) = -4$

Figure 2.1-6

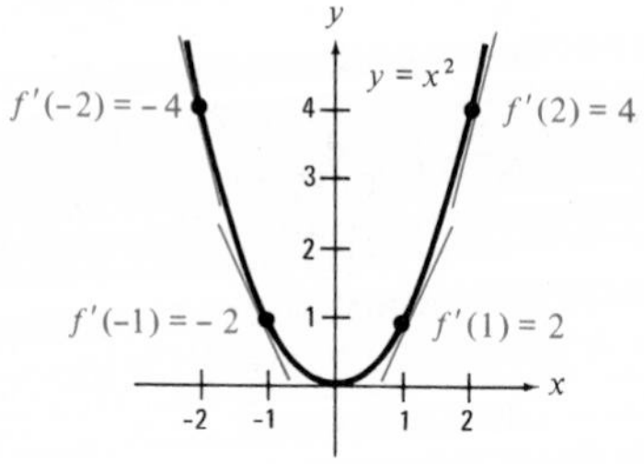

Pause for a moment now to consider the intuitive implications of the fact that the derivative of $f(x) = x^2$ is $f'(x) = 2x$. Look at Figure 2.1–6. Remember that the derivative is the slope of the tangent line to the curve at the given point. The calculations say that if x is positive, the derivative (being twice x) will also be positive; this corroborates the illustration.

Furthermore, as x gets larger, the derivative will get larger, too. On the other hand, if x is negative, the derivative (being 2x) will be negative, also; this indicates the fact that the function is decreasing when x is less than zero. And the only place where the tangent line is horizontal (the only place where the derivative is zero) is at the vertex, where $x = 0$. This last observation is an omen of what is to come; much of the remainder of this chapter will be devoted to calculating and interpreting the x's for which $f'(x) = 0$, that is, those x's for which the tangent line is horizontal.

Before giving more examples, we make two comments. First, if the limit that defines $f'(x)$ (the first derivative) exists at every x_1 in the domain of f, then $f'(x_1)$ defines another function over the same domain as the function $f(x)$. The function $f'(x)$ is also often written as $\dfrac{df}{dx}, \dfrac{dy}{dx}, f'$, and y'. Notice that the word "derivative" can mean either a *number* (the slope of the tangent line at a particular point) or a *function* [the correspondence from each x_1 in the domain of $f(x)$ to the slope of the tangent line to the graph of $f(x)$ at the point $(x_1, f(x_1))$].

Second, notice that we were careful to include the words "if the limit exists" in our definition of derivative. This is a necessary precaution, because for some functions this limit will not exist. (See problem 1 of Exercises 2.1.C.) In this book we shall avoid such difficulties by confining our discussion to "well-behaved" functions. That is, the limit will exist except in very obvious cases.

Equations of Tangent Lines

Derivatives can be used in many ways, but one of the most obvious is as a tool for finding the equations of the tangent lines to various curves. Before we explore the details of how this is done, let us formalize our ideas of what a "tangent line" is. Suppose that $f(x)$ has a derivative at x_1.

2.1.4   Definition

The straight line through the point $(x_1, f(x_1))$ with slope $f'(x_1)$ is called the tangent line to the graph of $y = f(x)$ at $(x_1, f(x_1))$.

You should now be able to convince yourself that this is a reasonable definition of the tangent line; doing so is equivalent to understanding the meaning of the derivative. Perhaps Figure 2.1–6 will help.

Suppose, now, that we are to find the equation of the tangent line to a given function at a given point where $x = x_1$. Definition 2.1.4 says that the slope of this tangent line will be the same as the derivative of the function (a number) at that point. This derivative, $f'(x_1)$, will therefore be the slope, m, of the tangent line. If we know the slope of any straight line and one point on it, we can find the equation of the straight line by using the formula

$$(y - y_1) = m(x - x_1)$$

where (x_1, y_1) is the given point. (See Definition 1.4.1.) Using function notation, this formula is

$$y - f(x_1) = f'(x_1)(x - x_1)$$

because we can find $y_1 = f(x_1)$ by substituting back into the original equation. We shall use this technique in Examples 2.1.5 and 2.1.7.

2.1.5   Example

Write the equations of the tangent lines to $f(x) = x^2$ at the points where
(a) $x = 1$ (b) $x = 2$
Graph these tangent lines.

Solution:

We know from the results of Example 2.1.3 that $f'(1) = 2$ and $f'(2) = 4$.
(a) We can use the point-slope formula for a straight line,

$$y - f(x_1) = m(x - x_1)$$

where $x_1 = 1$, $m = f'(x_1) = 2$, and $y_1 = f(x_1) = (1)^2 = 1$. We get

$$y - 1 = 2(x - 1)$$

or

$$y - 1 = 2x - 2$$

or

$$y = 2x - 1$$

Figure 2.1–7
(a)   (b)

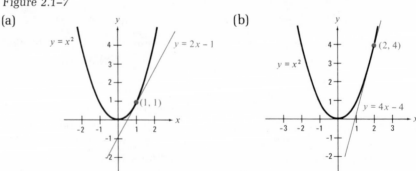

(b) Similarly, if $x_1 = 2$, $f'(x_1) = 4$, and $f(x_1) = (2)^2 = 4$, so we have $y - 4 = 4(x - 2)$, which is $y - 4 = 4x - 8$ or $y = 4x - 4$.

Instantaneous Change

We remind you again that

$$\frac{\Delta y}{\Delta x} = \frac{y_2 - y_1}{x_2 - x_1}$$

gives the *average* rate of change between two points, (x_1, y_1) and (x_2, y_2); and the derivative $f'(x_1)$ gives, in some sense, the *exact* or *instantaneous* rate of change at a single point (x_1, y_1).

2.1.6   **Example**

If $P(x) = 100x - x^2 - 500$, find
(a) $P'(x)$
(b) $P'(30)$
(c) $P'(40)$

Solution:

In Example 1.6.7 we found that

$$\frac{\Delta P}{\Delta x} = \frac{100(x + \Delta x) - (x + \Delta x)^2 - 500 - (100x - x^2 - 500)}{\Delta x}$$

$$= \frac{\boxed{100x} + 100\,\Delta x - \cancel{x^2} - 2x\,\Delta x - (\Delta x)^2 - \cancel{500} - \boxed{100x} + \cancel{x^2} + \cancel{500}}{\Delta x}$$

$$= \frac{100\,\Delta x - 2x\,\Delta x - (\Delta x)^2}{\Delta x} = 100 - 2x - \Delta x$$

Notice that we use "x" instead of "x_1" this time, only because we acknowledge at the outset that x can be any number in the domain of P, and it is easier to write a variable without subscripts. Using the above computation and Definition 2.1.1, we have

$$P'(x) = \lim_{\Delta x \to 0} \frac{P(x + \Delta x) - P(x)}{\Delta x} = \lim_{\Delta x \to 0} \frac{\Delta P}{\Delta x}$$

$$= \lim_{\Delta x \to 0} (100 - 2x - \Delta x) = 100 - 2x$$

Thus the answer to part (a) is $P'(x) = 100 - 2x$. We can use this to immediately compute the other two answers.

(b) $P'(30) = 100 - 2(30) = 40$
(c) $P'(40) = 100 - 2(40) = 20$

2.1.7   Example

(a) Find the slope of the tangent lines to the graph of the function $P(x) = 100x - x^2 - 500$ at the points where $x_1 = 30$ and $x_1 = 40$.

(b) Find the equations of these tangent lines. (Note that even though these lines are hard to graph because they are so steep, it is not particularly difficult to develop their equations.)

Solution:

(a) The slopes of the tangent lines were found in Example 2.1.6 to be $P'(30) = 40$ and $P'(40) = 20$.

(b) We again substitute into the formula $y - y_1 = m(x - x_1)$.
    If $x_1 = 30$, $P(30) = 100(30) - 30^2 - 500 = 3,000 - 900 - 500 = 1,600$, so the equation is

$$y - 1,600 = 40(x - 30)$$

or

$$y - 1,600 = 40x - 1,200$$

or

$$y = 40x + 400$$

If $x_1 = 40$, $P(40) = 100(40) - 40^2 - 500 = 4,000 - 1,600 - 500 = 1,900$ so the equation is

$$y - 1,900 = 20(x - 40)$$

or

$$y - 1,900 = 20x - 800$$

or

$$y = 20x + 1,100$$

2.1.8   Example

Find the instantaneous rate of change in profit (in other words, marginal profit) at a sales level of $x = 30$ if the profit is given by the function (you guessed it!) $P(x) = 100x - x^2 - 500$.

Solution:

Since $P'(x)$ is the instantaneous change in profit at sales level x, we use the result of Example 2.1.6(b).

$$P'(30) = 100 - 2(30) = 40$$

Hence there is an additional profit of *approximately* $40 for each additional unit sold. (*Caution:* This does not mean that exactly $40 will be made, since the tendency to make a profit decreases slightly at each point beyond $x = 30$. You can verify this.)

2.1.9   Example in Biology or Sociology

(*See also Example 1.6.8*) Suppose that $P(t)$ is the size of a certain (plant, animal, or human) population at time t. Let t_1 be any particular time and $\Delta t = t_2 - t_1$ be the change in time between times t_1 and t_2. Then

$$\lim_{\Delta t \to 0} \frac{\Delta P}{\Delta t} = \lim_{\Delta t \to 0} \frac{P(t_1 + \Delta t) - P(t_1)}{\Delta t} = P'(t_1)$$

is the *instantaneous rate of change* of population at time $t = t_1$.

2.1.10   Example in Physics

Remember (see Example 1.6.5) that if we drop an object from a high place, the distance in feet that it will fall in t seconds is given by the formula

$$y = f(t) = 16t^2$$

Use this to calculate how fast it is falling at time $t = t_1$.

Solution:

In Example 1.6.5(f) we calculated that the average speed of the object between time $t = t_1$ and another time, $t = t_1 + \Delta t$, will be $\frac{\Delta y}{\Delta t} = 32t_1 + 16 \, \Delta t$. As Δt becomes very small, the average speed over a very small interval with t_1 as one endpoint is precisely $32t_1 + 16 \, \Delta t$. Thus we say that the velocity (how fast it is falling) "at" time $t = t_1$ is

$$\frac{dy}{dt} = \lim_{\Delta t \to 0} (32t_1 + 16 \, \Delta t) = 32t_1$$

In the language of physics, the velocity of a freely falling object t seconds after it is dropped is $32t$ ft/sec.

SUMMARY

In this section we formally defined the "derivative" and discussed its meaning from several viewpoints. We calculated specific derivatives; if you are having trouble with this, you might want to try the following four-step rule:

1.   Substitute $(x + \Delta x)$ for x in $f(x)$, getting $f(x + \Delta x)$.
2.   Subtract $f(x)$ from $f(x + \Delta x)$, getting $f(x + \Delta x) - f(x) = \Delta y$.
3.   Divide your result by Δx, getting

$$\frac{f(x + \Delta x) - f(x)}{\Delta x} = \frac{\Delta y}{\Delta x}$$

4. Take the limit as $\Delta x \to 0$, getting

$$f'(x) = \lim_{\Delta x \to 0} \frac{f(x + \Delta x) - f(x)}{\Delta x}$$

Using the derivative to find the slope of tangent lines, we then wrote the equations of tangent lines to certain functions. Finally, we showed how the derivative can be interpreted as marginal profit, instantaneous rate of change of population, and velocity.

EXERCISES 2.1. A

1. (a) What is the slope of the tangent line to this parabola at its bottom point?

(b) What is the value of the derivative at this point?
(c) What can you say about the value of the derivative to the right of this point? (d) To its left?

HC 2. Refer to Example 2.1.3. Find the slope of the tangent line to the parabola in Figure 2.1–6 at the point $(2, f(2)) = (2, 4)$ by completing the following table for $f(x) = x^2$.

x	Δx	$x + \Delta x$	$f(x + \Delta x)$	$f(x + \Delta x) - f(x)$	$\dfrac{f(x + \Delta x) - f(x)}{\Delta x}$
2	0.5				
2	0.1				
2	0.03				
2	0.005				

3. For each of the following functions, $y = f(x)$, write an expression for $\dfrac{\Delta y}{\Delta x}$ in terms of x_1 and Δx. [See Examples 1.6.5(f) and 2.1.3(b).] Then take the limit as $\Delta x \to 0$ (using Definition 2.1.1) to find $f'(x)$ in terms of x. You may find the following table helpful.

$y = f(x)$	$f(x + \Delta x)$	$f(x + \Delta x) - f(x)$	$\dfrac{f(x + \Delta x) - f(x)}{\Delta x}$	$f'(x)$
(a) $y = 2$				
(b) $y = 2x$				
(c) $y = 2x^2$		.		
(d) $y = x^2 + 2x + 4$				
(e) $y = 3x^2 - 2$				
(f) $y = x^2$	$(x + \Delta x)^2$	$2x\,\Delta x + (\Delta x)^2$	$2x + \Delta x$	$2x$

4. Find the slope of the tangent line at the point where $x = 1$ for each of the functions in problem 3. Write the equation of the tangent line in each case. (See Example 2.1.5.)

5. If the cost of producing x items of a commodity is given by $C(x) = 100 + 3x - 0.05x^2$, find an expression for the marginal cost $C'(x)$. Then find $C'(10)$ and $C'(20)$.

6. A person's rate of respiration (in breaths per minute) is a linear function of the partial pressure p of carbon dioxide in the lungs (in torr) given by $f(p) = -10.35 + 0.59p$. Find the instantaneous rate of change in respiration when the partial pressure of carbon dioxide is at a level of 40 torr.

EXERCISES 2.1. B

1. (a) What is the slope of the tangent line to this parabola at its top point?

(b) What is the value of the derivative at this point?

(c) What can you say about the value of the derivative to the right of this point? (d) To the left?

HC 2. Refer to Example 2.1.3. Find the slope of the tangent line to the parabola in Figure 2.1-6 at the point $(-2, f(-2)) = (-2, 4)$ by completing the following table for $f(x) = x^2$.

x	Δx	$x + \Delta x$	$f(x + \Delta x)$	$f(x + \Delta x) - f(x)$	$\dfrac{f(x + \Delta x) - f(x)}{\Delta x}$
-2	-0.5				
-2	-0.1				
-2	-0.03				
-2	-0.005				

3. For each of the following functions $y = f(x)$, write an expression for $\dfrac{\Delta y}{\Delta x}$ in terms of x_1 and Δx. [See Examples 1.6.5(f) and 2.1.3(b).] Then take the limit as $\Delta x \to 0$ (using Definition 2.1.1) to find $f'(x)$ in terms of x. You may find the following table helpful.

$y = f(x)$	$f(x + \Delta x)$	$f(x + \Delta x) - f(x)$	$\dfrac{f(x + \Delta x) - f(x)}{\Delta x}$	$f'(x)$
(a) $y = 3$				
(b) $y = 3x$				
(c) $y = 3x^2$				
(d) $y = x^2 + 3x - 5$				
(e) $y = 2x^2 + 7$				
(f) $y = 4x^2$	$4(x + \Delta x)^2$	$8x \Delta x + 4(\Delta x)^2$	$8x + 4 \Delta x$	$8x$

4. Find the slope of the tangent line at the point where $x = 1$ for each of the functions in problem 3. Write the equation of the tangent line in each case. (See Example 2.1.5.)

5. If the profit from selling x items of a commodity is given by $P(x) = 5x - 100 - 0.1x^2$, find an expression for the marginal profit $P'(x)$. Then find $P'(10)$ and $P'(20)$.

6. An animal's pulmonary ventilation, in liters/minute, has been found to be a function of the excess hydrogen-ion concentration h measured in moles/liter. If the function is given by $V(h) = 5.4 + 1.5h$, find the instantaneous rate of change in pulmonary ventilation when the excess hydrogen-ion concentration is 2 moles/liter.

EXERCISES   2.1. C

1. Define a function $f(x)$ as follows (see the figure below):

$$f(x) = -x \quad \text{if x is less than zero}$$
$$f(x) = x \quad \text{if x is greater than or equal to zero}$$

Show that the limit

$$\lim_{\Delta x \to 0} \frac{f(0 + \Delta x) - f(0)}{\Delta x}$$

equals -1 if we let Δx approach zero through negative values, and the same limit equals 1 if we let Δx approach zero through positive values. [Hence we cannot define the derivative of $f(x)$ at $x = 0$ (see the second comment on page 69).]

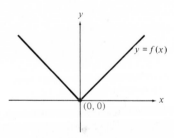

2. Find $f'(x)$ if $f(x) = x^3$.
3. Find $f'(x)$ if $f(x) = x^4$.
4. Find $f'(x)$ if $f(x) = x^5$.
5. If we throw an object downward from an airplane with a velocity of 10 ft/sec, the distance it will fall (measured in feet) in x seconds is given by $y = 10t + 16t^2$. How fast is it falling after 5 sec?

ANSWERS   2.1. A

1. (a) 0 (b) 0   (c) It is positive. (d) It is negative.

2.	x	Δx	x + Δx	f(x + Δx)	f(x + Δx) − f(x)	$\dfrac{f(x + Δx) − f(x)}{Δx}$
	2	0.5	2.5	6.25	2.25	4.5
	2	0.1	2.1	4.41	0.41	4.1
	2	0.03	2.03	4.1209	0.1209	4.03
	2	0.005	2.005	4.020025	0.020025	4.005

3. (a) $\dfrac{Δy}{Δx} = \dfrac{0}{Δx} = 0; f'(x) = 0$

(b) $\dfrac{Δy}{Δx} = \dfrac{2\,Δx}{Δx} = 2; f'(x) = 2$

(c) $\dfrac{Δy}{Δx} = \dfrac{4x\,Δx + 2(Δx)^2}{Δx} = 4x + 2\,Δx; f'(x) = 4x$

(d) $\dfrac{Δy}{Δx} = 2x + Δx + 2; f'(x) = 2x + 2$

(e) $\dfrac{Δy}{Δx} = 6x + 3\,Δx; f'(x) = 6x$

4. (a) $f'(1) = 0; y = 2$ (b) $f'(1) = 2; y = 2x$ (c) $f'(1) = 4; y = 4x − 2$
(d) $f'(1) = 4; y = 4x + 3$ (e) $f'(1) = 6; y = 6x − 5$
5. $C'(x) = 3 − 0.1x; C'(10) = 2; C'(20) = 1$
6. 0.59 breath/min/torr.

ANSWERS   2.1. C

1. When $0 + Δx$ is negative, $\dfrac{f(0 + Δx) − f(0)}{Δx} = \dfrac{−Δx − 0}{Δx} = −1$, and when $0 + Δx$
   is positive, $\dfrac{f(0 + Δx) − f(x)}{Δx} = \dfrac{Δx − 0}{Δx} = 1.$
2. $f'(x) = 3x^2$
3. $f'(x) = 4x^3$
4. $f'(x) = 5x^4$
5. 170 ft/sec

2.2   Techniques of Differentiation

If you have done most of the exercises in Section 2.1, you will probably agree that finding derivatives by using the definition is a tedious process. Fortunately there are a few simple formulas for finding the derivatives of complicated functions more easily. Some of these formulas are very easy; in fact, you may have guessed them already. For example, we know that the slope (that is, the derivative) of a horizontal line (one in the form of $y = b$) is zero. In other words:

2.2.1   Rule

If $f(x) = b$ [that is, if $f(x)$ is a *constant function*], then $f'(x) = 0$.

It is not hard to prove that this rule is true. Looking at it rigorously, we have for $f(x) = b$,

$$f'(x) = \lim_{\Delta x \to 0} \frac{f(x + \Delta x) - f(x)}{\Delta x} = \lim_{\Delta x \to 0} \frac{b - b}{\Delta x} = \lim_{\Delta x \to 0} 0 = 0$$

Next we turn to less trivial functions. If you did Exercises 2.1.C, you now have the following table for $f'(x)$ given $f(x)$:

$f(x)$	$f'(x)$	Reason
$f(x) = b$	$f'(x) = 0$	Rule 2.2.1
$f(x) = mx$	$f'(x) = m$	The derivative of any straight line is its slope
$f(x) = x^2$	$f'(x) = 2x$	Example 2.1.3
$f(x) = x^3$	$f'(x) = 3x^2$	Problem 2 of Exercises 2.1.C
$f(x) = x^4$	$f'(x) = 4x^3$	Problem 3 of Exercises 2.1.C
$f(x) = x^5$	$f'(x) = 5x^4$	Problem 4 of Exercises 2.1.C

Even if you didn't do Exercises 2.1.C, you probably know that you could, given sufficient time; the last three entries are calculated using the same techniques as were used for $y = x^2$, except that the algebra is more demanding. If you look at this table, you may yourself be able to guess an important rule in differential calculus.

2.2.2   Rule

If $f(x) = bx^n$, where b and n are any real numbers, then $f'(x) = bnx^{n-1}$.

This rule is proved in Appendix II for the special case where n is a positive integer. But it is always true.

2.2.3   Example

Find the derivatives of the functions
(a) $y = 23$
(b) $y = x^3$
(c) $y = x^{3/2}$
(d) $y = x^{-3/2}$
(e) $y = 4x^{3/2}$

Solution:

We use the notation $y' = f'(x)$ (mentioned on page 69) because it is simpler.
(a) $y' = 0$
(b) $y' = 3x^2$
(c) $y' = \frac{3}{2}x^{(3/2)-1} = \frac{3}{2}x^{1/2}$
(d) $y' = -\frac{3}{2}x^{-(3/2)-1} = -\frac{3}{2}x^{-5/2}$
(e) $y' = 4(\frac{3}{2})x^{(3/2)-1} = 6x^{1/2}$

Addition and Subtraction

In Example 2.1.6 we found that if $P(x) = 100x - x^2 - 500$, then $P'(x) = 100 - 2x$. This might lead you to suspect that we might have computed this by merely finding in turn the derivatives of $100x$, $-x^2$, and -500. Indeed, it is true that you can find the derivative of a sum of functions by merely taking the derivative of each term and adding the answers.

2.2.4 Rule

If $f(x) = g(x) \pm h(x)$, where both g and h have derivatives at x, then $f'(x) = g'(x) \pm h'(x)$.

Again, you can find the proof in Appendix II. Here we turn to examples showing you how these rules can be used. Finding derivatives is much easier than you might have suspected after completing Section 2.1! But the effort you expended in that section was worthwhile because it helped you understand what derivatives are and how to use them.

2.2.5 Example

Find the derivatives of the functions
(a) $y = x^2 + 3x + 2$
(b) $y = 3x^{-2} + x^{10} - x^{1/2}$
(c) $y = \sqrt{x} + 1/x$

Solution:

(a) Since $(x^2)' = 2x$, $(3x)' = 3$, and $(2)' = 0$, we have $y' = (x^2 + 3x + 2)' = (x^2)' + (3x)' + (2)' = 2x + 3$.

(b) $y' = 3(-2)x^{-2-1} + 10x^{10-1} - \frac{1}{2}x^{1/2-1} = -6x^{-3} + 10x^9 - \frac{1}{2}x^{-1/2}$

(c) To do this problem you must first write both terms in exponential notation. Then you can apply Rule 2.2.2 as before.

$$y = x^{1/2} + x^{-1}$$
$$y' = \frac{1}{2}x^{(1/2)-1} + (-1)x^{-1-1} = \frac{1}{2}x^{-1/2} - x^{-2}$$

One of the most common uses of derivatives in business and economics is in "marginal analysis." The derivative of total revenue, $R'(x)$ (that is, the instantaneous rate of change of total revenue) is often called the marginal revenue. The derivative of the cost, $C'(x)$, is often called the marginal cost. And the derivative of the profit function, $P'(x)$, can be called the marginal profit. Since a little reflection should convince you that

Profit = Revenue - Cost

(that is, the profit is the difference between the income and the expenses), we can write in mathematical terminology

$$P(x) = R(x) - C(x)$$

Applying Rule 2.2.4, we have

$$P'(x) = R'(x) - C'(x)$$

This is a useful formula in economic analysis such as that contained in the following problem.

2.2.6   Example

Suppose that a revenue function is defined by $R(x) = 3x$ and a cost function by $C(x) = 100 + 5x - 0.01x^2$. Using the fact that $P(x) = R(x) - C(x)$, find the marginal profit when $x = 10$.

Solution:

Since $R(x) = 3x$, $R'(x) = 3$.
Since $C(x) = 100 + 5x - 0.01x^2$, $C'(x) = 5 - 0.02x$.
Therefore, $P'(x) = R'(x) - C'(x) = 3 - (5 - 0.02x) = -2 + 0.02x$.
Thus $P'(10) = -2 + 0.02(10) = -2 + 0.2 = -1.8$. The significance of the minus sign before the marginal profit is that the profit is actually a loss (a negative profit). In this case, if an extra item is sold after the tenth item, the company will tend to *lose* $1.80.

SUMMARY

This section explained some rules that will enable you to compute derivatives much more easily than by using the definition, as in Section 2.1. The proofs of these rules are in the Appendixes, but it is essential here only that you be able to use them.

EXERCISES   2.2. A

Find y' for the following functions.

1. $y = 3$
2. $y = x^5$
3. $y = x^3$
4. $y = x^{-4}$
5. $y = 1/x^5$
6. $y = -4x^{2/3}$
7. $y = 7\sqrt[5]{x}$
8. $y = 3x^{1/3} - \frac{1}{5}x^{4/5}$
9. $y = x^{1/5} + x^5$

10. $y = 2x^4 + 4x^3 - 1$
11. $y = x^3 - x^2 + 3x$
12. $y = x^2(x^{1/2} - 2x)$
13. $y = x(x^3 - x^2 + 3x)$
14. If the revenue function is given by $R(x) = 5x$ and the cost function is given by $C(x) = 200 + 2x + 0.03x^2$, find the marginal profit for $x = 30$. [*Hint:* $P(x) = R(x) - C(x)$.]

EXERCISES 2.2. B

Find y' for the following functions.

1. $y = 7$
2. $y = x^6$
3. $y = x^4$
4. $y = x^{-3}$
5. $y = 1/x^4$
6. $y = -3x^{1/4}$
7. $y = 4x^{2/5}$
8. $y = 2x^{1/4} - \frac{1}{3}x^{3/2}$
9. $y = \sqrt[4]{x} + x^4$

10. $y = 3x^5 + 7x^2 - \pi$
11. $y = x^6 - x^3 + 4x$
12. $y = x^3(x^{1/3} - 4x)$
13. $y = x(x^5 - x^3 + 5x)$
14. If the revenue function is given by $R(x) = 7x$ and the cost function is given by $C(x) = 50 + 4x + 0.02x^2$, find the marginal profit when $x = 20$. [Hint: $P(x) = R(x) - C(x)$.]

EXERCISES 2.2. C

1. If $R(x) = (2x^2 - 2)/(x + 1)$ is a revenue function and $C(x) = (x^2 - 1)/(x + 1)$ is a cost function, find the marginal profit when $x = 2$.

ANSWERS 2.2. A

1. $y' = 0$
2. $y' = 5x^4$
3. $y' = 3x^2$
4. $y' = -4x^{-5}$
5. $y' = -5x^{-6}$
6. $y' = -\frac{8}{3}x^{-1/3}$
7. $y' = \frac{7}{5}x^{-4/5}$

8. $y' = x^{-2/3} - \frac{4}{25}x^{-1/5}$
9. $y' = \frac{1}{5}x^{-4/5} + 5x^4$
10. $y' = 8x^3 + 12x^2$
11. $y' = 3x^2 - 2x + 3$
12. $y = x^{5/2} - 2x^3$, so $y' = \frac{5}{2}x^{3/2} - 6x^2$
13. $y' = 4x^3 - 3x^2 + 6x$
14. $P'(x) = 3 - 0.06x$, so $P'(30) = 1.2$.

ANSWERS 2.2. C

1. $P'(2) = 1$

2.3 More Techniques and Higher-Order Derivatives

Multiplication and Division

Addition and subtraction follow such natural rules when you are differentiating that you might think similar rules follow for multiplication and division. Alas, such thoughts are in error, as the following examples show.

2.3.1 Example

Find the derivatives of the function

(a) $y = (3x)(2x + 1)$

(b) $y = \dfrac{x^3 + x}{x^2 + 1}$

Solution:

(a) First we multiply, and we get $y = (3x)(2x + 1) = 6x^2 + 3x$. Then we apply Rules 2.2.2 and 2.2.4, which gives us

$$y' = 12x + 3$$

[*Caution:* $y' \neq (3)(2) = 6$, which is what you get if you take the derivatives first and then multiply; you *must* multiply first and then take the derivative at this stage of your knowledge. See Rule 2.3.2.]

(b) First we divide, getting

$$y = \frac{x^3 + x}{x^2 + 1} = \frac{x(x^2 + 1)}{x^2 + 1} = x$$

Hence

$$y' = 1$$

[*Caution:* $y' \neq (3x^2 + 1)/2x$, which is what you get if you take the derivatives first and then divide; you *must* divide first and then take the derivative at this stage. See Rule 2.3.3.]

Now that we have shown that the "obvious" rules are false, we give the correct rules. Their ingenious proofs are given in Appendix II.

2.3.2   Rule

If $f(x) = g(x) \cdot h(x)$, where g' and h' both exist, then $f'(x) = g'(x)h(x) + g(x)h'(x)$.

2.3.3   Rule

If $f(x) = \dfrac{g(x)}{h(x)}$, where g' and h' both exist and $h(x) \neq 0$, then

$$f'(x) = \frac{h(x)g'(x) - g(x)h'(x)}{[h(x)]^2}$$

Both these rules have short forms, which may be easier to remember:

2.3.2   Rule
   If $y = uv$, then $y' = u'v + uv'$.

2.3.3   Rule
   If $y = \dfrac{u}{v}$, then $y' = \dfrac{vu' - uv'}{v^2}$.

When writing down the quotient formula, try writing the v^2 in the denominator of the derivative first; this may help you remember that v (without the prime!) is the first letter to write in the numerator. This is important; if you get the terms in the numerator backward, the minus sign will be in the wrong place, and your answer will be the negative of what it should be.

2.3.4   Example

Find the derivatives of
(a) $y = (x + 1)(3x^2 + 3)$
(b) $y = (x^2 + 1)(x^2 + x + 1)$

Solution:

(a) We let $u = x + 1$ and $v = 3x^2 + 3$. Then, using $y' = u'v + uv'$,

$$y' = (1)(3x^2 + 3) + (x + 1)(6x)$$

(b) We let $u = x^2 + 1$ and $v = x^2 + x + 1$. Using $y' = u'v + uv'$,

$$y' = (2x)(x^2 + x + 1) + (x^2 + 1)(2x + 1)$$

2.3.5   Example

Find the derivatives of
(a) $y = \dfrac{\sqrt{x}}{2x + 1}$
(b) $y = \dfrac{2x^2 - 1}{x + 1}$

Solution:

(a) We let $u = \sqrt{x} = x^{1/2}$ and $v = 2x + 1$. Then, using $y' = \dfrac{vu' - uv'}{v^2}$,

$$y' = \frac{(2x + 1)\frac{1}{2}x^{-1/2} - x^{1/2}(2)}{(2x + 1)^2}$$

(b) We let $u = 2x^2 - 1$ and $v = x + 1$. Using $y' = \dfrac{vu' - uv'}{v^2}$,

$$y' = \frac{(x + 1)(4x) - (2x^2 - 1)(1)}{(x + 1)^2}$$

Higher-Order Derivatives

We pointed out in Section 2.1 that the derivative of a function is often a new function, written $y' = f'(x)$. If the original function was $y = x^2$, for example, its derivative at each point is the new function $y' = 2x$. We can now often take the derivative of this new function and obtain what we call the second derivative of the original function. It is written $y'' = f''(x)$. When $y = x^2$, we obtain $y'' = (y')' = 2$.

$$y = x^2 \qquad y' = 2x \qquad y'' = 2$$

The second derivative is "the rate of change of the rate of change of $f(x)$" at each point. This concept has widespread significance in our modern world; population growth provides a typical example. It is important that the world population is increasing, that is, that the derivative (rate of change) of the population is positive. Even more ominous, however, is the fact that the rate of change of the increase in world population (its second derivative) is positive. The *rate* of population growth itself is increasing—and in many parts of the world at a faster rate than food supplies are increasing.

We postpone further discussion of the applications of the second derivative until Section 2.6. In this section we restrict our attention to definitions and computations.

2.3.6   Definition

Suppose that $y = f(x)$ and $f'(x)$ exists. Then if

$$\lim_{\Delta x \to 0} \frac{f'(x + \Delta x) - f'(x)}{\Delta x}$$

exists, it is called the second derivative of $f(x)$. It is denoted

$$f''(x) \qquad \text{or} \qquad \frac{d^2 y}{dx^2} \qquad \text{or} \qquad \frac{d^2 f}{dx^2} \qquad \text{or} \qquad y''$$

2.3.7   Example

Find the second derivatives of the following functions.
(a) $y = 2$
(b) $y = 2x$
(c) $y = x^2$
(d) $y = x^{-2}$
(e) $y = x/(x + 1)$

Solution:
(a) $y' = 0$ and $y'' = 0$
(b) $y' = 2$ and $y'' = 0$
(c) $y' = 2x$ and $y'' = 2$
(d) $y' = -2x^{-3}$ and $y'' = 6x^{-4}$
(e) $y' = \dfrac{(x + 1)(1) - x(1)}{(x + 1)^2} = \dfrac{1}{(x + 1)^2} = \dfrac{1}{x^2 + 2x + 1}$

$$y'' = \frac{(x^2 + 2x + 1)(0) - 1(2x + 2)}{(x^2 + 2x + 1)^2} = \frac{-2x - 2}{(x^2 + 2x + 1)^2} = \frac{-2}{(x + 1)^3}$$

Since the second derivative, y'', is a function itself, one may attempt to find its derivative. Often it is possible to differentiate repeatedly obtaining higher and higher derivatives.

$$y = x^4 \quad y' = 4x^3 \quad y'' = 12x^2 \quad y''' = 24x \quad y^{(iv)} = 24$$
$$y^{(v)} = 0 \quad y^{(vi)} = 0$$

2.3.8   Definition

If the function $y = f(x)$ can be differentiated n times, the result is called the nth derivative of $f(x)$. It is denoted

$$f^{(n)}(x) \quad \text{or} \quad \frac{d^n y}{dx^n} \quad \text{or} \quad \frac{d^n f}{dx^n} \quad \text{or} \quad y^{(n)}$$

[*Caution:* $f^{(n)}(x)$ and $y^{(n)}$ are different from the nth powers of $f(x)$ and y. The latter are written $[f(x)]^n$ and y^n, respectively.]

2.3.9   Example

Find the first four derivatives of $f(x) = y = x^6 + x^2 - 3x^{-1}$.

Solution:

$$y' = 6x^5 + 2x + 3x^{-2}$$
$$y'' = 6(5)x^4 + 2(1) + 3(-2)x^{-3} = 30x^4 + 2 - 6x^{-3}$$
$$y''' = 120x^3 + 0 + 18x^{-4}$$
$$y^{(iv)} = 360x^2 - 72x^{-5}$$

(*Note:* This function and its derivatives do not exist at $x = 0$. Why not?)

In stating subsequent rules, we shall always assume that the first two derivatives of the functions in question exist unless otherwise noted. (This assumption makes the rules easier to state and to remember even though it may be more than one normally needs.)

SUMMARY

In this section we stated and used rules that enable us to differentiate (calculate the derivative of) products and quotients of polynomials. With sufficient practice you can differentiate even the trickiest product and quotient with no more effort than you expend on long division.

The second derivative $f''(x)$ is the instantaneous rate of change (that is, derivative) of $f'(x)$. Similarly, the nth derivative, $f^{(n)}(x)$, of a function $y = f(x)$ is the derivative of its $(n - 1)$st derivative, $f^{(n-1)}(x)$. Although we have postponed discussion of the significance of higher-order derivatives, you should be able to calculate them, using the techniques of Section 2.2 and this section. Go to it!

EXERCISES 2.3. A

In problems 1–6, find y'.

1. $y = (x + 1)(x^3 - x^2 + 3x)$

2. $y = (3x^2 - 2)(x^7 - 5x^3 + 1)$

3. $y = \dfrac{x^3 - x^2 + 3x}{x^2}$

4. $y = \dfrac{x^3 - x^2 + 3x}{x}$

5. $y = \dfrac{5x^2 + 1}{x^2 + 2x - 3}$

6. $y = \dfrac{x^3 - x^2 + 3x}{x + 1}$

7. Find the first four derivatives of
   (a) $y = 5$
   (b) $y = 4x^3$
   (c) $y = \sqrt{x}$
   (d) $y = x^{7/2} + 1/x$

EXERCISES 2.3. B

In problems 1–6, find y'.

1. $y = (3x + 4)(x^4 - 3x^2 + 2x)$

2. $y = (4x^2 - 3)(x^8 - 4x^4 + 4)$

3. $y = \dfrac{x^4 - x^3 + \pi x}{x^3}$

4. $y = \dfrac{x^4 - x^3 + \pi x}{x}$

5. $y = \dfrac{3x^2 - 1}{x^2 + 3x - 5}$

6. $y = \dfrac{x^5 - x^3 + 8x}{2x + 4}$

7. Find the first four derivatives of
   (a) $y = \pi$
   (b) $y = 2x^2$
   (c) $y = \sqrt[3]{x}$
   (d) $y = x^{7/3} + 1/x^2$

EXERCISES 2.3. C

Find y' and y'' of

1. $y = (3x^2 + 5x)(4x^5 + x^7)$
2. $y = (x^2 + 3)/(x^3 + 4x)$

3. What can you say about the mth derivative of $y = x^n$?

ANSWERS 2.3. A

1. $y' = (1)(x^3 - x^2 + 3x) + (x + 1)(3x^2 - 2x + 3)$
2. $y' = (6x)(x^7 - 5x^3 + 1) + (3x^2 - 2)(7x^6 - 15x^2)$
3. $y = x - 1 + 3x^{-1}$, so $y' = 1 - 3x^{-2}$
4. $y = x^2 - x + 3$, so $y' = 2x - 1$
5. $y' = \dfrac{(x^2 + 2x - 3)(10x) - (5x^2 + 1)(2x + 2)}{(x^2 + 2x - 3)^2}$

6. $y' = \dfrac{(x + 1)(3x^2 - 2x + 3) - (x^3 - x^2 + 3x)}{(x + 1)^2}$

7. (a) $y' = 0$; $y'' = 0$; $y''' = 0$; $y^{(iv)} = 0$ (b) $y' = 12x^2$; $y'' = 24x$; $y''' = 24$; $y^{(iv)} = 0$
   (c) $y' = \tfrac{1}{2}x^{-1/2}$; $y'' = -\tfrac{1}{4}x^{-3/2}$; $y''' = \tfrac{3}{8}x^{-5/2}$; $y^{(iv)} = -\tfrac{15}{16}x^{-7/2}$ (d) $y' = \tfrac{7}{2}x^{5/2} - x^{-2}$;
   $y'' = \tfrac{35}{4}x^{3/2} + 2x^{-3}$; $y''' = \tfrac{105}{8}x^{1/2} - 6x^{-4}$; $y^{(iv)} = \tfrac{105}{16}x^{-1/2} + 24x^{-5}$

ANSWERS   2.3. C

1. $y' = (6x + 5)(4x^5 + x^7) + (3x^2 + 5x)(20x^4 + 7x^6)$
    $y'' = (6)(4x^5 + x^7) + (6x + 5)(20x^4 + 7x^6) + (6x + 5)(20x^4 + 7x^6)$
    $+ (3x^2 + 5x)(80x^3 + 42x^5)$

2. $y' = \dfrac{(x^3 + 4x)(2x) - (x^2 + 3)(3x^2 + 4)}{(x^3 + 4x)^2} = \dfrac{-x^4 - 5x^2 - 12}{x^6 + 8x^4 + 16x^2}$

    $y'' = \dfrac{(x^6 + 8x^4 + 16x^2)(-4x^3 - 10x) - (-x^4 - 5x^2 - 12)(6x^5 + 32x^3 + 32x)}{(x^6 + 8x^4 + 16x^2)^2}$

3. If n is a positive integer, then the mth derivative of $y = x^n$ for $m > n$ is always zero. Otherwise, the $(m + 1)$th derivative is always different from the mth derivative. In general, $y^{(m)} = n(n - 1)(n - 2) \cdots (n - m + 1)x^{n-m}$, $n \geq m$.

2.4   Using the First Derivative

Increasing and Decreasing Functions

Since the derivative was defined so that it would describe how a function is changing, it is just what we need to discover where particular functions are increasing, decreasing, or neither. Remember that a linear function has a positive slope if and only if it is increasing, and it has a negative slope if and only if it is decreasing. Similarly, if an arbitrary differentiable function (that is, a function that can be differentiated) has a positive derivative throughout some interval, then the function will be increasing at each point x_1 in that interval. By the same reasoning, if an arbitrary differentiable function has a negative derivative throughout some interval, then the function is decreasing at each point x_1 in that interval.

A straight line is horizontal if and only if its slope is zero; an arbitrary differentiable function has a horizontal tangent at a point if and only if its derivative is zero there.

2.4.1   Rule

(a) If $f'(x_1) > 0$ throughout some interval, then $f(x)$ is <u>increasing</u> at each point of the interval.

Figure 2.4–1

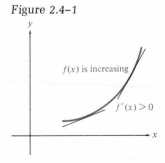

(b) If $f'(x_1) < 0$ throughout some interval, then $f(x)$ is <u>decreasing</u> at each point of the interval.

Figure 2.4–2

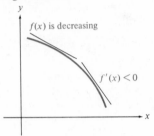

$f(x)$ is decreasing

$f'(x) < 0$

(c) If $f'(x_1) = 0$, then $f(x)$ is <u>stationary</u> (by which we mean it has a <u>horizontal tangent</u>) at the point $x = x_1$.

Figure 2.4–3

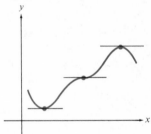

2.4.2   Example

Determine where the given functions are increasing and decreasing and where the tangent line is horizontal. Graph these functions.
(a) $f(x) = -2x + 1$
(b) $f(x) = x^2 + 2x + 1$
(c) $f(x) = x^2 + bx + c$

Solution:

(a) $f'(x) = -2$ for all x, so the function is always decreasing. [We can recognize immediately that f is a straight line. Thus its slope is constant; *linear functions are either always increasing or always decreasing or never changing (that is, constant).*]

Figure 2.4–4

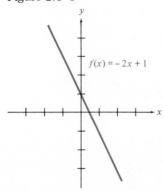

$f(x) = -2x + 1$

(b) $f'(x) = 2x + 2$. Setting $2x + 2 = 0$ and solving for x, we get $x = -1$. Thus we have a horizontal tangent at $x = -1$. For $x < -1$, $f'(x)$ is negative, so $f(x)$ is decreasing. For $x > -1$, $f'(x)$ is positive, so $f(x)$ is increasing. Since $f(x)$ is a quadratic function, its graph is a parabola; the previous statements should now make the graph clear.

Figure 2.4–5

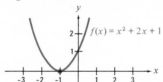

$f(x) = x^2 + 2x + 1$

(c) $f'(x) = 2x + b$. This is similar to the previous problem, but more general. If $f'(x) = 0$, then $2x + b = 0$, so $x = -b/2$; this will be the x-coordinate of the point where the function takes its minimum. To find the y-coordinate of this point, we substitute

$$f\left(-\frac{b}{2}\right) = \left(-\frac{b}{2}\right)^2 - \frac{b^2}{2} + c$$
$$= \frac{b^2}{4} - \frac{b^2}{2} + c$$
$$= -\frac{b^2}{4} + c$$

Figure 2.4–6

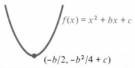

$f(x) = x^2 + bx + c$

$(-b/2, -b^2/4 + c)$

The placement of the axes of this graph depends on b and c. For $x > -b/2$, we have $f'(x) = 2x + b > 0$, so $f(x)$ increases to the right of $-b/2$ and its graph rises there. And if $x < -b/2, f'(x) = 2x + b < 0$, so $f(x)$ decreases to the left of $-b/2$; its graph is falling there. Thus $(-b/2, -(b^2/4) + c)$ is a minimum point of this function.

Local Maxima and Minima

2.4.3   Definition

If $f'(x_1) = 0$, then x_1 is called a <u>critical point</u> of f.

There are three types of pictures to keep in mind when $f'(x_1) = 0$ (that is, when x_1 is a critical point of f).

2.4.4   Illustrations

Figure 2.4–7
(a) A peak   (b) A trough   (c)

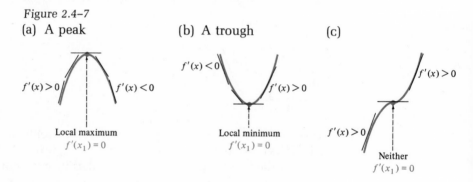

2.4.5   Definition

$f(x_1)$ is called a <u>local maximum</u> (or <u>relative maximum</u>) of $f(x)$ if $f(x_1) \geq f(x)$ for all x in some interval around x_1. [See Figure 2.4–7(a).]

2.4.6   Definition

$f(x_1)$ is called a <u>local minimum</u> (or <u>relative minimum</u>) of $f(x)$ if $f(x_1) \leq f(x)$ for all x in some interval around x_1. [See Figure 2.4–7(b).]

At a local maximum, the derivative of a well-behaved function changes from positive to negative; thus its derivative, if it exists, must be zero there. Similarly, at a local minimum the derivative changes from negative to positive, so it must be zero there. We postpone a discussion of Figure 2.4–7(c) until Section 2.6.

Figure 2.4–8

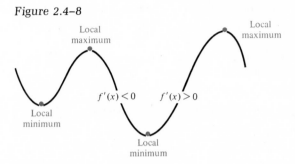

2.4.7   Rule

If $y = f(x)$ has a local maximum or a local minimum at $x = a$, then $f'(a) = 0$.

This is a very important rule. The following example may give you some idea of how it can be used. (If you want to review how to solve inequalities, see the Review of Algebra.)

2.4.8   Example

Suppose that the profit from making x items is given by $P(x) = 4x - 500 - 0.001x^2$.
(a) Tell for which values of x the profit is increasing and decreasing.
(b) For what value of x is the profit largest?
(c) What is the maximum profit?

Solution:

(a) The profit will be increasing when the derivative of $P(x)$ is positive and it will be decreasing when the derivative of $P(x)$ is negative. So we first take the derivative.

$$P'(x) = 4 - 0.002x$$

To see when $P'(x)$ is positive or negative, we first find when $P'(x)$ equals zero. To do this, we write

$$P'(x) = 4 - 0.002x = 0$$

and solve for x, getting

$$x = 2,000$$

It becomes clear that if $x < 2,000$, $P'(x) > 0$ and the profit is increasing. Also, if $x > 2,000$, $P'(x) < 0$ and the profit is decreasing.

(b) Since the profit increases until $x = 2{,}000$ and decreases afterward, it must have its maximum at $x = 2{,}000$. As Rule 2.4.7 predicted, $f'(2{,}000) = 0$.

(c) Since the profit is a maximum when $x = 2{,}000$, the maximum profit will be $P(2{,}000) = 4 \cdot 2{,}000 - 500 - 0.001(2{,}000)^2 = 3{,}500$.

Absolute Maxima and Minima

In applications of calculus there are usually restrictions on which x's are permissible. For example, we cannot produce fewer than zero items! If we want to maximize or minimize a function, we usually want to consider only certain values of x.

2.4.9   Definition

M is called the <u>absolute maximum</u> of a function $f(x)$ over a given domain if $f(x) \leqslant M$ for all x in the given domain, and there exists x_1 in the domain such that $f(x_1) = M$.

Figure 2.4–9

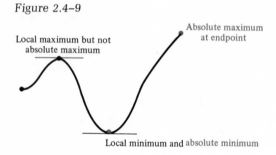

Local maximum but not absolute maximum

Absolute maximum at endpoint

Local minimum and absolute minimum

m is called the <u>absolute minimum</u> of a function $f(x)$ over a given domain if $f(x) \geqslant m$ for all x in the given domain and there exists x_1 in the domain such that $f(x_1) = m$.

2.4.10   Rule

If a function is continuous, then the absolute maximum (if it exists) over an interval occurs *either at an endpoint of the interval or at a local maximum.*

If a function is continuous, then the absolute minimum (if it exists) over an interval occurs *either at an endpoint of the interval or at a local minimum.*

It is not difficult to understand Rule 2.4.10 if you consider special cases. For example, in Figure 2.4–10, x_1 is the x-coordinate of our left endpoint and x_2 is the x-coordinate of an *adjacent* critical point, which in this case is a local minimum. Looking at the picture it becomes clear that if $x_1 \leqslant x \leqslant x_2$, then $f(x_1) \geqslant f(x) \geqslant f(x_2)$; $f(x)$ lies between $f(x_1)$ and $f(x_2)$ and therefore cannot be either a minimum or a maximum. Continuing to travel right from x_2, the function must increase until we either come to a local maximum (a peak) or an endpoint. If a local maximum, we then slide down again to a valley, or an endpoint, or we just keep sliding down.

Think of the curve as a one-way roller-coaster ride; if you don't get dizzy, you'll be convinced!

Figure 2.4–10

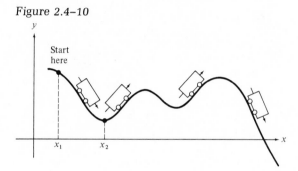

2.4.11   Example

Find the absolute maximum and the absolute minimum of $f(x) = x^2 + 2x + 1$ on the interval $-2 \leqslant x \leqslant 2$.

Solution:

In Example 2.4.2 we showed that this quadratic function has a local minimum at $x = -1$, its only critical point. Thus we need only check this point and the endpoints of the interval: $f(-2) = 1$, $f(-1) = 0$, and $f(2) = 9$. We see that the absolute minimum over this interval is the same as the local minimum; it is the value 0 and is taken at $x = -1$. The absolute maximum is 9; it is taken at $x = 2$.

2.4.12   Example

Suppose that the profit from making x items is given by $P(x) = 4x - 500 - 0.001x^2$. (See Example 2.4.8.)
(a) If the company is capable of producing between 0 and 1,500 items, how many should it produce for (absolute) maximum profit?
(b) If it can produce between 40 and 2,500, how many should it produce for maximum profit?

Solution:

(a) From Example 2.4.8 we know that the only critical point of the function is at $x = 2,000$. Since this is not in the interval $0 \leqslant x \leqslant 1,500$, we need only test the endpoints. Since the derivative $P' = 4 - 0.002x$ is always positive on the interval, the function is increasing there, so the maximum will be at the right endpoint, $x = 1,500$.

(b) Since the critical point occurs in the given interval ($x = 2,000$ is between $x = 40$ and $x = 2,500$), we must test three points—the critical point and the two endpoints.

$$P(40) = 4(40) - 500 - 0.001(40)^2 = 160 - 500 - 1.6 = -341.6$$
$$P(2,000) = 4(2,000) - 500 - 0.001(2,000)^2 = 8,000 - 500 - 4,000$$
$$= 3500$$
$$P(2,500) = 4(2,500) - 500 - 0.001(2,500)^2$$
$$= 10,000 - 500 - 6,250 = 3,250$$

Thus the maximum profit is \$3,500; it occurs when $x = 2,000$.

2.4.13   Example in Biology or Pre-Med

In a test for blood sugar metabolism, conducted over a time interval, the amount of sugar in the blood at time t was found to be first increasing and then decreasing, according to the formula

$$A = 4.5 + 0.1t - 0.1t^2$$

where t was measured in hours. Find the maximum amount of sugar in the blood.

Solution:

To find the critical point, we calculate the first derivative and set it equal to zero:

$$A' = 0.1 - 0.2t$$
$$0 = 0.1 - 0.2t$$
$$t = \frac{0.1}{0.2} = \frac{1}{2}$$

If $t < \frac{1}{2}$, we can see that A' is positive, so A is increasing. If $t > \frac{1}{2}$, A' is negative, so A is decreasing. It follows that A takes its maximum when $t = \frac{1}{2}$. Thus the maximal value of A is

$$4.5 + 0.1(\tfrac{1}{2}) - 0.1(\tfrac{1}{2})^2 = 4.5 + 0.05 - 0.025 = 4.525$$

SUMMARY

The first derivative $f'(x)$ can be used to find where $f(x)$ is increasing, decreasing, or stationary, according to Rule 2.4.1. Wherever $f(x)$ has a *local* maximum or a *local* minimum (a peak or a trough), $f'(x) = 0$. If we want to find the *absolute* maximum or minimum of a continuous function, we need to check only the critical points [that is, the places where $f'(x) = 0$] and the endpoints. We shall encounter and explore these ideas again in the next few sections. Master them here and you're well on your way!

EXERCISES 2.4. A

1. Use the derivative to find where the following functions are increasing and decreasing. Plot any critical points on a graph and sketch the function. Tell what the absolute minima (m) and maxima (M) are on the interval $-1 \leqslant x \leqslant 2$.

    (a) $y = 3x^2 - 6x + 1$ (d) $y = x^3$
    (b) $y = x^2 - 4$ (e) $y = x^4$
    (c) $y = -x^2 + 4x - 2$

2. Suppose that the profit from making x items is given by $P(x) = 6x - 400 - 0.01x^2$. What number should be manufactured for maximum profit? What is the maximum profit?

3. In problem 2, how many should be made if the company can manufacture between 10 and 200 of the items?

4. If the profit from producing x items is $P(x) = 12x - 0.03x^2 - 15$, how many items yield a maximum profit? What is this maximum profit?

EXERCISES 2.4. B

1. Use the derivative to find where the following functions are increasing and decreasing. Find the critical points. Use the above information to make a rough sketch of the functions. Tell what the absolute minima (m) and maxima (M) are on the interval $-2 \leqslant x \leqslant 1$.

    (a) $y = 2x^2 + 4x - 5$ (d) $y = x^3$
    (b) $y = x^2 + 1$ (e) $y = x^4$
    (c) $y = -x^2 + 6x - 7$

2. A manufacturer of widgets has daily production costs of $\$(1,600 - 40x + \frac{1}{2}x^2)$ for x widgets. How many widgets should he produce to minimize costs? What is the minimum possible cost?

3. In problem 2, how many widgets should he produce if the company cannot make more than 35?

4. If the profit from producing x items is $P(x) = 24x - 0.04x^2 - 20$, how many items yield a maximum profit? What is the maximum profit?

EXERCISES 2.4. C

1. The number of people (in hundreds) infected t days after a certain epidemic began is found to be fairly well described by the function $P(t) = -t^2 + 30t + 1$.
    (a) Find the day at which a maximum number of people will be infected.
    (b) How many people will be sick on the worst day?

2. Suppose the cost of producing x dinette tables is given by $C(x) = 500 + 14x + \frac{1}{4}x^2$. The price can be set at $p(x) = 650 - \frac{3}{2}x$, where x is the number sold, and the total revenue is given by $R(x) = xp(x)$.

(a) Find the value of x that maximizes revenue.

(b) Find the value of x that maximizes profit. (Again use Profit = Revenue − Cost.)

(c) Are the answers to (a) and (b) equal?

3. Suppose the cost of manufacturing an item is $C = 800 + 3x + 0.01x^2$ and the selling price is $10 each. What is the number that should be manufactured for maximum profit? How much will that maximum profit be? (*Hint:* Profit = Revenue − Cost.)

4. A car rental agency owns a fleet of 108 cars, all of which can be rented if the rate per car is $25 per day. But the agency leaves 4 cars unrented for each dollar that it raises the rate per car. In other words, if they set the price at $26 per day, only 104 cars will be rented; if they set the price at $27 per day, 100 cars will be rented; and so on. What rate per car will maximize income? What is the maximum possible income?

5. A car rental agency owns a fleet of 150 cars, all of which can be rented if the rate per car is $30 per day. But 3 cars will be left idle for each dollar that the agency raises its rate. In other words, if they charge $31 per day, only 147 cars will be rented; if they charge $32 per day, 144 cars will be rented; and so on. How much should they charge to maximize the total revenue received? What will this maximum revenue be?

6. Graph $y = \dfrac{1}{x} + x$, showing critical points.

HC 7. Convince yourself, by completing the following table for $A(t) = 4.5 + 0.1t - 0.1t^2$, that $A(t)$ in Example 2.4.13 does indeed take a local maximum near 0.5. Give answers that are accurate to three decimal places.

t	0.2	0.3	0.4	0.6	0.7	0.8
$A(t)$						

ANSWERS   2.4. A

1. (a) $y' = 6x - 6$, so the only critical point is at $x = 1$. For $x < 1$, the function decreases. For $x > 1$, the function increases. $M = 10$ at $x = -1$. $m = -2$ at $x = 1$.

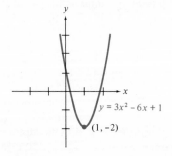
$y = 3x^2 - 6x + 1$
$(1, -2)$

(b) $y' = 2x$, so the only critical point is at $x = 0$. For $x < 0$, y decreases.
For $x > 0$, y increases. $M = 0$ at $x = 2$ and $m = -4$ at $x = 0$.

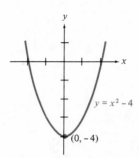

(c) $y' = -2x + 4$, so the only critical point is at $x = 2$. For $x > 2$, y is decreasing. For $x < 2$, y is increasing. $M = 2$ at $x = 2$ and $m = -7$ at $x = -1$.

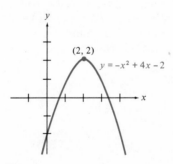

(d) $y' = 3x^2$, so the only critical point is at $x = 0$. For all x, $y' \geq 0$, so the
function is never decreasing. $M = 8$ at $x = 2$. $m = -1$ at $x = -1$.

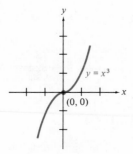

(e) $y' = 4x^3$, so the only critical point is at $x = 0$. y is increasing if $x > 0$ and y is decreasing if $x < 0$. $M = 16$ at $x = 2$. $m = 0$ at $x = 0$.

2. $x = 300$; $P(300) = \$500$
3.   200
4. $x = 200$; $P(200) = \$1{,}185$

ANSWERS 2.4. C

1.   (a) $t = 15$ (b) $P(15) = -225 + 450 + 1 = 226$, so 22,600 people will be infected on the worst (fifteenth) day.
2.   (a) Since $R(x) = x(650 - \frac{3}{2}x) = 650x - \frac{3}{2}x^2$, $R'(x) = 650 - 3x$.   Setting $R'(x) = 0$, we have that revenue is a maximum if $x = 216\frac{2}{3}$.
     (b) Since $P(x) = R(x) - C(x) = x(650 - \frac{3}{2}x) - (500 + 14x + \frac{1}{2}x^2)$, $P'(x) = 636 - 4x$ and profit is a maximum if $x = 159$.
     (c) The answers are not the same.
3. $x = 350$; $P(350) = \$425$
4. $\$26$; $\$2{,}704$
5. $\$40$; $\$4{,}800$
6. $y' = -x^{-2} + 1$, so the critical points are at $x = \pm 1$.

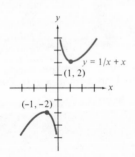

7.

t	0.2	0.3	0.4	0.6	0.7	0.8
$A(t)$	4.516	4.521	4.524	4.524	4.521	4.516

SAMPLE QUIZ   ## Sections 2.1, 2.2, 2.3, and 2.4

1.   Use the *definition* of derivative to find the derivative of $f(x) = 2x^2 - 4x + 7$.
2.   Find the first and second derivatives of $y = 4x^3 + \sqrt{x} + 1/x$.

3. Find y' if $y = (3x^2 + 7x)(4x^7 - 8)$.

4. Find $f'(x)$ if $f(x) = \dfrac{3x + 2}{x^2 - x}$. [*Hint:* If $y = u/v$, $y' = (vu' - uv')/v^2$.]

5. Suppose that the profit is given by the function

$$P(x) = 60x - 0.2x^2 - 10$$

(a) For what values of x is the profit increasing?
(b) For what values of x is the profit decreasing?
(c) For what x is the profit a maximum?
(d) What is the maximum profit?

The answers are at the back of the book.

2.5 First Derivative Test and the Inventory Model

You may wonder if it is possible to tell whether a critical point of a function [that is, a point x_1 such that $f'(x_1) = 0$] is a local maximum or a local minimum without going to all the trouble of making a graph. There are two ways of doing this; we discuss one in this section and the second in the next section.

If a function is decreasing just before it reaches a critical point and increasing just afterward, then clearly the function takes a local minimum at that critical point. That is, the function f has a local minimum at f if $f'(x) < 0$ just before x_1 and $f'(x) > 0$ just after x_1. Similar statements can be made for maxima. We now make both statements precisely.

2.5.1 First Derivative Test

If x_1 is a critical point of the function $f(x)$ and if $f'(x) < 0$ in some interval just to the left of x_1 and $f'(x) > 0$ in some interval just to the right of x_1, then f takes a local minimum at $x = x_1$.

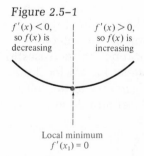

Figure 2.5–1

$f'(x) < 0$,
so $f(x)$ is
decreasing

$f'(x) > 0$,
so $f(x)$ is
increasing

Local minimum
$f'(x_1) = 0$

If x_1 is a critical point of the function $f(x)$ and if $f'(x) > 0$ in some interval just to the left of x_1 and $f'(x) < 0$ in some interval just to the right of x_1, then f takes a local maximum at $x = x_1$.

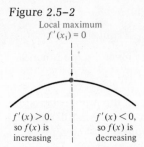

Figure 2.5–2

Local maximum
$f'(x_1) = 0$

$f'(x) > 0$,
so $f(x)$ is
increasing

$f'(x) < 0$,
so $f(x)$ is
decreasing

If the above ideas seem familiar to you, it is because we used them in the previous section. We have merely formalized them here; we now use them in a more complicated example.

2.5.2   Example

Find the local minima and maxima of $f(x) = \frac{1}{3}x^3 + \frac{1}{2}x^2$ without graphing.

Solution:

Taking the derivative, we get $f'(x) = x^2 + x = x(x + 1)$. Setting $x(x + 1) = 0$, we see that $f(x)$ has critical points at $x_1 = -1$ and $x_2 = 0$.

Figure 2.5–3

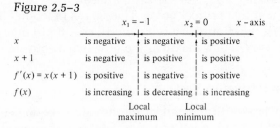

	$x_1 = -1$	$x_2 = 0$	x – axis
x	is negative	is negative	is positive
$x + 1$	is negative	is positive	is positive
$f'(x) = x(x + 1)$	is positive	is negative	is positive
$f(x)$	is increasing	is decreasing	is increasing
	Local maximum	Local minimum	

If $x > 0$, it is clear that $f'(x) = x(x + 1) > 0$, and with a little reflection you should realize that for $x < -1$, $f'(x) = x(x + 1)$ is also greater than zero because both of its factors are negative. If $-1 < x < 0$, then $x + 1 > 0$ and $x < 0$, so their product, $f'(x)$, is negative. Thus we conclude that $f(x)$ is decreasing on the interval $-1 < x < 0$ and increasing everywhere else. Since the function increases to the left of $x_1 = -1$ and decreases to its right, we must have a local maximum at $x_1 = -1$. Similarly, since the function decreases to the left of $x_2 = 0$ and increases to its right, we must have a local minimum at $x_2 = 0$.

The Inventory Problem

We now consider a well-known application of differential calculus to business. We begin with a particular example before we turn to the general pattern.

2.5.3   Example

Suppose that a company wants to make 10,000 gizmos this year. It costs $600 to get the plant ready for each run, the marginal cost per gizmo is $4, and it costs $12 per year to keep one gizmo in inventory. If the company wants to make a total of 10,000 this year, how many gizmos should be manufactured in each batch to minimize costs? What will the minimum cost be?

Solution:

Gizmos are some product manufactured in large quantities (called "batches" or "runs"); instead of a daily (or hourly) production process, the company occasionally "sets up" for a manufacturing run, which costs $600 no matter how many are manufactured at that time. Then the gizmos are stored in a rented inventory space and gradually sold until none remain and another batch must be manufactured. (For a graph that shows how many are available at a given time, look ahead to Figure 2.5–4.)

The idea of the solution is to write the total cost, $C(x)$, in terms of batch size, x; differentiate; set the derivative equal to zero; and solve for x, which is then the "best" batch size. To do this, we let

$$x = \text{number of gizmos produced per batch}$$

The basic equation is

$$\text{Cost} = \text{setup costs} + \text{variable costs} + \text{storage costs}$$

We consider each term of this equation in turn.

The annual batch setup cost is the cost per batch ($600) times the number of batches per year ($10{,}000/x$). Convince yourself that the number of batches per year will indeed be the total number produced divided by the number produced per batch.

The second term is the cost per item ($4) times the number of gizmos produced (10,000).

The third term is the cost of keeping one item in inventory ($12) times the average number in inventory ($x/2$). This last factor, $x/2$, is largely a matter of convention; we shall discuss it more later.

Substituting these mathematical expressions into the above word equation, we have

$$C = 600 \cdot \frac{10{,}000}{x} + 4 \cdot 10{,}000 + 12 \cdot \frac{x}{2}$$

This expression can be rewritten

$$C = 6{,}000{,}000x^{-1} + 40{,}000 + 6x$$

To find where this expression is a minimum, we differentiate:

$$C' = -6{,}000{,}000x^{-2} + 6$$

Setting this equal to zero, we solve for x:

$$0 = -6{,}000{,}000x^{-2} + 6$$
$$1{,}000{,}000x^{-2} = 1$$
$$1{,}000{,}000 = x^2$$
$$\pm 1{,}000 = x$$

Since we cannot produce a negative number of gizmos per batch, we know $x = -1{,}000$ is wrong, so the only critical point that we need examine is $x = 1{,}000$. To find whether this is a minimum or a maximum, we use the first derivative test:

$$C'(1{,}000) = \frac{-6{,}000{,}000}{(1{,}000)^2} + 6 = 0$$

as it should be. If we let x get a little bigger than 1,000, then the denominator of the fraction grows larger; so the fraction grows smaller and C' is positive. Thus $C(x)$ increases when x is just to the right of 1,000.

If we let x get a little smaller than 1,000, the fraction grows larger, so $C'(x)$ becomes negative. Thus $C(x)$ is a decreasing function to the left of 1,000. From these two facts we can conclude that $C(x)$ takes its minimum at $x = 1{,}000$. To find the minimum cost, we substitute into the original equation:

$$C(1{,}000) = \frac{6{,}000{,}000}{1{,}000} + 40{,}000 + 6 \cdot 1{,}000 = 52{,}000$$

Now we consider the previous inventory problem in greater generality. When making plans to manufacture an item, it is common for a company to be faced with the decision regarding how big a production batch is best. If the batches are too large, the inventory costs are large; if the batches are too small, the overhead costs of setting up the batches are high. In between there will be a minimum possible cost.

Suppose that x items are made per batch, all are immediately stored in inventory, and they are sold at a constant rate. Then the number available at any one time is as shown in Figure 2.5–4. We assume, therefore, there will be x/2 stored in inventory on the average during a year. We use this assumption in the following general equation for writing the total costs in terms of the batch size (compare with the word equation on page 101):

$$C(x) = b\,\frac{t}{x} + mt + v\,\frac{x}{2} = btx^{-1} + mt + \frac{v}{2}\,x$$

The variable x denotes batch size, and the other letters on the right side of the equation represent constants. b is the cost of setting up a batch; t is the total number of items desired; m is the marginal cost per item; and v is the cost of keeping one item in inventory for a year.

Figure 2.5–4

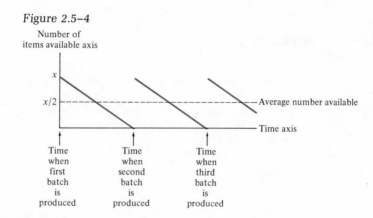

The exercises accompanying this section will give you further familiarity with this model.

SUMMARY

In this section we formally presented a test for discovering whether a critical point is a local maximum or minimum; this test is called the first derivative test. Then we used the first derivative in a popular model, the inventory model, which we explored in detail. In the next section we shall give another test for extrema, which is easier to apply. But although the second derivative test (as it will be called) is simpler, it does not always work; so it is a good idea to be reasonably familiar with the first derivative test.

EXERCISES 2.5. A

1. Use the first derivative test to tell, without graphing, whether the critical points in the functions of problem 1 of Exercises 2.4.A give local maxima or local minima.

2, 3. Use the first derivative test to tell whether the critical points of the following give local minima or maxima:

2. $y = \dfrac{1}{3} x^3 - x$

3. $y = \frac{1}{3}x^3 - x^2 - 3x$

4. Suppose that the cost of holding one item in inventory is $2, that the extra cost of producing one item is $4, and that the cost of readying the machinery for one run is $40. If the annual demand for the item is 4,000 units, how

many should be made at each run to minimize costs? What will the minimum cost be?

5. A company wants to make 100,000 items during the year. It costs $3,000 to get the plant ready for each run. If the marginal cost per item is $5 and it costs $6 per year to carry an item in inventory, how many should be made in each batch to minimize the cost? What will the minimum cost be?

EXERCISES 2.5. B

1. Use the first derivative test to tell whether the critical points in the functions of problem 1 of Exercises 2.4.B give local maxima or local minima without graphing.

2, 3. Use the first derivative test to tell whether the critical points of the following give local minima or maxima:

2. $y = \frac{1}{3}x^3 - 4x$
3. $y = 5x - 2x^2 - \frac{1}{3}x^3$
4. Suppose that the annual cost of holding one item in inventory is $10, that the extra cost of producing one item is $6, and that the cost of readying the machinery for one run is $200. If the annual demand for the item is 1,000 units, how many should be made at each run to minimize costs? What will the minimum cost be?

5. A company wants to make 1,000 items during the year. It costs $10 to get the plant ready for each run. If the marginal cost per item is $3 and it costs $2 per year to carry each item in inventory, how many should be made in each batch to minimize the cost? What will the minimum cost be?

EXERCISES 2.5. C

1. Using the general expression for $C(x)$ given on page 102, find the batch size that yields minimum cost if the cost of setting up a batch is b, the total number of items desired is t, the marginal cost per item is m, and the cost of keeping one item in inventory for one year is v.
2. Using the first derivative, find the local minima and maxima of $y = 2x^4 - 4x^2$.
HC 3. Verify the results you found in problem 4 in Exercises 2.5.A by completing the following table.

Lot size, x	Number of batches per year, $\frac{4,000}{x}$	Batch setup costs, $40\frac{4,000}{x}$	Marginal costs, $4 \cdot 4,000$	Average number in inventory, $\frac{x}{2}$	Total inventory costs, $2\frac{x}{2}$	Total costs
4,000	1	$ 40	$16,000	2,000	$4,000	$20,040
1,000	4	$160	$16,000	500		
500	8	$320	$16,000			
400	10					
200						
100						
40						

4. The inventory of the owner of a mink farm consists of a population of fur-bearing animals that is "harvested" each year in such a way as to (1) obtain the largest possible harvest without (2) depleting the initial population P_0. If the population of P mink grows to $f(P)$ animals in 1 year, the amount that may be harvested each year is $f(P) - P$. Let $h(P) = f(P) - P$ be the harvest function.

   (a) Show that $f'(P) = 1$ if P is the population level at which one obtains a maximum possible harvest. [Hint: $h'(P) = 0$.]

   (b) If the population of mink is given by the function $f(P) = P(21 - P)$, where P is measured in hundreds, find the population at which the mink farmer obtains the maximum sustainable harvest.

   (c) Find the maximum sustainable harvest of the mink farm.

   (d) Use the first derivative test to verify that you have a maximum.

ANSWERS 2.5. A

2. Local maximum at $x = -1$; local minimum at $x = 1$.
3. Local maximum at $x = -1$; local minimum at $x = 3$.
4. 400; $16,800
5. 10,000; $560,000

ANSWERS 2.5. C

1. $x = \sqrt{2bt/v}$
2. Local minima at $x = -1$ and $x = 1$; local maximum at $x = 0$.

3.

4,000	1	$ 40	$16,000	2,000	$4,000	$20,040
1,000	4	$ 160	$16,000	500	$1,000	$17,160
500	8	$ 320	$16,000	250	$ 500	$16,820
400	10	$ 400	$16,000	200	$ 400	$16,800
200	20	$ 800	$16,000	100	$ 200	$17,000
100	40	$1,600	$16,000	50	$ 100	$17,700
40	100	$4,000	$16,000	20	$ 40	$20,040

4. (a) $f'(P) - 1 = 0$ implies that $f'(P) = 1$.   (b) $P = 10$, or 1,000 mink.
   (c) $h(10) = 100$, or 10,000 mink.   (d) $h'(P) = 20 - 2P > 0$ if $P < 10$ and $h'(P) < 0$ if $P > 10$.

2.6   Using the Second Derivative

The second derivative is the derivative of the (first) derivative. Therefore, it tells how the derivative is changing. If the first derivative is increasing, then the second derivative must be positive, and conversely. If the first derivative is decreasing, then the second derivative is negative; the converse of this is true also.

2.6.1   Example

The costs of a certain operation have been observed to increase at a decreasing rate up to some number x_1, where the increase proceeds at an ever-increasing rate. If the cost is given by $C(x) = 0.1x^3 - 3x^2 + 250x + 1,000$, find the number x_1.

Solution:

Where the increase in cost is decreasing, the second derivative will be negative; where the increase in cost is increasing, the second derivative will be positive. To find where the second derivative changes sign, *we calculate the second derivative and find where it equals zero.*

$$C'(x) = 0.3x^2 - 6x + 250$$
$$C''(x) = 0.6x - 6$$
$$0 = 0.6x - 6$$

implies that $x = 10$, so $x_1 = 10$.

Looking more closely at the expression for $C''(x)$, we can see that if $x > 10$, $C''(x) > 0$, so the increase in costs is increasing for $x > 10$. If $x < 10$, the second derivative, $C''(x)$, is negative, so the costs are increasing at a decreasing rate.

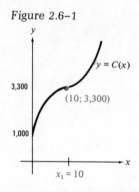

Figure 2.6–1

We summarize these ideas in the following rule.

2.6.2   Rule

(a) $f''(x) > 0$ if and only if $f'(x)$ is increasing.
(b) $f''(x) < 0$ if and only if $f'(x)$ is decreasing.
(c) $f''(x) = 0$ if and only if $f'(x)$ is stationary.

By interpreting this rule in various ways, we find that the second derivative can be very useful. We shall now explore the three most common applications of the second derivative: concavity, the second derivative test, and inflection points.

Concavity

Sometimes a picture is worth many words; this is certainly true with the previous rules. If you look at the picture below, you can see that the derivative (that is, slope of the tangent line to the curve) is increasing as x moves from a to b to c; this suggests the significance of having $f''(x) > 0$.

Figure 2.6-2

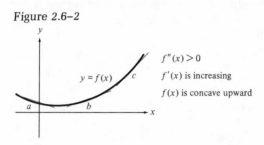

$f''(x) > 0$
$f'(x)$ is increasing
$f(x)$ is concave upward

2.6.3   Definition

If a curve lies above its tangent lines at each point, it will be shaped like a bowl and called <u>concave upward</u>.

As Figure 2.6-2 suggests, the fact that a curve is lying above its tangent lines is equivalent to its having a positive second derivative or an increasing first derivative. Similar (opposite!) reasoning holds for $f''(x) < 0$.

Figure 2.6-3

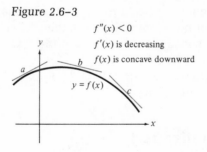

$f''(x) < 0$
$f'(x)$ is decreasing
$f(x)$ is concave downward

2.6.4   Definition

If a curve lies below its tangent lines at each point, it will be shaped like an upside-down bowl and called <u>concave downward</u>.

We see in Figure 2.6-3 that a concave-downward function will have a decreasing derivative (slope of its tangent lines) and therefore a negative second derivative. We summarize these thoughts in the following rule.

2.6.5   Rule

(a) $f''(x) > 0$ on an interval if and only if $f(x)$ is *concave upward* on that interval.
(b) $f''(x) < 0$ on an interval if and only if $f(x)$ is *concave downward* on that interval.

The standard device for remembering the relationship between the second derivative and concavity is that $f''(x) > 0$ corresponds to the bowl "holding water."

2.6.6   Example

(See Example 2.4.2 for graphs.) Investigate the concavity of the following functions:

(a) $y = mx + b$
(b) $y = x^2 + 2x + 1$
(c) $y = ax^2 + bx + c$

Solution:

(a) $y' = m$, so $y'' = 0$, which says that the graph of a straight line is neither concave upward nor concave downward anywhere. (In other words, the function $y = mx + b$ is indeed "straight"!)
(b) $y' = 2x + 2$, so $y'' = 2 > 0$; we conclude that this function is concave upward everywhere.
(c) $y' = 2ax + b$, so $y'' = 2a$; the concavity of a quadratic function (whose graph, we noted in Section 1.4, is a parabola) is determined by the coefficient of the x^2 term. If $a > 0$, then $f''(x) > 0$ everywhere, so the function is concave upward. If $a < 0$, $f''(x) < 0$ everywhere, so the function is concave downward.
   In Example 2.4.2, parts (b) and (c), we had $x = 1 > 0$, so the graph was concave upward.

Second Derivative Test

If we have a critical point x_1 [that is, a point such that $f'(x_1) = 0$], there are three possible figures, as we noted in Illustrations 2.4.4.
   In Figure 2.6–4(a) the function is concave upward in some interval around x_1, so $f''(x) > 0$ in that interval. In Figure 2.6–4(b) the function is concave downward in an interval around x_2, so $f''(x) < 0$ in that interval. In Figure 2.6–4(c) the function is concave downward to the left of x_3 and concave upward to the right of x_3, so $f''(x)$ must change sign at x_3. Thus $f''(x_3) = 0$. More about this comes later; now we turn to the conclusions from Figures 2.6–4(a) and 2.6–4(b).

Figure 2.6–4

(a) Local minimum   (b) Local maximum   (c) Inflection point

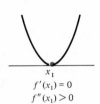

$f'(x_1) = 0$
$f''(x_1) > 0$

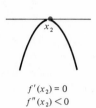

$f'(x_2) = 0$
$f''(x_2) < 0$

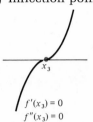

$f'(x_3) = 0$
$f''(x_3) = 0$

If a function has a horizontal tangent [that is, $f'(x) = 0$] at a point, we can find whether it has a peak (that is, a local maximum) there or a trough (that is, a local minimum), by examining the concavity via the second derivative.

2.6.7   Second Derivative Test

If $f(x)$ is a function that has a second derivative and x_1 is a critical point of f, then
(a) If $f''(x_1) > 0$, $f(x_1)$ is a local minimum (bowl holds water).
(b) If $f''(x_1) < 0$, $f(x_1)$ is a local maximum (bowl inverted).

Figure 2.6–5   Figure 2.6–6

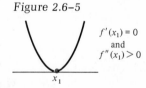

$f'(x_1) = 0$
and
$f''(x_1) > 0$

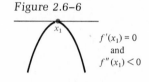

$f'(x_1) = 0$
and
$f''(x_1) < 0$

2.6.8   Example

Find the local maxima and minima of $y = x^2$.

Solution:

Since $f'(x) = 2x$, the critical point is at $x = 0$. Since $f''(x) = 2 > 0$ everywhere, by the second derivative test, (0, 0) is a local minimum.

Figure 2.6–7

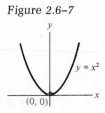

2.6.9   Example

Find the local maxima and minima of $y = \frac{1}{3}x^3 + \frac{1}{2}x^2$.

Solution:

$y' = x^2 + x = x(x + 1)$, so the critical points are at $x = 0$ and $x = -1$.
$y'' = 2x + 1$.
Since $2(0) + 1 > 0$, there is a local minimum at $x = 0$.
Since $2(-1) + 1 = -1 < 0$, there is a local maximum at $x = -1$.

Figure 2.6–8

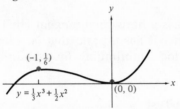

The second derivative test is often easy to use. Its drawback is that sometimes we have both $f'(x_1) = 0$ and $f''(x_1) = 0$, in which case the second derivative test gives no information.

2.6.10   Example

Find the local maxima and minima of $y = x^4$.

Solution:

$f'(x) = 4x^3$, so the only critical point is at $x = 0$. $f''(x) = 12x^2$, so $f''(0) = 0$, and the second derivative test fails. To answer the question, we must either graph the curve or use the first derivative test. $f'(x) < 0$ for $x < 0$, so the function is decreasing before the critical point. Since $f'(x) > 0$ when $x > 0$, the function increases to the right of the critical point. From these facts we can conclude, by the first derivative test, that f has a local minimum at $x = 0$.

Figure 2.6–9

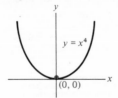

We list here the three methods for deciding whether a critical point gives a local minimum or a local maximum. It is often easiest to try them in this order.

1. Second derivative test (this section).
2. First derivative test (Section 2.5).
3. Graphing (Sections 2.4 and 2.7).

Inflection Points

We now return to an examination of Figure 2.6–4(c). Here the function was concave downward to the left of x_3 and concave upward on the right of x_3. Such a value gives an "inflection point." Inflection points do not have to be given by critical points (see below), although they may be, as we saw in Figure 2.6–4(c).

2.6.11   Definition

If the second derivative of $f(x)$ changes sign at a point $x = x_1$, then x_1 gives an inflection point of f.

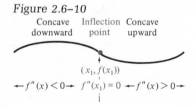

Figure 2.6–10

At any inflection point $(x_1, f(x_1))$, $f''(x_1) = 0$. These are the points where the curve changes from concave upward to concave downward, or vice versa.

2.6.12   Example

Find the inflection points of $C = 0.1x^3 - 3x^2 + 250x + 1{,}000$. (See Example 2.6.1 for the previous discussion of this function and its graph.)

Solution:
To find inflection points, we first find the points where the second derivative equals zero. Then we check these points to see at which of them the sign of the second derivative changes.

$$C'(x) = 0.3x^2 - 6x + 250$$
$$C''(x) = 0.6x - 6$$

If $C''(x) = 0.6x - 6 = 0$, then $x = 10$, so if there are any inflection points, there is only one, and it is where $x = 10$. Convince yourself that the sign of $C''(x)$ does indeed change at $x = 10$ (and, therefore, this is an inflection point); if you have trouble, look back at Example 2.6.1.

2.6.13   Example

Find the inflection points of the function $y = x^4 - 6x^2$.

Solution:

Inflection points can occur only where $f''(x) = 0$, so we first find the second derivative and set it equal to zero. After solving for x, we will then determine whether the second derivative does indeed change sign at these points.

$$y' = 4x^3 - 12x$$
$$y'' = 12x^2 - 12 = 12(x + 1)(x - 1) = 0$$
$$12 \neq 0 \qquad x + 1 = 0 \qquad x - 1 = 0$$
$$x = -1 \qquad x = 1$$

Thus there are possible inflection points where $x = -1$ and $x = 1$.

Figure 2.6–11

	$x = -1$	$x = 1$		x – axis
$(x + 1)$	is negative	is positive	is positive	
$(x - 1)$	is negative	is negative	is positive	
$f''(x) = 12(x + 1)(x - 1)$	is positive	is negative	is positive	

Since $f''(x) > 0$ for $x > 1$ and for $x < -1$, and $f''(x) < 0$ between these two values of x, the second derivative does change sign both where

Figure 2.6–12

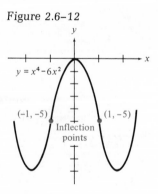

$y = x^4 - 6x^2$

$(-1, -5)$ $(1, -5)$

Inflection
points

$x = -1$ and where $x = 1$. Thus there are indeed inflection points at these values.

2.6.14   Example in Medicine or Sociology

When an epidemic first affects a given population, usually not only are an increasing number of people infected, but the increase is growing at an increasing rate; the second derivative of the number of sick people is positive as well as the first derivative. Public health officials who are trying to find remedies for the disease generally consider it "under control" when an inflection point is reached, in other words, when the number infected is increasing at a decreasing rate. After the concavity of the curve is downward, it is likely that the number infected will soon begin to decrease, too, in other words, that the first derivative will soon become negative. (See also Example 4.4.10.)

Figure 2.6–13

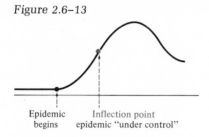

Epidemic   Inflection point
begins   epidemic "under control"

2.6.15   Example in Biology

The velocity (in cm/sec) at which an amphibian sartorius (a thigh muscle) is shortened when it is loaded with various weights of magnitude G (in grams) has been found by A. V. Hill to be

$$v(G) = \frac{87.9105}{G + 14.35} - 1.03$$

Sketch the graph of the function $y = v(G)$ for $G > 0$.

Solution:

It is easy to calculate $y' = -(87.9105)(G + 14.35)^{-2}$ and $y'' = 2(87.9105)(G + 14.35)^{-3}$. Since $y' \neq 0$ for all values of G, there are no relative extreme points. Furthermore, if $G > 0$, it is clear that $y'' > 0$, and the graph is concave upward. Indeed, the graph is given in Figure 2.6–14, where $v(G) > 0$ and $G > 0$. The plotted points are based upon Hill's experimental data; his data agree so closely with the values predicted by the theory that the thickness of the dots makes them appear to lie exactly on the curve.

Figure 2.6–14

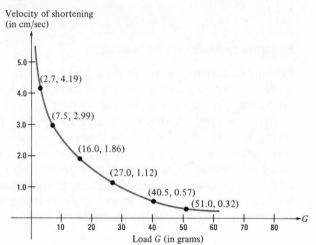

Velocity of shortening
(in cm/sec)

SUMMARY

The second derivative (the derivative of the derivative) tells how much the change of the original function is changing. It can be used to determine concavity [where $f''(x)$ is positive or negative], or inflection points [where $f''(x) = 0$], or the significance of critical points [where $f'(x) = 0$]. Both the first and second derivatives are very useful when graphing functions; the next section will use the material of the past four sections to draw good graphs of certain nonlinear functions.

EXERCISES 2.6. A

Find the critical points and inflection points of the following functions. Use the second derivative test to find whether the critical points give local maxima or minima. Use the second derivative to find where the functions are concave upward and concave downward. Save your answers, because the exercises for the next section will be to graph these functions.

1. $y = x^2 + 4x - 2$ 5. $y = x^3/3 - x^2$
2. $y = -x^2 + 4x - 2$ 6. $y = x^3/3 - x$
3. $y = 3x^2 - x + 1$ 7. $y = x^4/4 - x^2/2$
4. $y = -\frac{1}{2}x^2 - 3x + 2$ 8. $y = -x^3$

EXERCISES 2.6. B

Follow the same directions as above for the following functions.

1. $y = -x^2 + 6x + 10$ 5. $y = x^3/3 - 4x$
2. $y = 4x^2 - x + 1$ 6. $y = x^3/3 - x^2/2$
3. $y = \frac{1}{2}x^2 - 3x + 2$ 7. $y = x^2/2 - x^4/4$
4. $y = x^2 - 8x - 3$ 8. $y = x^3 + 1$

EXERCISES   2.6. C

Follow the same directions as above for the following functions.

1. $y = x^3 - 6x^2 + 9x + 1$
2. $y = (1 - 2x)^3$
3. $y = (1 - 2x)^4$
4. $y = \sqrt{1 - x^2}$
5.   Suppose that a flu epidemic hits a city in such a way that the healthy population seems to be following the function

$$P(t) = 100{,}000 - 60t^2 + t^3$$

where t is the number of days after the onset of the epidemic.
(a) For what t is the population decreasing?
(b) When is this decrease in population proceeding at an increasing rate?
(c) When is the decrease in population happening at a decreasing rate?
(d) When will the population be a minimum?
(e) What will this minimum population be?
(f) On what day will the number of healthy inhabitants begin to increase?
(g) What is the rate of change of this increase on the tth day?

ANSWERS AND SOME SOLUTIONS   2.6. A

1.   Since $y' = 2x + 4$, the only critical point is when $2x + 4 = 0$ and must therefore be at $x = -2$. At this point $y = (-2)^2 + 4(-2) - 2 = -6$. Thus the only
critical point gives $(-2, -6)$. Since $y'' = 2 > 0$, the function is always concave upward and there are no inflection points. The point $(-2, -6)$ must be a
local and absolute minimum.
2.   Relative maximum at $(2, 2)$; no inflection points; concave down everywhere.
3.   Relative minimum at $(\frac{1}{6}, \frac{11}{12})$; no inflection points; concave up everywhere.
4.   Relative maximum at $(-3, \frac{13}{2})$; no inflection points; concave down everywhere.
5. $y' = x^2 - 2x = x(x - 2)$, so there are critical points at

$$x = 0 \qquad \text{and} \qquad x = 2$$
$$f(0) = 0 - 0 = 0 \qquad \text{and} \qquad f(2) = \frac{2^3}{3} - 2^2 = -\frac{4}{3}$$

Since $f''(x) = 2x - 2$, $f''(0) = -2 < 0$, and $f''(2) = 2 \cdot 2 - 2 = 2 > 0$, we
have that $(0, 0)$ is a local maximum and $(2, -\frac{4}{3})$ is a local minimum. Setting
$f''(x) = 2x - 2 = 0$, we discover an inflection point at $x = 1$. Since $f(1) =$
$1^3/3 - 1^2 = -\frac{2}{3}$, the only inflection point is $(1, -\frac{2}{3})$. The function is concave
upward when $2x - 2 > 0$ (that is, when $x > 1$) and it is concave downward
when $2x - 2 < 0$ (that is, when $x < 1$).
6.   Critical points give $(-1, \frac{2}{3})$ and $(1, -\frac{2}{3})$; the first is a relative maximum and
the second a relative minimum. The only inflection point is $(0, 0)$. The func-
tion is concave upward for $x > 0$ and concave downward for $x < 0$.
7.   Local minima at $(-1, -\frac{1}{4})$ and $(1, -\frac{1}{4})$; local maximum at $(0, 0)$. Inflection
points at $(1/\sqrt{3}, -\frac{5}{36})$ and $(-1/\sqrt{3}, -\frac{5}{36})$; the function is concave downward
between these two points and concave upward elsewhere.
8.   Since $y' = -3x^2 \leq 0$, the function is decreasing everywhere except at the

point $(0, 0)$. Since $y'' = -6x$, the point $(0, 0)$ is also an inflection point and the function is concave upward for $x < 0$ and concave downward for $x > 0$.

ANSWERS 2.6. C

1. Relative maximum at $(1, 5)$; relative minimum at $(3, 1)$; inflection point at $(2, 3)$; concave upward if $x > 2$; concave downward if $x < 2$.
2. No relative maximum or minimum; $(\frac{1}{2}, 0)$ is an inflection point; concave upward if $x < \frac{1}{2}$; concave downward if $x > \frac{1}{2}$.
3. Relative minimum at $(\frac{1}{2}, 0)$; no relative maxima or inflection points; concave upward everywhere.
4. Relative maximum at $(0, 1)$; concave downward everywhere.
5. We use the facts that $P'(t) = -120t + 3t^2$ and $P''(t) = -120 + 6t$.
    (a) $0 \leqslant t \leqslant 40$ (b) $0 \leqslant t < 20$ (c) $20 < t \leqslant 40$ (d) At $t = 40$
    (e) $P(40) = 68{,}000$ (f) The fortieth day (g) $-120 + 6t$

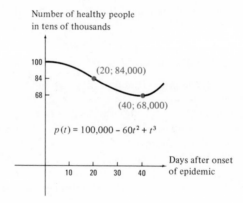

Number of healthy people
in tens of thousands

$(20; 84{,}000)$

$(40; 68{,}000)$

$p(t) = 100{,}000 - 60t^2 + t^3$

Days after onset
of epidemic

2.7 Sketching Graphs

It is often important to see a situation as a whole instead of piecemeal. When studying the behavior of a function, few methods are more useful than graphing. For example, a doctor looks at a graph of a patient's progress before giving a diagnosis; an economist examines a graph to analyze a company's or nation's economic prospects. Such graphs usually describe functions.

As we saw in Chapter 1, a function and its graph are often thought of as the same thing. Thus one can survey the overall behavior of a given function by looking at its graph. Since one cannot plot every point of most functions, it is valuable to pick out certain especially revealing points. The techniques of the last few sections are useful for this purpose. The first and second derivatives of a function provide the most essential information about the "shape" of a function.

2.7.1   **Example**

Sketch $y = x^3$ using first and second derivatives.

Solution:

The first derivative is $y' = 3x^2 \geqslant 0$, so the graph is never falling (decreasing). It is always rising (increasing) except when $x = 0$, its only critical point. Since $y'' = 6x$, we have $y'' < 0$ when $x < 0$ and the curve is concave downward there. If $x > 0$, then $y'' > 0$, so the curve is concave upward there. Thus $x = 0$ gives the inflection point $(0, 0)$.

Figure 2.7–1

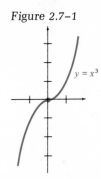

The steps used to sketch $y = x^3$ can be followed for most functions. Occasionally we are interested in some extra features, such as the intersections of the graph with the x and y axes. But usually the analysis given above will suffice, so we summarize it here.

2.7.2   **Curve-Sketching Rule**

To sketch the graph of $y = f(x)$:

1. Find $f'(x)$ and $f''(x)$.
2. Find critical points x by setting $f'(x) = 0$.
3. Classify critical points with either the first or second derivative tests. (See Sections 2.5 and 2.6.)
4. Find where the graph is rising $[f'(x) > 0]$ or falling $[f'(x) < 0]$.
5. Find where the graph is concave upward $[f''(x) > 0]$ or concave downward $[f''(x) < 0]$, and identify inflection points.
6. ·Plot all critical points, inflection points, and, possibly, a few others.
7. Use the information about the shape to fill in between.

2.7.3   **Example**

Graph $f(x) = \frac{1}{3}x^3 + \frac{1}{2}x^2$ using the first and second derivatives.

Solution:

$f'(x) = x^2 + x = x(x + 1)$, so the critical points are at $x = 0$ and $x = -1$.

Substituting into the original equation, we see that they give $(0, 0)$ and $(-1, \frac{1}{6})$.

$f''(x) = 2x + 1$, so the function is concave downward for $x < -\frac{1}{2}$ and concave upward for $x > -\frac{1}{2}$. There is an inflection point at $x = -\frac{1}{2}$. We note that $f(-\frac{1}{2}) = \frac{1}{12}$.

Since $f''(0) = 1 > 0$, $(0, 0)$ is a local minimum. Since $f''(-1) = -1 < 0$, $(-1, \frac{1}{6})$ is a local maximum.

Figure 2.7–2

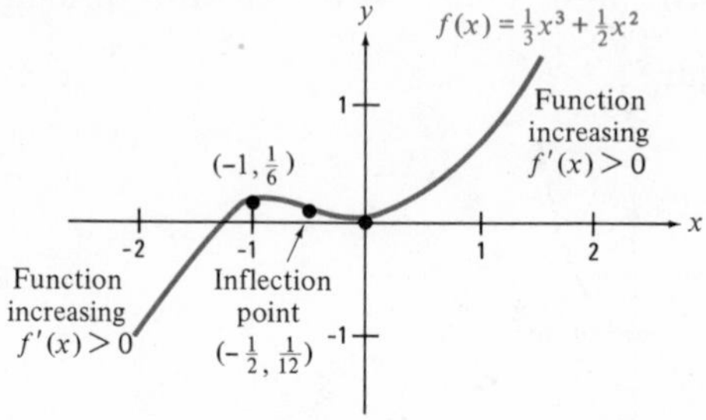

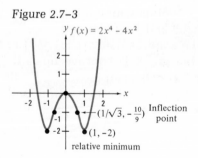

2.7.4   Example

Graph $f(x) = 2x^4 - 4x^2$.

Solution:

$f'(x) = 8x^3 - 8x = 8x(x^2 - 1)$, so the critical points are at $x = 0$ and $x = \pm1$. $f''(x) = 24x^2 - 8 = 8(3x^2 - 1)$, so the inflection points are at $x = \pm1/\sqrt{3}$. From the second derivative test, we can tell that $x = 0$ gives a local maximum and $x = \pm1$ give local minima. Factoring $f''(x) = 8(\sqrt{3}\,x + 1) \cdot (\sqrt{3}\,x - 1)$, we see that the second derivative is negative (and the curve concave downward) for $-1/\sqrt{3} < x < 1/\sqrt{3}$ and positive elsewhere.

Figure 2.7–3

Graphs can be used to give us a visual understanding of many practical functions. For our next example, let us sketch a function like the one we discussed in the inventory model (Section 2.5).

2.7.5   Example

Sketch the graph of $C(x) = 1/x + x$.

Solution:

$C = x^{-1} + x$, so $C' = -x^{-2} + 1 = -1/x^2 + 1$ and $C'' = 2x^{-3} = 2/x^3$. We first notice that C, C', and C'' are not defined at zero, since it is impossible to divide by zero.

Setting $C' = 0$, we find critical points at $x = \pm 1$. They give a local minimum and a local maximum, respectively, since $C''(\pm 1) = \pm 2$. By examining $C'(x)$, we see that if $x > 1$ or $x < -1$, the graph is rising. Elsewhere it is falling. By examining $C''(x)$, we find that $C(x)$ is concave upward if $x > 0$ and is concave downward if $x < 0$. The graph is shown in Figure 2.7–4.

Figure 2.7–4

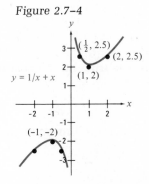

$y = 1/x + x$

$(\tfrac{1}{2}, 2.5)$ $(2, 2.5)$ $(1, 2)$ $(-1, -2)$

Sometimes it is instructive to sketch more than one curve using the same set of axes. (See Figure 2.7–5.) This is true with closely related functions such as a cost function, the corresponding average cost function, and the marginal cost function.

2.7.6   Example

If total cost of producing x items is given by $C(x) = x^2 - 4x + 8$ for $x > 0$, sketch the cost function, the average cost function $A = C/x$, and the marginal cost function, C', on the same set of axes.

Solution:

We have $C(x) = x^2 - 4x + 8$, so $C'(x) = 2x - 4$ and

$$A(x) = \frac{C(x)}{x} = x - 4 + \frac{8}{x} = \frac{8}{x} + x - 4$$

We recognize the graphs of these functions to be, respectively, a parabola, a straight line, and a case examined in Example 2.7.5. Using the information obtained in that example, we graph the three as in Figure 2.7–5.

Figure 2.7–5

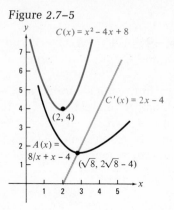

Sometimes such a graph reveals a fact that we might otherwise overlook. For example, the above graph clearly shows that the *marginal cost and average cost are equal where the average cost is a minimum*, in this case. We now prove that this is true for any cost function.

2.7.7   Example

Prove that when average cost is a minimum, the marginal cost equals the average cost.

Solution:

Let $C(x)$ be any cost function. Then $A(x) = C(x)/x$, so

$$A'(x) = \frac{xC'(x) - C(x)(1)}{x^2}$$

by the formula for differentiating quotients, Rule 2.3.3. Setting $A'(x) = 0$ to get the critical points of $A(x)$, we get

$$xC'(x) - C(x) = 0 \quad \text{or} \quad xC'(x) = C(x)$$

which gives

$$C'(x) = \frac{C(x)}{x}$$

where A is a minimum. Since $C'(x)$ is the marginal cost and $C(x)/x$ is the average cost, we have proved they are equal where A is a minimum.

SUMMARY

In this section we used the first and second derivatives to draw graphs of functions that show their critical and inflection points. We explored the inventory model again and showed how graphs can be used to reveal the subtle interplay among the total cost, average cost, and marginal cost functions. Then we proved that where the average cost is a minimum (for any cost function), the average cost equals the marginal cost. In the next section we shall use the techniques of this chapter to investigate several business functions.

EXERCISES   2.7. A

1–8. Sketch the functions in problems 1–8 of Exercises 2.6.A; use your information about the first and second derivatives.
9. If cost is given by the function $C(x) = x^2 - 2x + 10$, sketch the cost, average cost, and marginal cost on the same set of axes.
10. Verify in problem 9 that average cost is a minimum where average cost equals marginal cost.

EXERCISES   2.7. B

1–8. Sketch the functions in problems 1–8 of Exercises 2.6.B; use the facts about first and second derivatives.
9. If the cost function is $C(x) = x^2 - 2x + 9$, sketch the cost, average cost, and marginal cost functions on the same graph.
10. Verify in problem 9 that average cost is a minimum where average cost equals marginal cost.

EXERCISES   2.7. C

1–4. Sketch the functions in the problems of Exercises 2.6.C; use the facts about first and second derivatives.

ANSWERS   2.7. A

1.

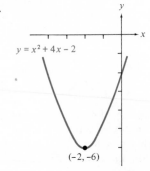

$y = x^2 + 4x - 2$

$(-2, -6)$

2.

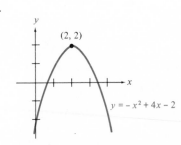

$(2, 2)$

$y = -x^2 + 4x - 2$

3.

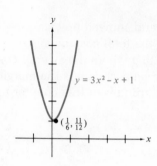

4.

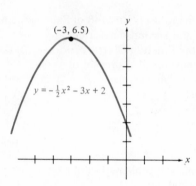

5.

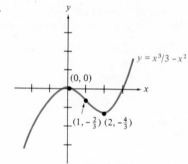

6.

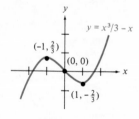

7.

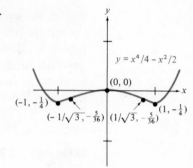

8.

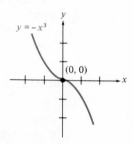

9.

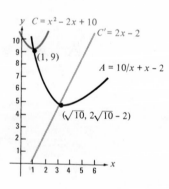

10. Average cost $= x - 2 + (10/x)$, so the derivative of the average cost is $1 - 10x^{-2}$. This equals zero when $x = \sqrt{10}$, and since the second derivative is positive when $x = \sqrt{10}$, this is a local minimum. When $x = \sqrt{10}$, both the average cost and the marginal cost are $2\sqrt{10} - 2$.

ANSWERS   2.7. C

1.

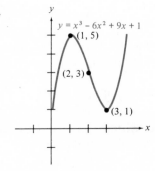

$y = x^3 - 6x^2 + 9x + 1$

(1, 5)

(2, 3)

(3, 1)

2.

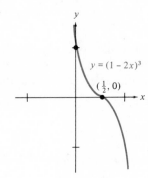

$y = (1 - 2x)^3$

$(\frac{1}{2}, 0)$

3.

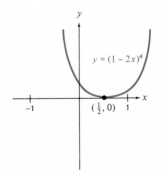

$y = (1 - 2x)^4$

$(\frac{1}{2}, 0)$

4.

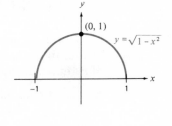

(0, 1)

$y = \sqrt{1 - x^2}$

2.8   More Marginal Analysis

In Section 1.1 we listed several common relationships among business variables, including the cost, revenue, profit, demand, and supply functions. In Section 1.6 we suggested how the marginal concept leads directly to that of the derivative as we pass to limits. Marginal analysis is concerned with the way in which changes in the independent variable (such as the quantity produced) affect the dependent variable of a business function. The problems in this chapter provide interesting applications and interpretations of the concept of a derivative in the business world.

In this section we shall let the independent variable be q instead of x because this notation is common in books about the applications where the independent variable designates the "quantity" produced. We shall use a different notation for derivatives because it explicitly shows both variables; as we consider more varied problems, this is desirable.

2.8.1   Example

(a) If $C = f(q)$ is the cost function, $\dfrac{dC}{dq} = f'(q)$ is the marginal cost.

(b) If $p = g(q)$ is the price function, then $R = p \cdot q = g(q) \cdot q$ is the revenue function, since the total revenue is the price per item times the quantity sold. The marginal revenue function is

$$\frac{dR}{dq} = g'(q) \cdot q + g(q) \cdot 1 = g'(q)q + g(q)$$

(using the product rule for differentiation, Rule 2.3.2).

(c) If $P = R - C$ is the profit function, then

$$\frac{dP}{dq} = \frac{dR}{dq} - \frac{dC}{dq}$$

is the marginal profit.

Before giving examples, we make two observations. Following the custom in business and economics books, we shall consistently let capital P represent the profit function. Lowercase p stands for price. Second, we want to emphasize that $\dfrac{dC}{dq}$ is *not* a fraction. It is merely another way of writing $C'(q)$. But $\dfrac{dC}{dq}$ indicates not only the cost (C' does this, too) but also shows that we are considering how the cost varies as the *quantity produced* varies. Later we might want to consider how it varies as the net sales vary $\left(\dfrac{dC}{dN} \text{ in Section 3.1}\right)$ or as time passes $\left(\dfrac{dC}{dt} \text{ in Section 3.6}\right)$. The derivative $\dfrac{dC}{dq}$ is the limit of fractions, $\dfrac{\Delta C}{\Delta q}$, but it is not itself a fraction. We repeat: $\dfrac{dC}{dq} = C'(q)$.

2.8.2   Example

Suppose that the cost function follows the linear model of Section 1.2 with fixed cost of $100 per day and marginal cost of $10 per item. Suppose, further, that price is related to quantity demanded by $p = 110 - 0.01q$. Find explicit formulas for cost, price, revenue, profit, marginal cost, the derivative of the price, marginal revenue, and marginal profit.

$$C = 100 + 10q \qquad \text{and} \qquad \frac{dC}{dq} = 10$$

$$p = 110 - 0.01q \qquad \text{and} \qquad \frac{dp}{dq} = -0.01$$

$$R = (110 - 0.01q)q = 110q - 0.01q^2 \quad \text{and} \quad \frac{dR}{dq} = 110 - 0.02q$$

$$P = (110q - 0.01q^2) - (100 + 10q) = 100q - 0.01q^2 - 100 \quad \text{and}$$

$$\frac{dP}{dq} = 100 - 0.02q$$

As we have observed before, where the profit is a maximum, $\frac{dP}{dq} = 0$. This happens precisely when the marginal cost equals marginal revenue, since

$$\frac{dP}{dq} = \frac{dR}{dq} - \frac{dC}{dq} = 0 \quad \text{if and only if} \quad \frac{dR}{dq} = \frac{dC}{dq}$$

We now explore in detail some mathematical models that involve marginal analysis.

Maxima and Minima

A business function, as we have pointed out, is merely a "model" of the real world and may be somewhat inaccurate. But if it is a good model, it will reflect the pattern of the total situation. Often, it can then be used to find local maxima and minima by setting its derivative equal to zero. To find absolute extrema, we must also check the endpoints.

2.8.3   **Example**

Find the maximum profit if $P(q) = 100q - 0.01q^2 - 100$.

Solution:

Since $\frac{dP}{dq} = 100 - 0.02q$, $q = 5,000$ is a critical point. Since $\frac{d^2P}{dq^2} = -0.02$, the parabola does not "hold water" and the critical point gives a maximum. Thus

$$P(5,000) = 500,000 - 0.01(5,000)^2 - 100 = \$249,900$$

is the maximum profit.

2.8.4   **Example**

Find the minimum average cost if total cost is $C(q) = 100q - 2q^2 + 0.02q^3$. [*Remember:* average cost $= A(q) = C(q)/q$.]

Solution:

$$A(q) = \frac{C(q)}{q} = 100 - 2q + 0.02q^2, \quad \text{so} \quad \frac{dA}{dq} = -2 + 0.04q, \quad \text{which is zero}$$

when $q = 50$. $\dfrac{d^2A}{dq^2} = 0.04 > 0$, so we have a local minimum at $q = 50$. Since $A(50) = 100 - 2(50) + 0.02(50)^2 = 50$, the minimum average cost is \$50.

Revenue and Elasticity of Demand

It might seem at first that the more an industry sells, the more revenue it receives. But this is not necessarily so. Consider our national farm problems; for some commodities, the *more* the farmers sell, the *less* their income! (This is because the price decreases proportionately more than the quantity increases.) This is a situation of <u>inelastic demand</u>; the greater the quantity sold, the less the revenue. We shall now examine this phenomenon in more detail.

When the price, p, of a commodity is considered as a function of the quantity sold, q (the demand function of Section 1.2), then a change, Δq, in q will yield a corresponding change, Δp, in p. If Δq is positive, Δp will be negative, because an increase in the quantity supplied will correspond to a lower price. Furthermore, the proportional change in quantity, $\dfrac{\Delta q}{q}$, may be different from the proportional change in price, $\dfrac{\Delta p}{p}$. When we want to compare these two proportional changes, we can consider their ratio.

$$\frac{\Delta q/q}{\Delta p/p} = \frac{p}{q \dfrac{\Delta p}{\Delta q}}$$

Since $\dfrac{\Delta p}{\Delta q}$ is always negative (because the demand function is decreasing), we put a minus sign in front of this expression to make it positive. Then we take a limit and define "elasticity."

2.8.5   **Definition**

If $p = f(q)$ is a differentiable price function, the <u>price elasticity of demand</u> is

$$E = -\lim_{\Delta q \to 0} \frac{p}{q \dfrac{\Delta p}{\Delta q}} = -\frac{p}{q \dfrac{dp}{dq}}$$

If $E > 1$:   elastic demand
If $E < 1$:   inelastic demand
If $E = 1$:   unit elasticity

Some books define elasticity of demand without a minus sign, but our definition seems to be more common and is certainly more convenient for the purposes of this text.

If the proportional change in quantity, $\dfrac{\Delta q}{q}$ is exactly equal to the proportional change in price, $\dfrac{\Delta p}{p}$, then their ratio is exactly 1 and we say we have unit elasticity; this is why we write "$E = 1$: unit elasticity" in Definition 2.8.5. If the proportional change in quantity is greater than the proportional change in price, then the elasticity, defined in Definition 2.8.5, is greater than 1 and we say we have elastic demand. In Example 2.8.7 we shall show that the demand is elastic if and only if an increase in the quantity sold brings an increase in revenue to the seller. In the opposite case — when $E < 1$ and an increase in quantity sold brings a decrease of total revenue — we say we have inelastic demand.

2.8.6   Example

If $p = 100 - q$, find expressions for (a) the revenue, (b) the marginal revenue, (c) the elasticity. (d) Then find the numerical value of the elasticity when $q = 40$. (e) For what q is there unit elasticity? (f) For what values of q is there elastic demand? (g) For what values of q is there inelastic demand? (h) For what q is the revenue a maximum?

Solution:

(a) Since Revenue = (price)(quantity),

$$R = (100 - q)q = 100q - q^2$$

(b) To find the marginal revenue, we differentiate:

$$\frac{dR}{dq} = 100 - 2q$$

(c) To find the elasticity, we find that $\dfrac{dp}{dq} = -1$ and substitute into the above expression for elasticity:

$$E = -\frac{p}{q\dfrac{dp}{dq}} = -\frac{100 - q}{q(-1)} = \frac{100 - q}{q}$$

(d) $E = (100 - 40)/40 = 60/40 = \frac{3}{2}$

(e) To find the point at which there is unit elasticity, we set $E = 1$ and solve for q:

$$\frac{100 - q}{q} = 1$$
$$100 - q = q$$
$$100 = 2q$$
$$50 = q$$

(f) The demand is elastic when $E > 1$, so now we solve an inequality:

$$\frac{100 - q}{q} > 1$$

$$100 - q > q$$
$$100 > 2q$$
$$50 > q$$

(g) The demand is inelastic when $E < 1$, which is when $E \neq 1$ and $E \not> 1$. In other words, the demand is inelastic when

$$q > 50$$

(h) To find the point at which the revenue is a maximum, we set the marginal revenue equal to zero and solve for q:

$$100 - 2q = 0$$
$$100 = 2q$$
$$50 = q$$

Notice that the revenue is a maximum when $E = 1$. This is true in general, as we shall show in Example 2.8.7.

2.8.7   Example

Prove that for any demand function, $p = f(q)$, revenue, $R = qp$, is a maximum precisely when $E = 1$. Prove that the revenue is increasing (with respect to quantity sold) if $E > 1$ and that the revenue is a decreasing function of quantity if $E < 1$.

Solution:

$$R(q) = q \cdot p$$

$$R'(q) = 1 \cdot p + q \frac{dp}{dq} = p \left[1 + \frac{q \frac{dp}{dq}}{p} \right] = p \left(1 - \frac{1}{E} \right)$$

Since p is never zero in the region of interest, we can have a critical point of $R(q)$ only when $1 - (1/E) = 0$, which happens precisely when $E = 1$.

Looking again at the expression for $R'(q)$, we see that if $E > 1$, $1/E < 1$, so $R' > 0$ and the revenue is increasing. When $E > 1$, we say that the demand is elastic.

If $0 < E < 1$, then $1/E > 1$, so $R' < 0$, which means that the revenue decreases as the quantity sold increases. This situation (having $E < 1$ for an industry) can have troublesome effects on an economy; the farm problems mentioned earlier are but one example. When $E < 1$, we say that the demand is inelastic.

Approximations

The alert reader may have noticed that we have introduced three slightly different meanings for the phrase "marginal cost":

1. "ΔC when $\Delta q = 1$" or "the additional cost incurred in producing an extra item" (Section 1.1).

2. "$\dfrac{\Delta C}{\Delta q}$," or "the change in cost per item when Δq more items are produced" (Section 1.6). This is especially useful when Δq is "small," where the meaning of "small" depends upon the problem.

3. "$\dfrac{dC}{dq} = C'(q)$" or "the derivative of the cost" (Section 2.1). This is often the easiest to calculate, so it is used to approximate the other two numbers. (See Example 2.8.8.)

In actual practice the values of these three numbers, if they are different, are so close that the slight difference does not matter.

The first two definitions above describe the "discrete case" and the third is the "continuous case." Notice that the second definition is exactly the same as the first if $\Delta q = 1$, and the third is the limit of the second as $\Delta q \to 0$. Similar comments can be made for marginal revenue and marginal profit.

2.8.8   Example

Using the functions of Example 2.8.2, evaluate for $q = 100$:
(a) marginal cost
(b) marginal revenue
(c) marginal profit
In each case use the first definition of the three above (for example, ΔC when $\Delta q = 1$) and then compute the quantity using the derivative. Notice how much easier it is to use the derivative! The answers may be slightly different in each case, but not much.

Solution:

(a) Since $C(q) = 100 + 10q$ (given on page 124), we have, for $q = 100$ and $\Delta q = 1$,

$$\Delta C = C(101) - C(100) = (100 + 1{,}010) - (100 + 1{,}000) = 10$$

Using the derivative definition of marginal cost, we have $\dfrac{dC}{dq} = 10$, so in this case the two answers are exactly the same; this is because the function is linear.

(b) Since $R(q) = 110q - 0.01q^2$ (given on page 125), we have, for $q = 100$ and $\Delta q = 1$,

$$\Delta R = R(101) - R(100) = [110(101) - 0.01(101)^2]$$
$$- [11{,}000 - 100]$$
$$= 107.99$$

If we use the derivative, we get

$$\frac{dR}{dq} = 110 - 0.02q = 110 - 0.02(100) = 110 - 2 = 108$$

This time the answers are not exactly the same (because the function is not linear), but they differ by only $108 - 107.99 = 0.01$.

(c) Since $P(q) = 100q - 0.01q^2 - 100$ (given on page 125), we have, for $q = 100$ and $\Delta q = 1$,

$$\begin{aligned} \Delta P &= P(101) - P(100) \\ &= [100(101) - 0.01(101)^2 - 100] \\ &\quad - [100(100) - 0.01(100)^2 - 100] \\ &= 97.99 \end{aligned}$$

If we use the derivative now, we get

$$\frac{dP}{dq} = 100 - 0.02q = 100 - 0.02(100) = 100 - 2 = 98$$

Again the difference between the two answers is only 0.01.

Notice that we could have calculated the marginal profit by subtracting marginal cost from marginal revenue (using either definition, as long as we are consistent).

SUMMARY

In this section we explored some of the more common interpretations of derivatives in the financial world. We began by shifting our notation from the simple C' to $\frac{dC}{dq}$, which displays the independent variable as well as the dependent variable; this is desirable both because it is common in books on applications and because it will be useful later in mathematics. After reviewing applications to minima and maxima using the "new" notation, we turned to the concept of elasticity. This was explored in detail for one demand function, and then we showed that in general the elasticity equals 1 where the revenue is a maximum, no matter what the demand function. Finally, we investigated the relationship between discrete and continuous marginal concepts.

EXERCISES   2.8. A

1. Suppose that $p = 5 - \frac{1}{2}q$ is the price per item when a given quantity q is demanded, and the costs of producing q items is given by $C = 3q - \frac{1}{4}q^2 + 10$. Use the notation of this section to write
    (a) The revenue function
    (b) The marginal revenue function
    (c) The marginal cost function
    (d) The profit function
    (e) The marginal profit function

2. Using the functions in problem 1, and the "one extra unit" definitions, find, when $q = 3$,
   (a) The marginal revenue
   (b) The marginal cost
   (c) The marginal profit

3. Using the functions in problem 1, and the "calculus" definitions, find, when $q = 3$,
   (a) The marginal revenue
   (b) The marginal cost
   (c) The marginal profit
      (Notice how much more time-consuming the arithmetic is in problem 2 than in problem 3. The latter, however, merely approximates the former.)

4. Using the functions in problem 1,
   (a) Find the value of q for which revenue is a maximum.
   (b) Find the value of q for which profit is a maximum.
   (c) Graph the demand function, the revenue function, and the marginal revenue function on one graph.

5. Using the functions in problem 1,
   (a) Find the elasticity function.
   (b) Find values of q that give unit elastic, elastic, and inelastic demand. (Note that revenue is a maximum when $E = 1$.)

6. If $p = 150 - 2q$, find
   (a) The total revenue function $R(q)$
   (b) The average revenue function R/q
   (c) The marginal revenue function $R'(q)$
   (d) The quantity that yields maximum revenue

7. If $p = 150 - 2q$, find
   (a) The elasticity function
   (b) The quantity that yields maximum revenue [Use $E = 1$ and compare with your answer to problem 6(d).]

8. If $p = -q^2 - q + 100$, find
   (a) The total revenue function $R(q)$
   (b) The average revenue function R/q
   (c) The marginal revenue function $R'(q)$
   (d) The quantity that yields a maximum revenue

9. If $p = -q^2 - q + 100$, find
   (a) The elasticity function
   (b) The quantity that yields a maximum revenue [Use $E = 1$ and compare with your answer to problem 8(d).]

EXERCISES   2.8. B

1. Suppose that $p = 4 - \frac{1}{3}q$ is the price per item when a given quantity q is demanded, and the cost of producing q items is given by $C = \frac{7}{3}q - \frac{1}{6}q^2 + 1$. Use the notation of this section to write
   (a) The revenue function
   (b) The marginal revenue function
   (c) The marginal cost function

      (d) The profit function

      (e) The marginal profit function

2.   Using the functions in problem 1, and the "one extra unit" definitions, find, when $q = 4$,

      (a) The marginal revenue

      (b) The marginal cost

      (c) The marginal profit

3.   Using the functions in problem 1, and the "calculus" definitions, find, when $q = 4$,

      (a) The marginal revenue

      (b) The marginal cost

      (c) The marginal profit

      (Notice how much more time-consuming the arithmetic is in problem 2 than in problem 3. The latter, however, merely approximates the former.)

4.   Using the functions in problem 1,

      (a) Find the value of q for which revenue is a maximum.

      (b) Find the value of q for which profit is a maximum.

      (c) Graph the demand function, the revenue function, and the marginal revenue function on one graph.

5.   Using the functions in problem 1,

      (a) Find the elasticity function.

      (b) Find the values of q that give unit elastic, elastic, and inelastic demand. (Note that revenue is a maximum when $E = 1$.)

6.   If $p = 120 - 3q$, find

      (a) The total revenue function $R(q)$

      (b) The average revenue function R/q

      (c) The marginal revenue function $R'(q)$

      (d) The quantity that yields maximum revenue

7.   If $p = 120 - 3q$, find

      (a) The elasticity function

      (b) The quantity that yields maximum revenue [Use $E = 1$ and compare with your answer to problem 6(d).]

8.   If $p = -q^3 + 500$, find

      (a) The total revenue function $R(q)$

      (b) The average revenue function R/q

      (c) The marginal revenue function $R'(q)$

      (d) The quantity that yields a maximum revenue

9.   If $p = -q^3 + 500$, find

      (a) The elasticity function

      (b) The quantity that yields a maximum revenue [Use $E = 1$ and compare with your answer to problem 8(d).]

ANSWERS   2.8. A

1.   (a) $R(q) = 5q - \tfrac{1}{2}q^2$ (b) $\dfrac{dR}{dq} = 5 - q$ (c) $\dfrac{dC}{dq} = 3 - \tfrac{1}{2}q$

      (d) $P(q) = 2q - \tfrac{1}{4}q^2 - 10$ (e) $\dfrac{dP}{dq} = 2 - \tfrac{1}{2}q$

2.   (a) 1.5 (b) 1.25 (c) 0.25

3. (a) 2 (b) $\frac{3}{2}$ (c) 0.5
4. (a) $q = 5$ (b) $q = 4$

(c)

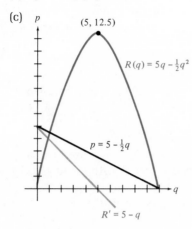

5. (a) $E = (5 - \frac{1}{2}q)/\frac{1}{2}q = (10 - q)/q$ (b) unit elastic at $q = 5$; elastic demand for $q < 5$; inelastic demand for $q > 5$
6. (a) $R(q) = 150q - 2q^2$ (b) $R/q = 150 - 2q$ (c) $R'(q) = 150 - 4q$ (d) 37.5
7. (a) $E = (150 - 2q)/2q$ (b) 37.5
8. (a) $R(q) = -q^3 - q^2 + 100q$ (b) $R/q = -q^2 - q + 100$
    (c) $R'(q) = -3q^2 - 2q + 100$ (d) $\dfrac{\sqrt{301} - 1}{3}$
9. (a) $E = (-q^2 - q + 100)/[q(2q + 1)]$ (b) $\dfrac{\sqrt{301} - 1}{3}$

2.9 Newton's Method and Finding the Zeros of a Function

How important it is to be able to find the points at which a function is equal to zero! To maximize a profit function, we find where the corresponding marginal profit (its derivative) is equal to zero. To minimize cost, we set the marginal cost function equal to zero.

It is easy to determine the single <u>zero</u> (that is, the value of x for which the function is zero) of a linear function $f(x) = mx + b$. If $m \neq 0$, it is just $x = -b/m$, the x-intercept. To find the zeros of a quadratic function $g(x) = ax^2 + bx + c$, we can use the quadratic formula:

$$x = \frac{-b \pm \sqrt{b^2 - 4ac}}{2a}$$

But for a function more complicated than a linear or quadratic function, there is usually no simple formula for finding the zeros. To approximate the zeros of such a function, we might first use the <u>graphical method</u>:

1. Graph the function, using the techniques explained earlier in this chapter.
2. See where the graph of the function crosses the x-axis; the x-coordinates of such points are the zeros of the function.

2.9.1   Example

Approximate the zeros of the function $f(x) = x^3/3 - x^2 + 1$ using the graphical method.

Solution:

Since $f'(x) = x^2 - 2x = x(x - 2)$, there are critical points of the function at $x = 0$ and $x = 2$. Testing each of these in $f''(x) = 2x - 2$, we see that the function has a local maximum at $x = 0$ and a local minimum at $x = 2$. We use these facts to graph the function as in Figure 2.9–1. We

Figure 2.9–1

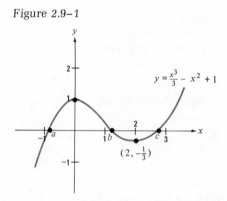

$$y = \frac{x^3}{3} - x^2 + 1$$

$$\left(2, -\tfrac{1}{3}\right)$$

look carefully at the graph to get estimates for the three zeros of this cubic; we guess that $b \approx 1.2$, $c \approx 2.6$, and $a \approx -0.9$. A quick check reveals that $f(1.2) = 0.136 \neq 0$. Hence, our estimate for b is close, but not exact. You can check that the same is true for our estimates of a and c.

Example 2.9.1 shows the limitations of the graphical method. A more precise graph would yield better answers, but a mere picture can never be regarded with complete confidence. It is more satisfactory, when precision is needed, to use a graph as a *first* approximation and then to use the following technique (named after Sir Isaac Newton, one of the inventors of the calculus) to find better and better results.

2.9.2   Newton's Method [for Approximating the Zeros of a Function $f(x)$]

1. Approximate a zero, x_0, of a function $f(x)$ using graphing.
2. Find the derivative f' of f and compute $x_1 = x_0 - [f(x_0)/f'(x_0)]$.

3. Using x_1 as the new approximation, compute $x_2 = x_1 - [f(x_1)/f'(x_1)]$.
4. Continue to get better approximations by letting, at each step,

$$x_{n+1} = x_n - \frac{f(x_n)}{f'(x_n)}$$

Important: At each stage it is essential that $f'(x_n) \neq 0$.

2.9.3   Example

Approximate the smaller positive zero of the function $f(x) = x^3/3 - x^2 + 1$ using three iterations of Newton's method.

Solution:

Using Figure 2.9–1, we guessed that $b = 1.2$. Therefore, we set $x_0 = 1.2$. Next, we find that $f'(x) = x^2 - 2x$, so $f'(1.2) = -0.96$ and $f(1.2) \approx 0.136$ gives

$$x_1 = 1.2 - \frac{f(1.2)}{f'(1.2)} \approx 1.342$$

Since $f(1.342) \approx 0.005$ and $f'(1.342) \approx -0.883$,

$$x_2 = 1.342 - \frac{f(1.342)}{f'(1.342)} \approx 1.348$$

And since $f(1.348) \approx -0.0006$ and $f'(1.348) \approx -0.879$,

$$x_3 = 1.348 - \frac{f(1.348)}{f'(1.348)} \approx 1.3473$$

You can check that $f(1.3473)$ is zero to five decimal places.

2.9.4   Example

Approximate the negative zero of the function $f(x) = x^3/3 - x^2 + 1$ using two iterations of Newton's method.

Solution:

Again referring to Example 2.9.1, we set $x_0 = -0.9$ (although in Example 2.9.1, we called it a). Since $f(-0.9) = -0.053$ and $f'(-0.9) = 2.61$,

$$x_1 = -0.9 - \frac{f(-0.9)}{f'(-0.9)} \approx -0.880$$

And since $f(-0.880) \approx -0.0016$ and $f'(-0.880) \approx 2.534$,

$$x_2 = -0.880 - \frac{f(-0.880)}{f'(-0.880)} \approx -0.8794$$

You can check that $f(-0.8794)$ is zero, to four decimal places.

Let us now see why Newton's method often works. The tangent line to the graph of $y = f(x)$ at the point $(x_0, f(x_0))$ has the equation $y - f(x_0) = f'(x_0)(x - x_0)$. This line crosses the x-axis where $x = x_0 - f(x_0)/f'(x_0)$; this is the point that we call x_1. The value of x_1 will often be closer to the place where $f(x)$ crosses the x-axis than x_0 was. We then repeat this process, with x_1 playing the role of x_0 above to obtain x_2; each x_{n+1} is obtained from x_n in a similar fashion. (See Figure 2.9–2.)

Figure 2.9–2

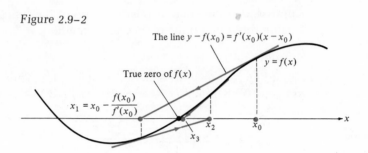

If at any point $f'(x_n) = 0$, Newton's method fails, because we cannot divide by zero. It is also possible that the sequence does not converge to the zero of the function, but this is rare; usually the x_n's, get closer and closer to the desired zero very rapidly.

2.9.5   Example

The cost of manufacturing x tons of a certain product is given by the function $C(x) = 250 + 2x - 0.1x^3$ and the revenue obtained by selling the x tons is $R(x) = 3x$. Approximate the break-even point, the point where $C(x) = R(x)$, for this product.

Solution:

If $R(x) = C(x)$, then $P(x) = R(x) - C(x) = 0$. Hence we must find a zero of the profit function, $3x - (250 + 2x - 0.1x^3)$. It is easy to see that $P'(x) = 1 + 0.3x^2 > 0$, so the profit is always increasing as we produce more. Since $P(0) = -250$, it follows that if $P(x) = 0$, $x > 0$.

Using our hand calculator to quickly compute a few values for P, we discover that $P(13) = -17.3$ and $P(14) = 38.4$. We decide, therefore, to set $x_0 = 13.3$. Using Newton's method, we obtain the following:

$$x_0 = 13.3 \qquad P(x_0) = -1.436 \qquad P'(x_0) = 54.067$$
$$x_1 = 13.327 \qquad P(x_1) = 0.026 \qquad P'(x_1) = 54.283$$
$$x_2 = 13.3265 \qquad P(x_2) = -0.0007 \text{ (which is very close to zero)}$$

EXERCISES 2.9. A

1. Find the zeros of the functions
    (a) $f(x) = 3x - 20$ (b) $f(x) - 5 = 3(x + 2)$
2. Use the quadratic formula to find the zeros of the function $f(x) = 2x^2 + 5x - 1$.

HC 3. Let $f(x) = 2x^3 - 1$. Since $f(x)$ is always increasing (how do we know this?), there is exactly one zero.
    (a) Use your calculator to find this zero to the nearest tenth by trial and error.
    (b) Use Newton's method, beginning with the result of part (a), until you find an x such that $|f(x)| < 0.005$.

HC 4. Suppose that we want to find the zeros of $f(x) = x^3 + x + 1$. Since $f(0) = 1$ and $f(-1) = -1$, let $x_0 = -0.5$. Use Newton's method to improve this value until you find x such that $|f(x)| < 0.005$.

EXERCISES 2.9. B

1. Find the zeros of the functions
    (a) $f(x) = 2x + 30$ (b) $f(x) + 5 = 3(x - 2)$
2. Use the quadratic formula to find the zeros of the function $f(x) = -x^2 + 3x + 1$.

HC 3. Let $f(x) = 3x^3 - 1$. Since $f(x)$ is always increasing (how do we know this?), there is exactly one zero.
    (a) Use your calculator to find this zero to the nearest tenth by trial and error.
    (b) Use Newton's method, beginning with the result of part (a), until you find an x such that $|f(x)| < 0.005$.

HC 4. Setting $x_0 = 1$ as a first approximation to a zero of $f(x) = x^3 + x^2 + x - 1$, use Newton's method to approximate this zero until you find x such that $|f(x)| < 0.005$.

EXERCISES 2.9. C

1. Convince yourself that $f(x) = x^3 + x^2 - 2x$ has a zero at $x = 0$ and at $x = 1$. Let $x_0 = 0.5$ be an initial approximation to either zero. Does Newton's method give an increasingly better approximation to either zero? What happens?

ANSWERS 2.9. A

1. (a) $x = \frac{20}{3}$. (b) $x = -\frac{11}{3}$.
2. $x = (-5 \pm \sqrt{33})/4$
3. The function is always increasing because $f'(x) = 6x^2 \geqslant 0$.
    (a) $x = 0.8$ (b) $x_1 = 0.794$ and $f(x_1) \approx 0.001$
4. $x_2 = -0.683$ and $f(x_2) \approx -0.002$

ANSWERS   2.9. C

1.   One obtains $x_1 = -2$. While $f(-2) = 0$, $x = -2$ is not the zero we sought. See the accompanying figure.

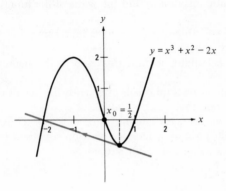

SAMPLE QUIZ   **Sections 2.5, 2.6, 2.7, and 2.8**

Do problem 1 and any three of the remaining four problems. Each problem counts 25 points.

Find the critical points and inflection points of each of the following functions. Use the second derivative to find whether the critical points are local maxima or minima. Then graph the functions.

1. $y = 4x - 2x^2 - 1$

2. $y = x^3/3 - x + 1$

3.   Suppose that a price–demand relationship is given by $p = 20 - 2q$.
   (a) Write an expression for the revenue.
   (b) Write an expression for the marginal revenue.
   (c) Write an expression for the elasticity.
   (d) For what q is there unit elasticity?
   (e) For what values of q is the demand elastic?

4.   Suppose that the price is given by $p = 20 - 2q$ and the cost is given by $C = 8q + q^2 - 1$. For what q is the profit a maximum? What is the maximum profit?

5.   Suppose that a company wants to make 1,000 of a certain item each year. It costs $10 to get ready for each batch, $8 to keep an item in inventory, and $4 per item for the additional costs per item.
   (a) What is the best batch size if costs are to be minimized?
   (b) What is the minimum total cost?
   (c) How often should batches be made?

The answers are at the back of the book.

SAMPLE TEST   **Chapter 2**

1. (24 pts) Find the first, second, and third derivatives of $y = 2x + (1/x) + 4\sqrt{x}$.
2. (6 pts) Find y' if $y = (x^2 + 3)(4x^3 + 2x)$.
3. (25 pts) Let $y = 2x^3 - 3x^2$.
    (a) Find the critical point(s) and inflection point(s) of this curve.
    (b) Sketch the graph.
    (c) Write the equation of the tangent line to this curve where $x = -1$.
4. (20 pts) Suppose that the profit when x items are produced is given by $P(x) = 6x - 200 - 0.03x^2$.
    (a) Write the expression for the marginal profit.
    (b) For what x is the profit a maximum?
    (c) How can you tell it is a maximum, not a minimum, at this point?
    (d) What is the maximum profit?
5. (15 pts) Suppose that the price is given by $p = 100 - 0.2q$, where q is the quantity demanded.
    (a) What is the price when 200 items are wanted?
    (b) Write an equation for the revenue.
    (c) Write an equation for the marginal revenue.
    (d) For what quantity q is the revenue a maximum?
    (e) Write an equation for the elasticity of demand for this function, given that

$$E = \frac{p}{-q\,\dfrac{dp}{dq}}$$

    (f) Is the demand elastic or inelastic when $q = 200$?
    (g) For what q is the elasticity equal to 1?
6. (10 pts) Use the *definition* of the derivative to find the derivative of $y = 3x^2 + 5x$.

The answers are at the back of the book.

Chapter 3

DERIVATIVES: PART II

3.1 The Chain Rule

In Section 2.2 we noticed that some functions had to be manipulated algebraically before they could be differentiated using the techniques of that section. For example, to differentiate $y = (x + 1)^4$ we had first to compute $y = x^4 + 4x^3 + 6x^2 + 4x + 1$ and then differentiate to get $y' = 4x^3 + 12x^2 + 12x + 4 = 4(x + 1)^3$. But sometimes algebraic computations cannot put a function into a form that can be differentiated using the skills of Sections 2.2 and 2.3. Even such an innocent function as $f(x) = (x + 1)^{1/2}$ cannot be differentiated with present skills unless we turn back to the definition. To differentiate such functions, we need another rule, which is the subject of this section.

We can often use a process called "composition" of functions to write more complicated functions in terms of simpler ones. Then we can use the techniques involving the simpler functions and the "chain rule" (pages 141–142) to find the derivative of the more complicated function.

3.1.1 Definition

If the domain of a function f includes the range of a function g, then $h(x) = f(g(x))$ is called the composition of f with g.

Figure 3.1–1

$$x \longrightarrow \boxed{g} \longrightarrow g(x) \longrightarrow \boxed{f} \longrightarrow f(g(x))$$

3.1.2 Example

Let $f(x) = 3x + 2$ and $g(x) = x^2$. Find
(a) $f(g(2))$
(b) $g(f(2))$
(c) $f(g(x))$
(d) $g(f(x))$

Solution:

(a) $f(g(2)) = f(2^2)$ [substituting 2 for x in $g(x) = x^2$]
          $= f(4)$ (since $2^2 = 4$)
          $= 3(4) + 2$ [substituting 4 for x in $f(x) = 3x + 2$]
          $= 12 + 2 = 14$

140

(b) $g(f(2)) = g(3(2) + 2)$ [substituting 2 for x in $f(x) = 3x + 2$]
    $= g(6 + 2) = g(8)$ (by arithmetic)
    $= 8^2$ [substituting 8 for x in $g(x) = x^2$]
    $= 64$

Thus we conclude that $f(g(2)) = 14$ and $g(f(2)) = 64$. The two answers are very different. In other words, $f(g(2)) \neq g(f(2))$. Look closely at how we obtain these two answers because now we shall perform the same computations for a general x instead of the specific number 2.
(c) $f(g(x)) = f(x^2) = 3x^2 + 2$

Figure 3.1-2
$$x \longrightarrow \boxed{g} \longrightarrow x^2 \longrightarrow \boxed{f} \longrightarrow 3x^2 + 2$$

(d) $g(f(x)) = g(3x + 2) = (3x + 2)^2 = 9x^2 + 12x + 4$

Figure 3.1-3
$$x \longrightarrow \boxed{f} \longrightarrow 3x + 2 \longrightarrow \boxed{g} \longrightarrow (3x + 2)^2$$

Notice again that $f(g(x)) \neq g(f(x))$; the answer will change if we change the order in which the functions work.

3.1.3   Example

Let $f(x) = x^2 + 1$ and $g(x) = \sqrt{x} = x^{1/2}$. Find
(a) $f(g(2))$
(b) $g(f(2))$
(c) $f(g(x))$
(d) $g(f(x))$

Solution:
(a) $f(g(2)) = f(\sqrt{2}) = (\sqrt{2})^2 + 1 = 2 + 1 = 3$
(b) $g(f(2)) = g(2^2 + 1) = g(5) = \sqrt{5}$ [since $3 \neq \sqrt{5}$, we see that $f(g(2)) \neq g(f(2))$]
(c) $f(g(x)) = f(\sqrt{x}) = (\sqrt{x})^2 + 1 = x + 1$
(d) $g(f(x)) = g(x^2 + 1) = \sqrt{x^2 + 1}$, which is the innocent function that we could not differentiate at the beginning of this section. [Again, $f(g(x)) \neq g(f(x))$.]

3.1.4   Chain Rule

If $y = f(u)$ and $u = g(x)$ are differentiable functions that define the composite function $y = f(g(x)) = h(x)$, then
$$h'(x) = f'(g(x)) \cdot g'(x)$$

In other words,

$$\frac{dy}{dx} = \frac{dy}{du} \cdot \frac{du}{dx}$$

Warning! The preceding expression looks as if it contains three fractions, but it doesn't. It contains three derivatives, arranged to state a relationship among them that is not particularly obvious. But the notation is suggestive, so it is popular.

The proof of the chain rule is complicated and therefore is not included in this book. But the chain rule itself is not difficult to use once you get the knack of it. Often it is helpful to let u equal the expression in parentheses, as in the following examples.

3.1.5   Example

Find the derivatives, where they exist, of
(a) $y = (x + 1)^4$
(b) $y = (7x + 1)^4$
(c) $y = (x^2 + 1)^4$
(d) $y = (x^2 + 1)^{1/2}$

Solution:

(a) Let $u = x + 1$. Then $y = u^4$ and we have

$$\frac{du}{dx} = 1 \qquad \text{and} \qquad \frac{dy}{du} = 4u^3$$

so

$$\frac{dy}{dx} = \frac{dy}{du}\frac{du}{dx} \qquad \text{(by the chain rule)}$$

$$\begin{aligned} &= 4u^3 \cdot 1 && \text{(by substituting the above expressions)} \\ &= 4(x + 1)^3 && \text{(by substituting } u = x + 1) \end{aligned}$$

(b) Let $u = 7x + 1$. Then $y = u^4$ and we have

$$\frac{du}{dx} = 7 \qquad \text{and} \qquad \frac{dy}{du} = 4u^3$$

so

$$\frac{dy}{dx} = \frac{dy}{du}\frac{du}{dx} \qquad \text{(by the chain rule)}$$

$$\begin{aligned} &= 4u^3 \cdot 7 && \text{(by substituting the above expressions)} \\ &= 28(7x + 1)^3 && \text{(by multiplying } 4 \cdot 7 = 28 \text{ and substituting} \\ &&& u = 7x + 1) \end{aligned}$$

(c) Let $u = x^2 + 1$. Then $y = u^4$ and we have

$$\frac{du}{dx} = 2x \quad \text{and} \quad \frac{dy}{du} = 4u^3$$

so

$$\frac{dy}{dx} = \frac{dy}{du}\frac{du}{dx} \qquad \text{(by the chain rule)}$$

$$= 4u^3 \cdot 2x \qquad \text{(by substituting the above expressions)}$$
$$= 4(x^2 + 1)^3 \cdot 2x \qquad \text{(by substituting } u = x^2 + 1\text{)}$$
$$= 8x(x^2 + 1)^3 \qquad \text{(by algebra)}$$

(d) Let $u = x^2 + 1$. Then $y = u^{1/2}$ and we have

$$\frac{du}{dx} = 2x \quad \text{and} \quad \frac{dy}{du} = \tfrac{1}{2}u^{-1/2}$$

so

$$\frac{dy}{dx} = \frac{dy}{du}\frac{du}{dx} \qquad \text{(by the chain rule)}$$

$$= \tfrac{1}{2}u^{-1/2} \cdot 2x \qquad \text{(by substituting the above expressions)}$$
$$= x(x^2 + 1)^{-1/2} \qquad \text{(by canceling the 2s and substituting}$$
$$u = x^2 + 1\text{)}$$

Each of the examples above illustrated the following rule, which is a corollary of the chain rule.

3.1.6 Rule

If $y = f(x) = (u(x))^n$, then $\dfrac{dy}{dx} = n(u(x))^{n-1} \cdot \dfrac{du}{dx}$, provided that everything in the expression on the right exists.

We shall use this rule in some more complicated examples.

3.1.7 Example

Find the derivatives, where they exist, of
(a) $y = (2x^3 + 4)^{10}$
(b) $y = (x^7 + 3x^5 - 2x^3)^4$
(c) $y = \left(\dfrac{x + 1}{x + 2}\right)^{3/2}$

Solution:

(a) Let $u = 2x^3 + 4$. Then $\dfrac{du}{dx} = 6x^2$ and $y = u^{10}$, so $\dfrac{dy}{du} = 10u^9$. Thus

$$\frac{dy}{dx} = \frac{dy}{du} \cdot \frac{du}{dx} = 10(2x^3 + 4)^9(6x^2) = 60x^2(2x^3 + 4)^9$$

(b) Let $u = x^7 + 3x^5 - 2x^3$. Then $\dfrac{du}{dx} = 7x^6 + 15x^4 - 6x^2$ and $y = u^4$, so

$\dfrac{dy}{du} = 4u^3$. Thus

$$\frac{dy}{dx} = \frac{dy}{du}\frac{du}{dx} = 4(x^7 + 3x^5 - 2x^3)^3(7x^6 + 15x^4 - 6x^2)$$

(c) Let $u = (x+1)/(x+2)$. Then

$$\frac{du}{dx} = \frac{(x+2)\cdot 1 - (x+1)\cdot 1}{(x+2)^2} = \frac{1}{(x+2)^2} \qquad \text{and} \qquad y = u^{3/2}$$

so $\dfrac{dy}{du} = \tfrac{3}{2}u^{1/2}$. Thus

$$\frac{dy}{dx} = \frac{dy}{du}\frac{du}{dx} = \frac{3}{2}\left(\frac{x+1}{x+2}\right)^{1/2}\frac{1}{(x+2)^2}$$

3.1.8   Example

Suppose that net sales are 70 percent of gross sales, and that profits are 10 percent of net sales. What is the rate of change of profits with respect to gross sales?

Solution:

We can write $N = 0.70G$ and $P = 0.10N$, where G, N, and P denote gross sales, net sales, and profits, respectively. Then we have $\dfrac{dN}{dG} = 0.7$ and $\dfrac{dP}{dN} = 0.1$. It follows that

$$\frac{dP}{dG} = \frac{dP}{dN}\frac{dN}{dG} = (0.1)(0.7) = 0.07$$

so the rate of change of profits with respect to gross sales is 0.07.

SUMMARY

This section tells how to differentiate the composition of two functions, both of whose derivatives are known. If $y = f(u)$ and $u = g(x)$, the chain rule can be easily remembered by the equation

$$\frac{dy}{dx} = \frac{dy}{du}\cdot\frac{du}{dx}$$

Learn to use this rule well. It will be useful throughout the remainder of this course.

EXERCISES   3.1. A

Find the derivatives of the following functions, where they exist.

1. $y = (3x + 2)^3$

2. $y = \sqrt{3x + 2}$

3. $y = \dfrac{1}{3x + 2}$

4. $y = (x^2 + 4x - 7)^4$

5. $y = \sqrt{6x^2 + 2}$

6. $y = \dfrac{1}{x^2 + x}$

7. $y = (3x^4 - 2x^7)^{1/3}$

8. $y = (2x + x^3)^{5/2}$

9. $y = \dfrac{1}{\sqrt{3x^2 + 7}}$

10. $y = (3x^7 + \pi)^{2/3}$

11. Suppose that a man discovers that his net monthly income is $P = 0.8x - 40$ dollars when his gross monthly income is x. He resolves to save $\sqrt{P}$ dollars per month. What is his marginal saving if his gross pay is $550? (That is, about how much extra will he save if he is paid an extra dollar?)

12. Suppose that net sales are 80 percent of gross sales and profit is 25 percent of net sales.
    (a) What is the rate of change of the profit with respect to net sales?
    (b) What is the rate of change of the net sales with respect to the gross sales?
    (c) What is the rate of change of the profit with respect to gross sales?
    (d) Complete the statement: The rate of change of profit with respect to gross sales is the rate of change of _____ times the rate of change of _____.

13. Repeat parts (a), (b), and (c) of problem 12 assuming that net sales are 75 percent of gross sales and profit $P = 10x - 100 - 0.01x^2$, where x is the net sales.

EXERCISES   3.1. B

Find the derivatives of the following functions, where they exist.

1. $y = (2x + 4)^2$

2. $y = \sqrt{2x + 4}$

3. $y = \dfrac{1}{2x + 4}$

4. $y = (3x^2 + 5x - 7)^5$

5. $y = \sqrt{4x^2 - 5}$

6. $y = \dfrac{1}{x^3 - x^2}$

7. $y = (5x^3 - 8x^9)^{6/5}$

8. $y = (3x + x^5)^{1/4}$

9. $y = (4x^2 - \pi)^{-1/2}$

10. $y = (5x^8 + 17)^{3/4}$

11. Suppose that net sales are 60 percent of gross sales and profit is 15 percent of net sales. Answer the questions of problem 12 in Exercises 3.1.A.

12. Repeat parts (a), (b), and (c) of problem 12 in Exercises 3.1.A assuming that net sales are 70 percent of gross sales and that the profit is $P = 5x - 50 - 0.02x^2$, where x is the net sales.

EXERCISES   3.1. C

Find the derivatives of the following functions, where they exist.

1. $y = [(x - 3x^2)(\sqrt{x} + 7)]^3$

2. $y = \left(\dfrac{x^2 + 1}{2x + 3x^3}\right)^2$

HC 3. (a) Find the derivative of the function $f(x) = (2x + 1)^3$ at the point where $x = 1$ using the chain rule. (b) Then approximate this derivative by finding the slope of the straight line through $(1, f(1)) = (1, 27)$ and $(x, f(x))$ as follows. Make your answers accurate to four decimal places.

x	1.1	1.01	1.001	1.0001
$\dfrac{f(x) - 27}{x - 1}$				

ANSWERS 3.1. A

1. $y' = 9(3x + 2)^2$
2. $y' = \frac{3}{2}(3x + 2)^{-1/2}$
3. $y' = -3(3x + 2)^{-2}$
4. $y' = 4(x^2 + 4x - 7)^3(2x + 4)$
5. $y' = 6x(6x^2 + 2)^{-1/2}$
6. $y' = -(x^2 + x)^{-2}(2x + 1)$

7. $y' = \frac{1}{3}(3x^4 - 2x^7)^{-2/3}(12x^3 - 14x^6)$
8. $y' = \frac{5}{2}(2x + x^3)^{3/2}(2 + 3x^2)$
9. $y' = -\frac{1}{2}(3x^2 + 7)^{-3/2}(6x)$
10. $y' = 14x^6(3x^7 + \pi)^{-1/3}$
11. \$0.02

12. (a) 0.25 (b) 0.8 (c) 0.20 (d) the net sales with respect to the gross sales; the profit with respect to the net sales
13. (a) $10 - 0.02x$ (b) 0.75 (c) $7.5 - 0.015x$

ANSWERS 3.1. C

1. $3[(x - 3x^2)(\sqrt{x} + 7)]^2[(1 - 6x)(\sqrt{x} + 7) + (x - 3x^2)(\frac{1}{2}x^{-1/2})]$

2. $2 \cdot \left[\dfrac{x^2 + 1}{2x + 3x^3} \right]\left[\dfrac{(2x + 3x^3)2x - (x^2 + 1)(2 + 9x^2)}{(2x + 3x^3)^2} \right]$

3. (a) 54 (b)

x	1.1	1.01	1.001	1.0001
$\dfrac{f(x) - 27}{x - 1}$	57.6800	54.3608	54.0361	54.0032

3.2   Inverse and Implicit Functions

In many relationships between two variables, either variable may be viewed as the independent variable, depending on the context. In Section 1.2 we saw how a linear demand curve can be viewed with price as a function of quantity, $p = f(q)$, or with quantity as a function of price, $q = g(p)$. Either relationship can be obtained from the other by solving for the original independent variable in terms of the dependent variable. In this case we say that $f(q)$ and $g(p)$ are "inverse functions."

3.2.1   Definition

If $y = f(x)$ defines y as a function of x, then a function $x = g(y)$ such that $x_0 = g(f(x_0))$ for each x_0 in the domain of $f(x)$ is called an inverse function of $f(x)$. (Sometimes we restrict the values of y.)

3.2.2 **Example**

Find inverses for the following functions.

(a) $y = 3x - 2$
(b) $y = 2x + 4$

Solution:
(a) If $y = 3x - 2$, then solving for x, we get $x = (y + 2)/3$, so $g(y) = (y + 2)/3$ is the only possible inverse of f. Checking, we see that

$$g(f(x)) = g(3x - 2) = \frac{(3x - 2) + 2}{3} = \frac{3x}{3} = x \qquad \text{for all } x$$

(b) Solving for x, we get that $x = (y - 4)/2$ is the inverse. Checking, we get that

$$g(f(x)) = g(2x + 4) = \frac{(2x + 4) - 4}{2} = \frac{2x}{2} = x$$

Next we take derivatives of these functions. If we look at part (a) of Example 3.2.2 we see that the derivative of the original function is 3 and the derivative of the inverse is $\frac{1}{3}$. In part (b) we note that the derivative of the original function is 2 and the derivative of the inverse is $\frac{1}{2}$. These observations suggest something that is, in fact, true: The derivative of the inverse of a function is the reciprocal of the derivative of the original function.

Speaking more formally, we have the following rule.

3.2.3 **Rule**

If $y = f(x)$ is a differentiable function with a differentiable inverse, $x = g(y)$ and $y_1 = f(x_1)$, $f'(x_1) \neq 0$, then

$$g'(y_1) = \frac{1}{f'(x_1)}$$

In other words,

$$\frac{dx}{dy} = \frac{1}{\dfrac{dy}{dx}} \qquad \text{at the point } (x_1, y_1)$$

At the end of Section 1.2, where we discussed inverse functions (price as a function of quantity, and quantity as a function of price), we noted that the graphs of two functions that are inverses of each other are reflections across the line at 45° to the axes through the origin from the lower left to the upper right. Similarly, if we express the inverse function of $y = f(x)$ in the rule above as $y = g(x)$ [instead of as $x = g(y)$], then the graph of g will be the reflection of the graph of f over the line $y = x$. This is because (a, b) will be on the graph of a function if and only if (b, a) is on the graph of its inverse. From Figure 3.2–1 you may be able to see that the slope (derivative) of the original function at (a, b) is the reciprocal of the slope (derivative) of its inverse at (b, a). Lack of this insight, however, need not hinder you from using Rule 3.2.3 accurately and effectively.

Figure 3.2–1

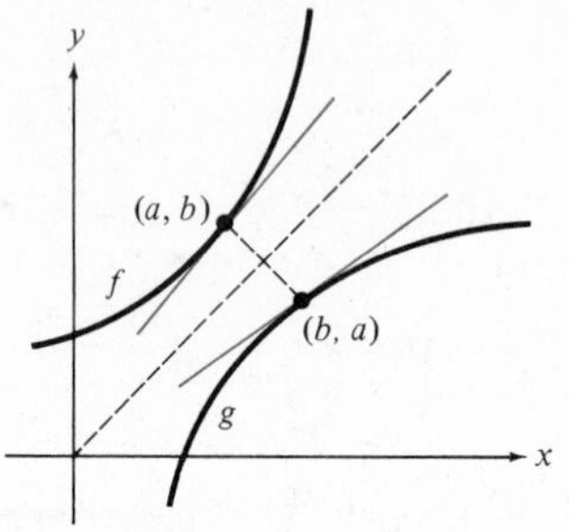

3.2.4   Example

Find $\dfrac{dx}{dy}$ at $y = 1$ if

(a) $y = 3x - 2$
(b) $y = 3x^3 - 23$
(c) $y = \sqrt{x}$

Solution:

(a) $\dfrac{dy}{dx} = 3$ implies that $\dfrac{dx}{dy} = \tfrac{1}{3}$ for all x and y.

(b) Since $\dfrac{dy}{dx} = 9x^2$, $\dfrac{dx}{dy} = 1/9x^2$ whenever $x \neq 0$. But, alas, x is not given — we know only that $y = 1$. So we solve for x in the original expression:

$$1 = 3x^3 - 23$$
$$24 = 3x^3$$
$$8 = x^3$$
$$2 = x$$

Now substituting back into the expression for $\dfrac{dx}{dy}$ we obtain

$$\frac{dx}{dy} = \frac{1}{9(2)^2} = \frac{1}{36}$$

(c) Since $\dfrac{dy}{dx} = \tfrac{1}{2}x^{-1/2}$, $\dfrac{dx}{dy} = 1/\tfrac{1}{2}x^{-1/2} = 2\sqrt{x}$ when the value of x is given.

Again we do not know x immediately, since we were given $y = 1$; but this time it is easy to substitute back into the original equation, $1 = \sqrt{x}$, and see that $x = 1$. Therefore,

$$\frac{dx}{dy} = 2\sqrt{1} = 2$$

Implicit Differentiation

Often we have an equation that involves both x and y which does not express a function, so that our arguments and proofs in Chapter 2 and Appendix II are not valid. However, we still might want to find how the change in one variable affects the change in the other. Or we might want to find the slope of a line tangent to the graph of this equation. Even though we do not have an explicit function, we can still use the chain rule to find $\frac{dy}{dx}$ and $\frac{dx}{dy}$. It is very difficult to *prove* that this method always works, but it is within the scope of this book to use the fact that it does.

3.2.5   Example

Find $\frac{dy}{dx}$ at the point $(1, -\sqrt{2})$ on the graph of $x^2 + y^2 = 3$.

Solution:

To find $\frac{dy}{dx}$ we differentiate each term of the given equation with respect to x. You know that the derivative of x^2 is $2x$ and the derivative of 3 is 0.

To find the derivative of y^2 with respect to x (not y!) we let $u = y^2$. Then we notice that

$$\frac{du}{dx} = \frac{du}{dy} \cdot \frac{dy}{dx}$$

This changes the notation of Section 3.1, but it means the same thing as Rule 3.1.4. Since $\frac{du}{dy}$ is clearly 2y, $\frac{du}{dx} = 2y\frac{dy}{dx}$. Thus differentiating $x^2 + y^2 = 3$ with respect to x, we get

$$2x + 2y\frac{dy}{dx} = 0$$

Solving this for $\frac{dy}{dx}$, we get $\frac{dy}{dx} = -\frac{x}{y}$. At the point $(1, -\sqrt{2})$ this has the value $-1/-\sqrt{2} = 1/\sqrt{2}$.

The argument used in Example 3.2.5 is needed so often in implicit differentiation that we summarize it here. It is merely the chain rule, but the notation is slightly different.

3.2.6   Rule

If $u = f(y)$ is a function of y, and y is a function (possibly an implicit function) of x, then

$$\frac{du}{dx} = f'(y)\frac{dy}{dx}$$

where $f'(y)$ indicates differentiation with respect to y.

Using Rule 3.2.6 involves differentiating both sides of an equation in x and y as if each side were a function of x. In so doing, (1) every expression in x is treated as usual; (2) every y is treated as a function $y = f(x)$; and (3) the derivatives of both sides of the given equation are set equal to each other and will contain the symbol $\frac{dy}{dx}$. (4) Then we may solve for $\frac{dy}{dx}$ in terms of x and y.

If both x and y appear in the same term of the equation, we must use the product rule, as in the following example.

3.2.7   Example

Find $\frac{dy}{dx}$ at the point (1, 2) on the graph of $xy = 2$.

Solution:

We use the product formula $u'v + uv'$, where $u = x$ and $v = y$. Differentiating both sides of $xy = 2$, we get

$$1 \cdot y + x\frac{dy}{dx} = 0$$

Solving for $\frac{dy}{dx}$, we have

$$x\frac{dy}{dx} = -y$$

$$\frac{dy}{dx} = -\frac{y}{x} = -\frac{2}{1} = -2$$

We can use the results of Examples 3.2.5 and 3.2.7 to find the equation of lines tangent to graphs that are not necessarily functions. Notice that although we use the words "circle," "ellipse," and "hyperbola" to describe the graphs of some of the equations in this section, it is not necessary to graph these curves (or even to know what they are) to be able to use implicit differentiation to find the slope of lines tangent to their graphs. (This is because you can use algebra and calculus without knowing the specific geometric interpretation.)

    Warning! In the following examples the only pairs of numbers (x, y) that can legitimately be substituted into the expression for $\dfrac{dy}{dx}$ are those which satisfy the original equation (and therefore lie on the graph).

3.2.8   **Example**

Find the equation of the line tangent to the circle $x^2 + y^2 = 3$ at the point where $x = 1$, $y = -\sqrt{2}$.

Solution:

If we knew $\dfrac{dy}{dx}$ at that point, we could use the point-slope equation for a straight line with $m = \dfrac{dy}{dx}$. But this is precisely the quantity we calculated in Example 3.2.5! Thus we can substitute

$$m = \frac{1}{\sqrt{2}}$$

and the point $(1, -\sqrt{2})$ into the point-slope form for a linear equation, which gives us

$$y - (-\sqrt{2}) = \frac{1}{\sqrt{2}}\,(x - 1) \qquad \text{or} \qquad y = \frac{1}{\sqrt{2}}\,x - \sqrt{2} - \frac{1}{\sqrt{2}}$$

as the equation of the tangent line at the given point.

Figure 3.2–2

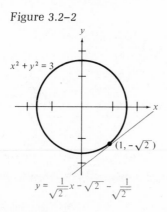

$$y = \frac{1}{\sqrt{2}}x - \sqrt{2} - \frac{1}{\sqrt{2}}$$

3.2.9   Example

Find the equation of the line tangent to the hyperbola $xy = 2$ at the point $(1, 2)$.

Solution:

Using the solution for Example 3.2.7, we have $m = -2$. Thus the equation of the tangent line is

$$y - 2 = -2(x - 1) \qquad \text{or} \qquad y = -2x + 4$$

Figure 3.2–3

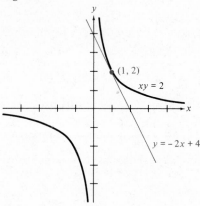

3.2.10   Example

If a price–demand law for a commodity is given by $p = -q^2 + 15q + 115$, find $\dfrac{dq}{dp}$ when $p = 15$.

Solution:

We differentiate the given expression implicitly, and then solve for $\dfrac{dq}{dp}$.

$$1 = -2q \frac{dq}{dp} + 15 \frac{dq}{dp}$$

$$1 = (-2q + 15) \frac{dq}{dp}$$

$$\frac{1}{-2q + 15} = \frac{dq}{dp}$$

This gives us an expression for $\dfrac{dq}{dp}$ in terms of q. In our problem we must use the fact $p = 15$ and the original equation to find q:

$$15 = -q^2 + 15q + 115$$
$$0 = -q^2 + 15q + 100 = (-q + 20)(q + 5)$$

Thus

$$q = 20 \quad \text{or} \quad q = -5$$

But we never have a negative quantity, so q must equal 20. Thus

$$\frac{dq}{dp} = \frac{1}{-2(20) + 15} = \frac{1}{-40 + 15} = -\frac{1}{25}$$

SUMMARY

In this section we showed how to find the derivative of an inverse and how to take derivatives of relationships that are not functions. Although the proofs behind these techniques are deep, the obvious steps "work."

The notation $\frac{dx}{dy}$ (not a fraction!) is a help in remembering that

$$\frac{dx}{dy} = \frac{1}{\dfrac{dy}{dx}}$$

EXERCISES   3.2. A

1. Find $\frac{dx}{dy}$ when

    (a) $y = (x + 3)^4$ (c) $y = \sqrt{x} + x^{-7} - \pi x^{2/3}$
    (b) $y = \sqrt{3x + 2}$ (d) $y = (3x^4 + x^{-4})^5$

2. Find $\frac{dy}{dx}$ when

    (a) $y^2 = 3x$ (g) $x^2 + y^2 = 36$
    (b) $y^2 = 5x + 7$ (h) $x^2 + y^2 = 49$
    (c) $y^3 = 8x^2 - x$ (i) $9x^2 + 16y^2 = 144$
    (d) $2y^3 = x^2 - 4x$ (j) $xy = 4$
    (e) $x^2 - y^2 = 2$ (k) $xy = 16$
    (f) $x^2 - y^2 = 5$ (l) $3xy = 7$

3. (a) Find the slope of the line tangent to $x^2 + y^2 = 25$ at $(4, 3)$.
    (b) Write the equation of this straight line.

4. (a) Find $\frac{dq}{dp}$ when the demand function $p = 80 - 0.2q$ is given.

    (b) Find $\frac{dq}{dp}$ when the demand function $p = 60 - 0.3q$ is given.

    (c) Find $\frac{dq}{dp}$ when the demand function $p = 100 - 2q$ is given.

5. (a) Find $\frac{dq}{dp}$ when the demand function $p = 200 - 8q^{1/4}$ is given.

(b) Find the elasticity of demand when the demand level is 10,000 units.
$$\left(\text{Recall that } E = -\frac{p}{q\,\dfrac{dp}{dq}}. \right)$$

6. Find the equation of the line tangent to the ellipse $3x^2 + y^2 = 12$ at the point $(1, 3)$.

7. Find the equation of the line tangent to the curve $x^{2/3} + y^{2/3} = 5$ at the point $(8, 1)$.

EXERCISES   3.2. B

1. Find $\dfrac{dx}{dy}$ when

   (a) $y = (x + 4)^5$
   (b) $y = \sqrt{2x + 3}$

   (c) $y = \pi x^{-3} + \sqrt{x} + 3x^{-2}$
   (d) $y = (2x^3 + x^{-3})^4$

2. Find $\dfrac{dy}{dx}$ when

   (a) $y^2 = 6x$
   (b) $y^2 = 4x + 7$
   (c) $y^3 = 4x^2 - x$
   (d) $3y^2 = x^3 + 3x$
   (e) $x^2 - y^2 = 4$
   (f) $2x^2 - y^2 = 9$

   (g) $x^2 + y^2 = 16$
   (h) $x^2 + y^2 = 25$
   (i) $3x^2 + 4y^2 = 25$
   (j) $xy = 8$
   (k) $xy = 11$
   (l) $5xy = \pi$

3. Write the equation of the line tangent to the ellipse $x^2 + 4y^2 = 13$ at the point $(3, -1)$.

4. Write the equation of the line tangent to the hyperbola $3x^2 - y^2 = 3$ at the point $(2, 3)$.

5. Write the equation of the line tangent to the circle $x^2 + y^2 = 4$ at the point $(-1, \sqrt{3})$.

6. (a) Find $\dfrac{dq}{dp}$ when the demand function is $p = 60 - 0.5q$.

   (b) Find $\dfrac{dq}{dp}$ when the demand function $p = 40 - 0.4q$ is given.

   (c) Find $\dfrac{dq}{dp}$ when the demand function $p = 50 - 2q$ is given.

7. (a) Find $\dfrac{dq}{dp}$ when the demand function is $p = 300 - 3q^{1/3}$.

   (b) Find the elasticity when $q = 1,000$. $\left(\text{Recall that } E = -\dfrac{p}{q\,\dfrac{dp}{dq}}. \right)$

EXERCISES   3.2. C

1. Find $\dfrac{dy}{dx}$ when

   (a) $y^2 + 3x = y$
   (b) $xy + x^2 = 7$

   (c) $\sqrt{y} = x^2 + \pi x$

2. Find $\dfrac{dx}{dy}$ when

(a) $y = (x^2 + 3x)\sqrt{2x - \pi}$

(b) $y = (x^2 + 2x^7)\sqrt[3]{2x - x^3}$

HC 3. See Examples 3.2.7 and 3.2.9. Verify that the slope of the tangent line to the curve $xy = 2$ at the point $(1, 2)$ is indeed -2 by calculating the slope between the point $(1, 2)$ and each of the following points on the curve. Give answers that are accurate to four decimal places.

(x, y)	$\left(1.1, \dfrac{2}{1.1}\right)$	$\left(1.01, \dfrac{2}{1.01}\right)$	$\left(1.003, \dfrac{2}{1.003}\right)$
$\dfrac{y - 2}{x - 1}$			

ANSWERS 3.2. A

1. (a) $\dfrac{1}{4(x + 3)^3}$ (b) $\frac{2}{3}\sqrt{3x + 2}$ (c) $\dfrac{1}{\frac{1}{2}x^{-1/2} - 7x^{-8} - 2\pi x^{-1/3}/3}$

(d) $\dfrac{1}{5(3x^4 + x^{-4})^4(12x^3 - 4x^{-5})}$

2. (a) $3/2y$ (b) $5/2y$ (c) $(16x - 1)/3y^2$ (d) $(x - 2)/3y^2$ (e) x/y (f) x/y
(g) $-x/y$ (h) $-x/y$ (i) $-9x/16y$ (j) $-y/x$ (k) $-y/x$ (l) $-y/x$

3. (a) $-\frac{4}{3}$ (b) $y = -\frac{4}{3}x + \frac{25}{3}$

4. (a) -5 (b) $-\frac{10}{3}$ (c) -0.5

5. (a) $-\dfrac{q^{3/4}}{2}$ (b) $E = \dfrac{p}{2q^{1/4}} = 6$

6. $y = -x + 4$

7. $y = -\frac{1}{2}x + 5$

ANSWERS 3.2. C

1. (a) $3/(1 - 2y)$ (b) $-(2x + y)/x$ (c) $(4x + 2\pi)\sqrt{y}$

2. (a) $\dfrac{1}{(2x + 3)\sqrt{2x - \pi} + (x^2 + 3x)(2x - \pi)^{-1/2}}$

(b) $\dfrac{1}{(2x + 14x^6)(2x - x^3)^{1/3} + (x^2 + 2x^7)(\frac{1}{3})(2x - x^3)^{-2/3}(2 - 3x^2)}$

3.

(x, y)	$\left(1.1, \dfrac{2}{1.1}\right)$	$\left(1.01, \dfrac{2}{1.01}\right)$	$\left(1.003, \dfrac{2}{1.003}\right)$
$\dfrac{y - 2}{x - 1}$	-1.8182	-1.9802	-1.9940

3.3 Exponential Functions

Functions of the form $f(x) = c \cdot a^{kx}$, where c, k, and $a \geqslant 0$ are constants, are called <u>exponential functions</u>. Such functions have many applications in economic analysis and elsewhere. One of their uses is to model the growth of capital under compound interest. If P dollars are invested at $100r$ percent per period, the value at the end of one period becomes

$$A(1) = P + Pr = P(1 + r)$$

The value at the end of two periods is

$$A(2) = P(1 + r) + [P(1 + r)]r = P(1 + r)(1 + r) = P(1 + r)^2$$

The value at the end of three periods is

$$A(3) = P(1 + r)^2 + [P(1 + r)^2]r = P(1 + r)^2(1 + r) = P(1 + r)^3$$

And the value at the end of n periods is

$$A(n) = P(1 + r)^n \qquad \text{where } n \text{ is any positive whole number}$$

3.3.1   Example

Find the amount to which \$100 will grow by the end of 2 years when it is invested at 4 percent and compounded
(a) annually
(b) semiannually
(c) quarterly

Solution:

(a) $A(2) = 100(1 + 0.04)^2 = 108.16$ (using Table I, at the back of the book) since the number of periods is 2 and $r = 0.04$.

(b) Compounding semiannually at 4 percent for 2 years is the same as compounding annually at 2 percent for 4 years; we have twice as many periods at one-half the rate. Thus we compute

$$A(4) = 100 \left(1 + \frac{0.04}{2}\right)^4 = 100(1 + 0.02)^4$$
$$= 108.24 \qquad \text{(using Table I)}$$

(c) This time we have four times as many periods at one-fourth the rate:

$$A(8) = 100 \left(1 + \frac{0.04}{4}\right)^8 = 100(1 + 0.01)^8$$
$$= 108.28 \qquad \text{(using Table I)}$$

If the number of periods k within a certain fixed time interval x (usually years) grows without bound, it is as if we were compounding continuously. (Banks call this compounding "daily.") Then, keeping $n = x$ fixed and varying the number of periods, k, we have

$$A(x) = \lim_{k \to \infty} \left[P \left(1 + \frac{r}{k}\right)^{kx}\right] = P \left[\lim_{k \to \infty} \left(1 + \frac{1}{k/r}\right)^{k/r}\right]^{rx} \qquad *$$

The limit on the right appears so often in mathematics and its applications that it has been given a special name, e. The limit can be shown to exist and to be equal to a number between 2 and 3. Like π, it cannot be written as a fraction or as a finite decimal, so we simply call it e or use an approximate value, such as $e \approx 2.7$ or 3. Computers have calculated e to many decimal places. (See also page 49.)

$$e = \lim_{m \to \infty} \left(1 + \frac{1}{m}\right)^m \approx 2.7182818284 \qquad \text{to the nearest 10 places}$$

The starred equation can now be written

$$A(x) = Pe^{rx}$$

In this expression $A(x)$ gives the value of the principal P compounded <u>continuously</u> at 100r percent per year (or month, or day, etc.) for x years (or months, or days, etc.).

3.3.2   Example

Find the value of \$100 compounded continuously for 2 years at 4 percent. (See Example 3.3.1.)

Solution:

$A(x) = Pe^{rx}$, where $P = 100$, $r = 0.04$, and $x = 2$, so

$$\begin{aligned}
A = 100e^{0.04(2)} &= 100e^{0.08} \\
&= 100(1.0833) \qquad \text{(by Table III, at the back of the book)} \\
&= \$108.33
\end{aligned}$$

Notice that this is just a bit more than we would have received if we had received quarterly compounding, as in Example 3.3.1.

Present Value

Thus far we have always regarded the amount to be invested, P, as given in the equation

$$A = Pe^{rx}$$

where A is the unknown quantity to be calculated. But there are situations in which we know how much we want to have after a certain period ot time, A, and we want to calculate how much, P, we must invest at a certain rate, compounded in a given way, so that it will be worth A after the given period. In that case we want P as a function of A instead of A as a function of P. To obtain P as a function of A we solve the equation above by multiplying both sides by e^{-rx}. This gives

$$Ae^{-rx} = P \qquad \text{or} \qquad P = Ae^{-rx}$$

an equation in which A is regarded as given, and we want to find P. In this case we say that P is the present value of A dollars due x years from now invested at 100r percent compounded continuously.

Similarly, the equation $A = P(1 + r)^n$ can be viewed differently if we multiply through by $(1 + r)^{-n}$. Then we obtain

$$A(1 + r)^{-n} = P \quad \text{or} \quad P = A(1 + r)^{-n}$$

In this case we say that P is the present value of A dollars due n years from now invested at 100r percent compounded annually. This means that the A dollars due n years from now is worth P dollars today, since P dollars invested now will get you A dollars n years hence, if invested at the given rate. Suitable modifications can be made for compounding semiannually or quarterly.

3.3.3   Example

Find the present value of $100 due 2 years from now when the interest rate is 4 percent compounded
(a) annually
(b) semiannually
(c) quarterly
(d) continuously

Solution:
(a) $P = 100(1 + 0.04)^{-2} = 92.46$ (using Table II)
(b) $P = 100 \left(1 + \dfrac{0.04}{2}\right)^{-4} = 92.38$

(c) $P = 100 \left(1 + \dfrac{0.04}{4}\right)^{-8} = 92.35$
(d) $P = 100e^{-0.04(2)} = 100e^{-0.08} = 100(0.9231) = \$92.31$ (using Table III)

Notice that in each case we need to invest a little bit less money than in the preceding case to get our $100 two years from now.

Other Exponential Functions

As was mentioned on page 155, any function that has its independent variable in the exponent is called an exponential function. The expression $A = Pe^{rx}$, where P and r are fixed numbers and x is the independent variable, is, therefore, an exponential function. There are many more exponential functions with a wide variety of applications.

3.3.4   Example

Suppose that a certain bacterium reproduces by dividing itself in half every minute. If there is one such bacterium at time $x = 0$, graph the number that exist by time x.

Solution:

Figure 3.3–1

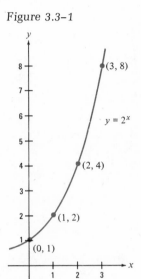

Now you see why you can get so sick so quickly once you catch "the bug"! When $x = 4$, the number of these bacteria is 16, and after 5 minutes there are 32 – the graph is off the page.

In the previous example, only the points on the graph have an applied interpretation. But the curve connecting the points can be obtained mathematically by remembering that for any rational number a/b, $2^{a/b} = \sqrt[b]{2^a}$ = and then approximating the rational points; it also looks reasonable to the eye. If you want to plot one point between the integers, try $2^{1/2} = \sqrt{2} \approx 1.414$, and you will see that it lies on the curve. Other fractional exponents can be graphed in a similar fashion.

The values of y for negative x are found by remembering that $2^{-x} = 1/2^x$. Thus, if $x = -1$, $y = 2^{-1} = \frac{1}{2}$. If $x = -2$, $y = 2^{-2} = \frac{1}{4}$.

Other useful exponential functions include $y = 3^x$, $y = 4^x$, and $y = 10^x$. Since $2 < e < 3$, it is interesting to see $y = 2^x$, $y = e^x$, and $y = 3^x$ graphed on the same coordinate system. This is shown in Figure 3.3–2 on page 160.

3.3.5   Example in Biology

The growth of a bacteria culture is often modeled by an exponential function. Suppose that the size of a particular culture is $f(x) = 200 \cdot 2^{x/20}$ when x is the number of minutes after the experiment has begun. What will be the size of the culture after an hour has elapsed?

Figure 3.3–2

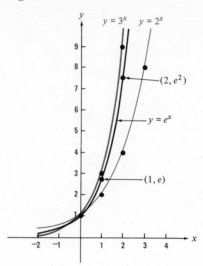

Solution:

An hour is 60 minutes, so we want to find $f(60)$.

$$f(60) = 200 \cdot 2^{60/20} = 200 \cdot 2^3 = 1{,}600$$

SUMMARY

We began this section by developing the formula $A = P(1 + r)^n$, which gives the amount A that a principal P will grow to if invested for n periods at $100r$ percent per period. We used this to derive the formula $A = Pe^{rx}$, which gives the amount that the principal will grow to in x years if invested at $100r$ percent compounded *continuously*. Using these two equations, we obtained two formulas for calculating *present value* in the discrete and continuous case, respectively.

In the second part of the section, we considered other exponential functions – any function with its independent variable in the exponent. The function $y = 2^x$ is often used when modeling the growth of a bacteria culture.

EXERCISES   3.3. A

1. Use Table III to find the values of e^x when $x = -1, -0.5, 0, 0.5, 1, 1.5$, and 2. Sketch $y = e^x$, including these points.
2. Find the value to which \$1,000 will grow if invested at 2 percent per year for 3 years and compounded
    (a) annually   (c) quarterly
    (b) semiannually   (d) continuously

3.   Find the value to which $100 will grow if invested at 6 percent for 2 years compounded
     (a) annually   (c) quarterly
     (b) semiannually   (d) continuously

4.   Find the present value of $1,000 due 3 years from now if the present rate of interest is 4 percent per year compounded
     (a) annually   (c) quarterly
     (b) semiannually   (d) continuously

5.   Find the present value of $100 due 4 years from now if the interest rate is 8 percent per year compounded
     (a) annually   (c) quarterly
     (b) semiannually   (d) continuously

6.   Suppose that the size of a bacteria culture is given by $f(x) = 50 \cdot 2^{x/15}$, where x is the number of minutes after the experiment began. What will be the size of the culture after an hour has elapsed?

7.   Humans, like bacteria, tend to double their number in some fixed period of time. The population of the earth in 1930 was about 2 billion and in 1975 was about 4 billion. Barring a catastrophe or an effective conscious effort to control population growth, when will the earth's population reach 8 billion? 16 billion?

EXERCISES   3.3. B

1.   Use Table III to find the values of e^x when $x = -1, -0.5, 0, 0.5, 1, 1.5$, and 2. Sketch $y = e^x$ including these points.

2.   Find the value to which $100 will grow if invested at 4 percent for 3 years and compounded
     (a) annually   (c) quarterly
     (b) semiannually   (d) continuously

3.   Find the value to which $100 will grow if invested at 8 percent for 4 years and compounded
     (a) annually   (c) quarterly
     (b) semiannually   (d) continuously

4.   Find the present value of $1,000 due 5 years from now if the interest rate is 8 percent compounded
     (a) annually   (c) quarterly
     (b) semiannually   (d) continuously

5.   Find the present value of $100 due 2 years from now if it is invested at 6 percent per year compounded
     (a) annually   (c) quarterly
     (b) semiannually   (d) continuously

6.   Suppose that the size of a bacteria culture is given by $f(x) = 40 \cdot 2^{x/30}$, where x is the number of minutes after the experiment began. What will be the size of the culture after an hour has elapsed?

7.   Humans, like bacteria, tend to double their number in some fixed period of time. The population of the United States in 1900 was about 75 million and in 1950 was about 150 million. Assuming there are no interruptions to typical growth, at what year will the population reach 300 million? 600 million?

EXERCISES   3.3. C

1. If payments of P are made into a fixed fund that gathers interest at the rate of $100\, r$ percent compounded per period, the value of the fund after k periods is

$$A = P(1+r)^{k-1} + P(1+r)^{k-2} + \cdots + P(1+r)^2 + P(1+r) + P$$

Show that

$$A = P\,\frac{(1+r)^k - 1}{r}$$

[*Hint:* Multiply A by $(1+r)$ and subtract.]

Since the periods are often taken to be 1 year in length, A is called the <u>value of an annuity</u>.

2. What is the value of an annuity if 100 dollars is deposited annually into a fund paying
    (a) 3 percent per year after 5 years?
    (b) 3 percent per year after 20 years?
    (c) 8 percent per year after 5 years?
    (d) 8 percent per year after 20 years?

HC 3. If you have a y^x key on your hand calculator, show how the function $y = 2^x$ can be approximated for the irrational number $x = \sqrt{2}$ by filling in the following table (accurate to four decimal places).

x	1.4	1.41	1.414	1.4142	1.41421
2^x					

HC 4. A population of *Escherichia coli* bacteria, a very common species that normally exists in human intestines, has an initial size P_0 and, if unchecked, doubles in size every 20 minutes according to the formula $P(t) = P_0 \cdot 2^{t/20}$, where t is given in minutes. If your hand calculator has a y^x key, you can complete the following table (accurate to two decimal places).

t (minutes)	0	5	10	15	20	30	40
$P = P_0 \cdot 2^{t/20} =$							

ANSWERS   3.3. A

1.

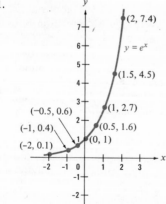

2. (a) $1,061.20 (b) $1,061.50 (c) $1,061.70 (d) $1,061.80
3. (a) $112.36 (b) $112.55 (c) $112.65 (d) $112.75
4. (a) $889.00 (b) $888.00 (c) $887.40 (d) $886.90
5. (a) $73.50 (b) $73.07 (c) $72.84 (d) $72.61
6. 800
7. 8 billion in 2020; 16 billion in 2065

ANSWERS   3.3. C

2. (a) $531 (b) $2,687 (c) $586.63 (d) $4,576.25

3.

x	1.4	1.41	1.414	1.4142	1.41421
2^x	2.6390	2.6574	2.6647	2.6651	2.6651

4.

t	0	5	10	15	20	30	40
$P_0 \cdot 2^{t/20}$	P_0	$1.19P_0$	$1.41P_0$	$1.68P_0$	$2P_0$	$2.83P_0$	$4P_0$

3.4   Logarithmic Functions and Derivatives

Natural Logarithms

The function

$$y = e^x$$

is defined for all x. Table III gives some values, most of them only approximations, of course. If we wish to express the value of x that yields the value y, we write

$$x = \log_e (y) \quad \text{or} \quad x = \ln (y)$$

In either case we call x the *natural logarithm* of y. Since it is conventional to let x be the independent variable when expressing functions, we shall generally write

$$y = \ln (x)$$

to express the logarithm function. (Refer to Sections 1.2 and 3.2.) We summarize these facts in the following definition.

3.4.1   **Definition**

The natural logarithm function, $y = \ln (x)$, is the inverse of the function $y = e^x$. That is,

$$y = e^x \quad \text{if and only if } x = \ln (y)$$

Notice that $y = \ln (x)$ is defined only when x is positive; this is because e^x is positive for all x. (See Figure 3.4–1.)

Figure 3.4–1

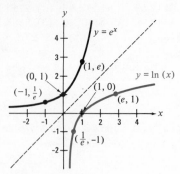

The fact that the exponential function, $y = e^x$, and the natural logarithm are inverse functions of each other means that

$$e^{\ln (x)} = x \qquad \text{and} \qquad \ln (e^x) = x$$

These are two expressions worth memorizing; they will be useful in our future work. Except for some special values of x, we must usually find the values of e^x and $\ln (x)$ by consulting a table, such as Table III or IV.

You probably remember studying *common logarithms* in high school. The common logarithm is the inverse of the function $y = 10^x$ and is often written $x = \log (y)$ or $x = \log_{10} (y)$. Incidentally,

$$\log_{10} (x) = \frac{\ln (x)}{\ln (10)} \approx \frac{\ln (x)}{2.3026} \qquad \text{and} \qquad \ln (x) = \frac{\log_{10} (x)}{\log_{10} (e)} \approx \frac{\log_{10} (x)}{0.4343}$$

You may remember seeing the following relationships for common logs. They are also true for natural logs, for the same reason. Each statement on the left is a direct consequence of the corresponding statement on the right, with $A = \ln (a)$ and $B = \ln (b)$.

3.4.2   Rule

(a) $\ln (ab) = \ln (a) + \ln (b)$ (since $e^A e^B = e^{A+B}$)
(b) $\ln (a^b) = b \ln (a)$ (since $(e^A)^b = e^{Ab}$)

(c) $\ln \left(\dfrac{a}{b}\right) = \ln (a) - \ln (b)$ $\left(\text{since } \dfrac{e^A}{e^B} = e^{A-B}\right)$

(d) $\ln (e) = 1$ (since $e^1 = e$)
(e) $\ln (1) = 0$ (since $e^0 = 1$)

The five statements in Rule 3.4.2 will be very useful in the remainder of the book; it is advisable to become thoroughly familiar with them now. The next example suggests some ways that they can be used.

3.4.3   Example

If ln (3) = 1.0986, find (without using tables)
(a) ln (3e)
(b) ln (e/3)
(c) ln (9)
(d) ln ($\sqrt{3}$)

Solution:
(a) ln (3e) = ln (3) + ln (e)   [by Rule 3.4.2 (a)]
           = 1.0986 + 1 = 2.0986   [by Rule 3.4.2 (d)]
(b) ln (e/3) = ln (e) − ln (3)   [by Rule 3.4.2 (c)]
             = 1 − 1.0986 = −0.0986   [by Rule 3.4.2 (d)]
(c) ln (9) = ln (3^2) = 2 ln (3)   [by Rule 3.4.2 (b)]
           = 2(1.0986) = 2.1972
(d) ln ($\sqrt{3}$) = ln ($3^{1/2}$) = $\frac{1}{2}$ ln (3)   [by Rule 3.4.2 (b)]
                   = $\frac{1}{2}$(1.0986) = 0.5493

Table IV at the back of the book gives the value of the natural logarithm function for some values of x. Reading down the left column of Table IV, you can verify that indeed ln (3) = 1.0986, as was given in Example 3.4.3. Reading across that same line, notice that ln (3.1) = 1.1314 and ln (3.2) = 1.1632; the numbers across the top of the table give the values of the independent variable to the tenth place.

3.4.4   Example

Compute ln (250) using Table IV.

Solution:

We see that 250 is not in the table, so we factor 250 into 250 = (2.5)(10^2). Hence

ln (250) = ln [(2.5)(10^2)]
         = ln (2.5) +   ln (10^2)   [Rule 3.4.2(a)]
         = ln (2.5) + 2 · ln (10)   [Rule 3.4.2(b)]
         = 0.9163 + 2(2.3026)   (Table IV)
         = 5.5215

Similarly, the logarithm of other numbers can be found by factoring them (perhaps only approximately) and then using Rule 3.4.2, part (a).
On the line with a "2" on the left (still in Table IV), you can see that ln (2) = 0.6931. This number is useful in many applied problems, including the following example.

3.4.5   Example

For how many years would you have to invest $100 in order to double its value if the interest rate is 4 percent per year compounded continuously?

Solution:

When we substitute $A = 200$, $P = 100$, and $r = 0.04$ into the formula $A = Pe^{rx}$, we have

$$200 = 100e^{0.04x}$$

$2 = e^{0.04x}$ (dividing by 100)

$\ln(2) = \ln(e^{0.04x}) = 0.04x$ [taking the logarithm of both sides and remembering that $\ln(e^x) = x$]

$0.6931 \approx 0.04x$ (using Table IV)

$x \approx \dfrac{0.6931}{0.04} \approx 17.33$ years (dividing)

Logarithms are also used in many fields to model important relationships in the real world. The following example shows an important application that uses common logarithms.

3.4.6   Example in Psychology or Biology

It has been found that the subjective impression of loudness is not a linear function of the physical intensity (power) of sound; it is a logarithmic function of the physical intensity. To define this function we let $I_0 = 10^{-12}$ watt per square meter; I_0 thus defined is the lowest intensity at which the subjective pitch corresponding to the physical frequency of 1,000 hertz (cycles per second) can be experienced by an average person.

Then if we let I be the physical intensity of any given sound, the subjective experience of loudness L can be given (and often is) by the formula

$$L = 10 \log_{10}\left(\frac{I}{I_0}\right) = 10[\log_{10}(I) - \log_{10}(I_0)]$$

One unit of L, thus given, is known as a decibel; it is named after Alexander Graham Bell (1847–1922). Some typical sounds are shown in Table 3.4–1.

Table 3.4–1 Intensity of Common Sounds

Physical intensity, I (watts/m²)	Loudness, L (decibels)	
10^{-12}	0	Barely audible
10^{-11}	10	Whisper
10^{-8}	40	Quiet automobile
10^{-6}	60	Ordinary conversation
10^{-4}	80	Loud orchestra
$10^0 = 1$	120	Thunder

When a logarithm is used in expressing a function, the function is often graphed using a logarithmic scale along one of the axes. This results in such a function being easier to picture. In Figure 3.4–2, $\log_{10}(I)$ is treated as the independent variable.

Figure 3.4–2

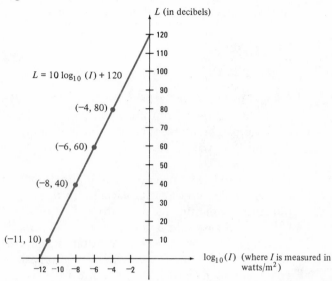

Differentiation of Logarithmic and Exponential Functions

Both exponential and logarithmic functions are used extensively in many subjects because of their special "growth" properties, which are reflected in the simplicity of their derivatives. Next we state the surprising derivatives of these two functions; the proof of Rule 3.4.7 can be found in Appendix IV.

3.4.7 Rule

(a) If $y = e^x$, then $y' = e^x$ for all x.
(b) If $y = \ln(x)$, then $y' = 1/x = x^{-1}$ for x > 0.

Notice that the derivative of the exponential function is exactly the function itself. The exponential function is the only function other than $f(x) = 0$ with this property. This is why the exponential function and functions closely related to it are useful in modeling growth (or decay) where the rate of growth (or decay) is proportional to the amount of the quantity already existing. Such applications include the growth of money in an account, the growth of populations of animals or humans (the more you have, the more you get), and the decay of radioactive substances.

3.4.8 Example

Find y' if
(a) $y = e^{3x}$
(b) $y = \ln(3x)$

Solution:

In both cases, let u = 3x and use the chain rule.

(a) $y' = \dfrac{dy}{du} \cdot \dfrac{du}{dx} = e^u \cdot 3 = 3e^{3x}$

(b) $y' = \dfrac{dy}{du} \cdot \dfrac{du}{dx} = \dfrac{1}{u} \cdot 3 = \dfrac{3}{3x} = \dfrac{1}{x}.$ [This is the same as the derivative of $y = \ln (x)$. That is because $\ln (3x) = \ln (3) + \ln (x)$, so the function $\ln (3x)$ differs from the function $\ln (x)$ by a constant [$\ln (3)$] everywhere. Therefore, their graphs will be, in a certain sense, "parallel," and they will have the same derivative.]

Warning! Do not confuse the function $f(x) = e^x$ with the function $g(x) = x^n$. The first has its variable in the exponent and is called an *exponential function*. Its derivative is $f'(x) = e^x$. The second has its variable in the base and a constant exponent. Its derivative is $g'(x) = nx^{n-1}$. The formulas are clearly very different; choose the appropriate one according to whether the exponent or the base is varying in the function you want to differentiate. (If both are varying, you need "logarithmic differentiation," which is explained at the end of Section 3.5.)

3.4.9   Example

Find y' and y'' when
(a) $y = e^{(x^2)}$
(b) $y = \ln (x^2)$

Solution:

(a) We use the chain rule, letting $u = x^2$.

$$y' = \frac{dy}{du} \cdot \frac{du}{dx} = e^u \cdot 2x = 2xe^{(x^2)}$$

Now we use both the product and chain rules, letting $u = 2x$ and $v = e^{x^2}$.

$$y'' = 2e^{x^2} + 2x(2xe^{x^2}) = 2e^{x^2} + 4x^2e^{x^2}$$

(b) Remembering that $\ln (x^2) = 2 \ln (x)$ [by Rule 3.4.2 (b)], we have

$$y' = 2\left(\frac{1}{x}\right) = 2x^{-1} \qquad \text{and} \qquad y'' = -2x^{-2}$$

SUMMARY

This section introduced the inverse of the natural exponential function $y = e^x$; its inverse is written $y = \ln (x)$, and is called the natural logarithm. A number of important properties of the logarithm function were mentioned, and then some applications to investments and perceptual theory. Finally, the derivatives of both the exponential and logarithmic functions were introduced; in the next section the derivatives will be used to show further properties and applications of these functions.

EXERCISES   3.4. A

1. Use Table IV to find the values of ln (x) when x = 0.25, 0.5, 1, 2, 3, 4, and 5. Sketch y = ln (x), including these points.
2. How long will it take $100 to double in value if it is invested at 8 percent per year compounded continuously? (*Hint:* See Example 3.4.5.)
3. Human populations grow similarly to money (compounded continuously, of course!) where r is the difference between the birth and death rates of a population. The accompanying table shows the annual birth and death rates of various countries in 1971. Find about how long it takes the population of each country to double. (*Hint:* See Example 3.4.5.)

Country	Birth rate	Death rate
Kenya	0.050	0.020
Chile	0.034	0.011
The Netherlands	0.0186	0.0082
United States	0.0177	0.0095
Sweden	0.0143	0.0104

4. At what interest rate must we invest $100, compounded continuously, if we want it to be worth $150 after 10 years?
5. A common rule of thumb used by accountants to discover how long it will take money to double if invested at a given rate r is to divide r into 72. Use the formulas of this chapter to show why this works. (Actually the number should be 69, but 72 is easier to divide into, and the answer is generally accurate enough for rough estimates.)
6. Find the derivatives of

(a) $y = e^{5x}$
(b) $y = e^{3x}$
(c) $y = e^{-2x}$
(d) $y = e^{-x^2}$
(e) $y = e^{x^2+1}$
(f) $y = e^{(x^2+1)^3}$
(g) $y = 3 \ln (x)$
(h) $y = \ln (x + 3)$

(i) $y = \ln (x + \pi)$
(j) $y = \ln (4x)$
(k) $y = \ln (-2x)$
(l) $y = \ln (x^3)$
(m) $y = \ln (x^{-2})$
(n) $y = \ln (\sqrt{x})$
(o) $y = \ln [(x^2 + 1)^3]$
(p) $y = \ln [(2x^2 - x)^{-1}]$

EXERCISES   3.4. B

1. Use Table IV to find the values of ln (x) when x = 0.25, 0.5, 1, 2, 3, 4, and 5. Sketch y = ln (x), including these points.
2. How long will it take $50 to double in value if it is invested at 6 percent per year compounded continuously?
3. Human populations grow similarly to money (compounded continuously, of course!), where r is the difference between the birth and death rates of a

population. The accompanying table shows the annual birth and death rates of various countries in 1971. Find about how long it takes the population of each country to double. (*Hint:* See Example 3.4.5.)

Country	Birth rate	Death rate
Liberia	0.044	0.025
Venezuela	0.046	0.010
New Zealand	0.0224	0.0084
United States	0.0177	0.0095
Sweden	0.0143	0.0104

4. At what interest rate must we invest $100, compounded continuously, if we want it to be worth $200 after 10 years?

5. A common rule of thumb used by accountants to discover how long it will take money to double if invested at a given rate r is to divide r into 72. Use the formulas of this chapter to show why this works. (Actually the number should be 69, but 72 is easier to divide into, and the answer is generally accurate enough for rough estimates.)

6. Find the derivatives of

(a) $y = e^{4x}$

(b) $y = e^{7x}$

(c) $y = e^{-3x}$

(d) $y = e^{2x^2}$

(e) $y = e^{x^3+1}$

(f) $y = e^{(x^3+1)^2}$

(g) $y = 4 \ln (x)$

(h) $y = \ln (x + 2)$

(i) $y = \ln (x + e)$

(j) $y = \ln (5x)$

(k) $y = \ln (-3x)$

(l) $y = \ln (x^4)$

(m) $y = \ln (x^{-3})$

(n) $y = \ln (\sqrt[3]{x})$

(o) $y = \ln [(x^3 + 1)^2]$

(p) $y = \ln [(3x^2 + x)^{-2}]$

EXERCISES 3.4. C

1. The half-life of potassium ^{42}K is 12.5 hours, so the amount of potassium remaining if A_0 units are allowed to decay for t hours is given by the formula $A = A_0 e^{-\alpha t}$, where $\alpha = [\ln (2)]/12.5$. Find the amount remaining after 25 hours (that is, $t = 25$).

2. The concentration of a drug in the body fluids depends on the time elapsed after administration. Often the concentration is given by a function in the form $C(t) = C_0 e^{-t}$, where C_0 is the initial dosage. How much time elapses until only half the drug remains?

3. Prove that $\log_{10} (x) = \ln (x)/\ln (10)$ by writing $y = \log_{10} (x)$ exponentially and then taking the natural log of both sides.

4. Rewrite the formula for loudness (Example 3.4.6) in terms of natural logarithms.

ANSWERS   3.4. A

1.

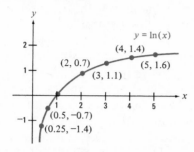

2.   8.7 years
3.   Kenya, 23 years; Chile, 30 years; The Netherlands, 67 years; United States, 85 years; Sweden, 178 years
4.   4 percent

5.   Consider the equation $2P = Pe^{rt}$; $2 = e^{rt}$; $\ln (2) = rt$; $\dfrac{\ln (2)}{r} = t \approx \dfrac{0.69}{r}$

6.   (a) $y' = 5e^{5x}$ (b) $y' = 3e^{3x}$ (c) $y' = -2e^{-2x}$ (d) $y' = -2xe^{-x^2}$
     (e) $y' = 2xe^{x^2+1}$ (f) $y' = 6x(x^2 + 1)^2e^{(x^2+1)^3}$ (g) $y' = 3/x$
     (h) $y' = 1/(x + 3)$ (i) $y' = 1/(x + \pi)$ (j) $y' = 1/x$ (k) $y' = 1/x$

     (l) $y' = 3/x$ (m) $y' = -2/x$ (n) $y' = \dfrac{1}{2x}$

     (o) $y' = 6x/(x^2 + 1)$ (p) $y' = -(4x - 1)/(2x^2 - x)$

ANSWERS   3.4. C

1. $\frac{1}{4}A_0$
2. $t = \ln (2)$; 0.69
4. $L = [10/\ln (10)] \ln (I/I_0)$

3.5   Using Derivatives of e^x and $\ln (x)$

The simplicity of the derivatives of $y = e^x$ and $y = \ln (x)$ can be used in many ways. We first turn to a discussion of graphs.

3.5.1   Example

Discuss the properties of the graphs of the functions
(a) $f(x) = e^x$ and
(b) $g(x) = \ln (x)$, using their first and second derivatives.

Solution:

(a) We observe that $f'(x) = e^x$ and $f''(x) = e^x$. Since $e^x > 0$ for all x, there are no relative extrema. Moreover, the graph is both increasing and concave upward everywhere, since $f' > 0$ and $f'' > 0$, respectively. The values of e^x increase very rapidly if $x > 1$ and more slowly if $x < 1$. The rate of growth depends directly on the value of the original function. This is what makes the exponential function an ideal model for many rapid-growth situations, such as compound interest and the "population explosion."

(b) We observe that $g'(x) = 1/x$ and $g''(x) = -1/x^2$. Since the logarithm function is only defined when x is positive, $1/x > 0$, and we conclude that the function is always increasing; the increase is very rapid for $0 < x < 1$ and slower for $x > 1$. Actually, multiplicative changes in x result in additive changes in ln (x), since ln (kx) = ln (k) + ln (x). The fact that the second derivative is negative implies that the graph is concave downward everywhere. The features of the logarithm function make it useful for the description of certain slow-growth situations, such as intensity of perception versus physical change in stimuli.

Figure 3.5–1

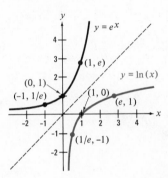

The given features of f and g can be seen in Figure 3.5–1, together with their relationship to each other as inverse functions; they are mirror images of each other over the line $y = x$.

3.5.2   Example in Demography

Figure 3.5–2

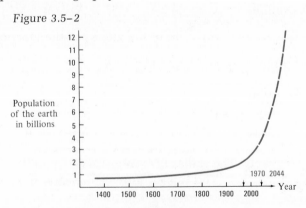

The population of the earth has been a function of the form $y = e^{rx}$, where x is the date and y is the number of people alive at that date; r is a constant, which you can calculate in the exercises. Figure 3.5–2 describes the population of the world, where the dashed part is the projected population if conditions stay the same and we do not run out of resources. This projection says that the population of the world will quadruple between 1970 and 2044, a period of only 74 years.

For any $a > 1$, the function $y = a^x$ can be defined in a manner similar to $y = e^x$ and will be everywhere increasing and positive. Therefore, $y = a^x$ will (like $y = e^x$) have an inverse defined over the set of positive numbers. This inverse function is called the <u>logarithm to the base a</u> and is written $y = \log_a(x)$.

Figure 3.5–3

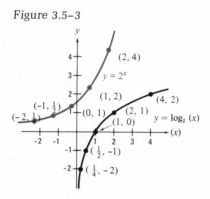

Exponential and logarithmic functions with bases different from e have many of the same properties as $y = e^x$ and $y = \ln(x)$ (see Figure 3.5–3 for $a = 2$), but the scales are different. Therefore, they are useful in many of the same types of models.

Logarithmic Differentiation

Instead of stating more rules for the differentiation of general exponential and logarithmic functions, we shall now introduce a computational technique called <u>logarithmic differentiation</u>. This long expression merely means that it is sometimes easier to take the natural logarithm of both sides of an equation before differentiating.

Expanding this idea, there are three steps in logarithmic differentiation:

1. Take the natural logarithm of both sides of the equation and simplify.
2. Differentiate both sides of the new equation.
3. Solve for y' in the new expression.

The only difficult part of logarithmic differentiation is recognizing when to use it. There are two types of functions for which it is often useful:

1. An exponential or logarithmic function with a (constant or varying) base other than e.
2. A complicated product or quotient.

3.5.3   **Example**

Find y' when
(a) $y = 10^x$
(b) $y = \log_{10}(x)$
(c) $y = \dfrac{(x+1)(x-2)}{3x(x^2-5)}$

Solution:

(a) $y = 10^x$
          $\ln(y) = \ln(10)^x = x\ln(10)$ [taking the logarithm of both sides and using Rule 3 4.2(b)]

          $\dfrac{y'}{y} = \ln(10)$ [differentiating both sides; use the chain rule on the left side and the fact that $x\ln(10) = mx$, where $m = \ln(10)$ is a constant, on the right side]

          $y' = y\ln(10) = 10^x \cdot \ln(10)$ (multiplying through by y and then substituting $y = 10^x$)

(b) If $y = \log_{10}(x)$, we soon notice that we cannot differentiate $\ln(\log_{10}(x))$ any more easily than the original function. However, if we first rewrite the original expression as $10^y = x$, then taking the log before differentiating does help.

          $10^y = x$ (by definition of $y = \log_{10}(x)$)
          $\ln(10^y) = \ln(x)$ (taking the natural logarithm of both sides)

          $y\ln(10) = \ln(x)$ [Rule 3.4.2(b)]

$$y' \ln (10) = \frac{1}{x}$$ [taking the derivative of both sides and recognizing that $y \ln (10) = mx$, where $m = \ln (10)$ is a constant]

$$y' = \frac{1}{x \ln (10)}$$ [dividing by $\ln (10)$]

(c) When we take the natural logarithm of both sides and use Rule 3.4.2(a) and (c), we have

$$\ln (y) = \ln \left[\frac{(x + 1)(x - 2)}{3x(x^2 - 5)} \right] = \ln (x + 1) + \ln (x - 2) - \ln (3x)$$
$$- \ln (x^2 - 5)$$

Differentiating the left and right members of this equation,

$$\frac{y'}{y} = \left[\frac{1}{x + 1} + \frac{1}{x - 2} - \frac{3}{3x} - \frac{2x}{x^2 - 5} \right]$$

Multiplying through this equation by $y = [(x + 1)(x - 2)]/[3x(x^2 - 5)]$, we obtain

$$y' = \frac{(x + 1)(x - 2)}{3x(x^2 - 5)} \left(\frac{1}{x + 1} + \frac{1}{x - 2} - \frac{1}{x} - \frac{2x}{x^2 - 5} \right)$$

If you now solve this problem using the product and quotient formulas of Section 2.3, you will see that logarithmic differentiation is much easier than the alternative.

If you plan to use exponential and logarithmic functions to bases other than e frequently, the following example should be considered a "Rule" and memorized. Otherwise, it is enough to notice that logarithmic differentiation is helpful when differentiating any exponential or logarithmic function with a base other than e.

3.5.4   Example

(a) Show that if $y = a^x$, $y' = a^x \ln (a)$.

(b) Show that if $y = \log_a (x)$, $y' = \dfrac{1}{x \ln (a)}$.

Solution:

(a) Taking the logarithm of both sides of $y = a^x$, we get $\ln (y) = \ln (a^x) = x \ln (a)$.

Differentiating both sides: $\dfrac{y'}{y} = \ln (a)$

Multiplying through by y: $y' = a^x \ln (a)$

(b) This time we first put both sides of the equation into the exponent, using a for the base:

$$y = \log_a (x)$$

$$a^y = a^{\log_a(x)} = x \qquad \text{(using the inverse property)}$$

$$\ln (a^y) = \ln (x) \qquad \text{(taking the log of both sides)}$$

$$y \ln (a) = \ln (x) \qquad \text{[using Rule 3.4.2(b)]}$$

$$y' \ln (a) = 1/x \qquad \text{(differentiating)}$$

$$y' = \frac{1}{x \ln (a)}$$

3.5.5   Example in Biology

It is generally accepted that the size of the aorta could be smaller in most animals if it were not for certain turbulences in the blood flow. The flow of blood becomes turbulent when the Reynolds number, R, exceeds a certain quantity. In many organisms the Reynolds number depends on the radius of the aorta, r, according to the function

$$R(r) = A \ln (r) - Br$$

where A and B are positive constants depending on the organism. If $A = 472$ and $B = 1{,}100$ for a certain breed of dog, find the maximum value of the Reynolds number, R.

Solution:

To find the maximum value of $R(r) = 472 \ln (r) - 1{,}100r$, we find R' and set it equal to zero; then we use the second derivative test to show that R takes a maximum at that point.

$$R' = 472 \frac{1}{r} - 1{,}100$$

so $R' = 0$ if, and only if, $r = 472/1{,}100 \approx 0.43$.

$$R'' = -\frac{472}{r^2}$$

which is clearly negative, so the critical point indeed gives a maximum. In this particular breed of dog, the maximum is taken when $r = 0.43$ cm and is $R(0.43) = -871$.

SUMMARY

The derivatives of $f(x) = e^x$ and $g(x) = \ln (x)$ are $f'(x) = e^x$ and $g'(x) = 1/x$, respectively. We use these to graph the exponential and logarithm functions, and to compute the derivatives of more complicated functions (using the chain rule). Logarithmic differentiation, the process of first taking the logarithm of a function and then differentiating it, is useful for exponential and logarithm functions to bases other than e and for complicated products.

EXERCISES   3.5. A

1. Sketch the graphs of the following functions, using derivatives. Plot the points where $x = -1, 0$, and 1.
   (a) $y = 2e^x$
   (b) $y = e^{2x}$
   (c) $y = e^{-x}$
   (d) $y = e^{-2x}$
   (e) $y = e^{0.05x}$

2. Sketch the graphs of the following functions, using derivatives. In each case plot either the points where $x = 0.5, 1$, and 2 or the points where $x = -0.5$, -1, and -2.
   (a) $y = 2 \ln(x)$
   (b) $y = -\ln(x)$
   (c) $y = \ln(-x)$
   (d) $y = \ln(2x)$

3. What is the marginal revenue if the revenue obtained for a certain commodity is $R = 100e^{-x}$?

4. Find the equation of the tangent line to the curve $y = e^x$ at the point where $x = 1$.

5. Graph the function $y = e^{-x^2}$ using the first and second derivatives. This is a very important function in statistics. It is called the "normal curve" or the "bell curve" or the "Gaussian curve" [after Karl Friedrich Gauss (1777–1855)].

6. Find the derivatives of the following functions, using logarithmic differentiation.
   (a) $y = 5^x$
   (b) $y = 8^x$
   (c) $y = 2^{4x}$
   (d) $y = 10^{3x}$
   (e) $y = 5^{(x^2)}$
   (f) $y = 5^{(x^2+1)^3}$
   (g) $\log_5(x) = y$
   (h) $\log_3(x) = y$
   (i) $\log_5(x^2) = y$
   (j) $\log_3(x^4) = y$
   (k) $y = \log_5[(x^2 + 1)^3]$

7. Prove that if $y = a^{f(x)}$, then $y' = a^{f(x)}f'(x) \ln(a)$.

EXERCISES   3.5. B

1. Sketch the graphs of the following functions, using derivatives. Plot the points where $x = -1, 0$, and 1.
   (a) $y = 2e^x$
   (b) $y = e^{2x}$
   (c) $y = e^{-x}$
   (d) $y = e^{-2x}$
   (e) $y = e^{0.05x}$

2. Sketch the graphs of the following functions, using derivatives. In each case plot either the points where $x = 0.5, 1$, and 2 or the points where $x = -0.5$, -1, and -2.
   (a) $y = 2 \ln(x)$
   (b) $y = -\ln(x)$
   (c) $y = \ln(-x)$
   (d) $y = \ln(2x)$

3 What is the marginal revenue if the revenue obtained for a certain commodity is $R = 50e^{-2x}$?

4. Find the equation of the tangent line to the curve $y = \ln(x)$ at the point where $x = 1$.

5. Graph the function $y = e^{-x^2}$ using the first and second derivatives. This is a very important function in statistics. It is called the "normal curve" or the "bell curve" or the "Gaussian curve" [after Karl Friedrich Gauss (1777–1855)].

6. Find the derivatives of the following, using logarithmic differentiation.
   (a) $y = 2^x$
   (b) $y = 7^x$
   (c) $y = 5^{4x}$
   (d) $y = 10^{4x}$
   (e) $y = 2^{(x^2)}$
   (f) $y = 2^{(x^3+1)^2}$
   (g) $y = \log_2 (x)$
   (h) $y = \log_4 (x)$
   (i) $y = \log_2 (x^3)$
   (j) $y = \log_5 (x^3)$
   (k) $y = \log_2 [(x^3 + 1)^2]$

7. Prove that if $y = \log_a (f(x))$, then

$$y' = \frac{f'(x)}{f(x)} \log_a (e) = \frac{f'(x)}{f(x) \ln (a)}$$

EXERCISES   3.5. C

1. If the concentration of a drug in the body fluids is given by the formula $C(t) = C_0 e^{-t}$, where C_0 is the initial dosage and t is the time elapsed after the drug is taken, what is the rate of change in the concentration when
   (a) $t = 1$?
   (b) $t = 2$?
   (c) $t = 3$?

2. Find the maximum value of the function $f(x) = xe^{-x^2}$.

3. What is the rate of change of the value $A(x) = Pe^{rx}$ of $100 after 10 years if it is invested at 6 percent compounded continuously?

4. Find the derivatives of the following, using logarithmic differentiation.
   (a) $y = \dfrac{(x^2 + 1)^3(x^3 - 3x)}{5x^4(x + 1)^2}$
   (b) $y = 2^{t^{2/3}}e^{-t/10}$
   (c) $y = 2^{\sqrt{t}}e^{-t}$

5. A population of *Escherichia coli* bacteria, a very common species that normally exists in human intestines, has an initial size P_0 and, if unchecked, doubles in size every 20 minutes according to the formula $P(t) = P_0 \cdot 2^{t/20}$, where t is given in minutes. What is the rate of change of this population when $t = 20$ minutes?

ANSWERS   3.5. A

1. (a)

(b)

(c)

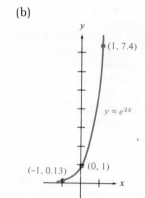

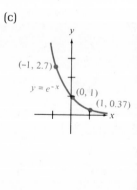

(d)   (e)

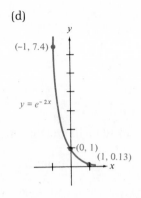

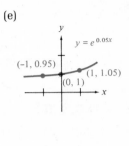

2.   (a)   (b)   (c)   (d)

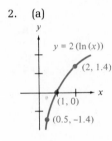

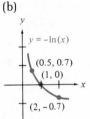

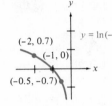

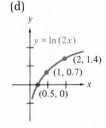

3. $R' = -100e^{-x}$
4. $y = ex$
5.

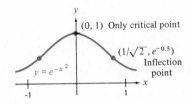

6.   (a) $y' = 5^x \ln(5)$ (b) $y' = 8^x \ln(8)$ (c) $y' = 2^{4x}(4) \ln(2)$
     (d) $y' = 10^{3x}(3) \ln(10)$ (e) $y' = 5^{(x^2)}(2x) \ln(5)$
     (f) $y' = 5^{(x^2+1)^3}6x(x^2 + 1)^2 \ln(5)$ (g) $y' = 1/[x \ln(5)]$ (h) $y' = 1/[x \ln(3)]$
     (i) $y' = 2/[x \ln(5)]$ (j) $y' = 4/[x \ln(3)]$ (k) $y' = \dfrac{6x}{(x^2 + 1) \ln 5}$
7. $\ln(y) = f(x) \ln(a)$; so $y'/y = f'(x) \ln(a)$; so $y' = a^{f(x)}f'(x) \ln(a)$

ANSWERS   3.5. C

1.   (a) $-0.37C_0$ (b) $-0.14C_0$ (c) $-0.05C_0$
2. $e^{-1/2}/\sqrt{2}$
3. $6e^{0.6}$

4. (a) $y' = \dfrac{(x^2 + 1)^3(x^3 - 3x)}{5x^4(x + 1)^2} \left(\dfrac{6x}{x^2 + 1} + \dfrac{3x^2 - 3}{x^3 - 3x} - \dfrac{4}{x} - \dfrac{2}{x + 1} \right)$

   (b) $y' = 2^{t^{2/3}} e^{-t/10} [\frac{2}{3} t^{-1/3} \ln (2) - \frac{1}{10}]$ (c) $y' = 2^{\sqrt{t}} e^{-t} [\frac{1}{2} t^{-1/2} \ln (2) - 1]$

5. $\frac{1}{10} P_0 \cdot \ln (2)$ units per minute

SAMPLE QUIZ   **Sections 3.1, 3.2, 3.3, 3.4, and 3.5**

1. (15 pts) Find the value to which $100 will grow if invested at 4 percent for 3 years and compounded
    (a) annually   (c) continuously
    (b) quarterly

2. (10 pts) At what interest rate must we invest $100 so that it will grow to $200 in 10 years if compounded continuously?

3. (10 pts) Sketch the graph of $y = e^x$.

4. (20 pts) Find the derivatives of
    (a) $y = e^{9x}$ (c) $y = x \ln (x)$
    (b) $y = e^{3x^2 - 4x}$ (d) $y = \ln (x^3 + 5x)$

5. (5 pts) Find the marginal revenue if the revenue is $R = x^2 \ln (x)$.

6. (20 pts) Find the derivatives of
    (a) $y = (x^3 - 5x^2 + 6)^7$ (b) $y = \sqrt{2x^4 - 6x}$

7. (10 pts) Find $\dfrac{dx}{dy}$ if $y = (2x^2 + 3x)^7$.

8. (10 pts) Find the equation of the line tangent to the ellipse $3x^2 + y^2 = 12$ at the point (1, 3).

The answers are at the back of the book.

3.6   Velocity, Acceleration, and Related Rates of Change

If the distance that a moving body has traveled is given by a function $s = f(t)$, then its <u>velocity at time t</u> is given by $f'(t) = \dfrac{ds}{dt}$. The velocity is the instantaneous rate of change of distance with respect to time. (See Section 1.6.) If the motion of the body is not constant, its velocity, as well as its position, will be changing. The rate of change of the velocity is called <u>acceleration</u>. Therefore, the acceleration is given by $f''(t) = \dfrac{d^2s}{dt^2}$.

3.6.1   Example

Suppose that a ball is thrown straight up into the air and is $s = 48t - 16t^2$ feet high after t seconds.
(a) What is its velocity and acceleration after 1 second (that is, when $t = 1$)?
(b) How long will it be until it reaches its maximum height?
(c) What is its maximum height?

Solution:

(a) $\dfrac{ds}{dt} = 48 - 32t$, so when $t = 1$, $\dfrac{ds}{dt} = 48 - 32(1) = 16$. Since $\dfrac{ds}{dt}$ is the velocity, the answer is 16 ft/sec. $\dfrac{d^2s}{dt^2} = -32$, so the acceleration is always 32 ft/sec^2 in a downward (negative) direction; it is due to gravity.

(b) The maximum height will be achieved when the derivative $\dfrac{ds}{dt}$ is zero. Hence we take the derivative of the height s, and set it equal to zero.

When $\dfrac{ds}{dt} = 48 - 32t = 0$, $t = \tfrac{3}{2}$. Since $\dfrac{d^2s}{dt^2} = -32 < 0$, the height is a maximum at $t = \tfrac{3}{2}$.

(c) $s(\tfrac{3}{2}) = 48(\tfrac{3}{2}) - 16(\tfrac{3}{2})^2 = 72 - 36 = 36$, so the maximum height is 36 feet.

3.6.2 Example

An arrow is shot straight upward at a branch of a tree that is 300 feet high. The initial velocity of the arrow is 128 ft/sec, so it will be $s = 128t - 16t^2$ high at time t. Will the arrow hit the branch if aimed properly?

Solution:

The question can be restated: Will the maximum height achieved by the arrow be at least 300 feet? To answer this, we take the derivative of the distance and set it equal to zero. $s' = 128 - 32t$, so $t = \tfrac{128}{32} = 4$ seconds. Thus the arrow will be as high as it gets after 4 seconds. To see how high this is, we substitute $s = 128(4) - 16(4)^2 = 256$ feet. Since 256 is the maximum height, the arrow will not hit the branch.

Rates of Change over Time

In physical problems, distance is often given explicitly as a function of time, so the calculation of velocity requires merely taking the derivative. In business applications, however, functions are often given in terms of the number of units produced instead of the time involved in their production. But the number of units produced may well vary as a function of time; this would result in the cost also being a function of time. Problems of this type can be handled by taking the derivative with respect to a variable (for example, time t) that is not visible in the equation. This involves using the chain rule.

3.6.3 Example

If the cost of producing x units of a commodity is given by $C(x) = 3x - 4[\ln (x)]$ and the constant production rate is 10 items per hour, at what rate is the cost changing (with respect to time t) when 50 units have been produced?

Solution:

Since both C and x are functions of time t, we can differentiate each term in the given equation with respect to t before making any numerical substitutions. We use the chain rule in a manner similar to that of Section 3.2 when doing implicit differentiation:

$$\frac{dC}{dt} = 3\frac{dx}{dt} - 4\frac{1}{x}\left(\frac{dx}{dt}\right)$$

Into this expression we now substitute $\frac{dx}{dt} = 10$ and $x = 50$.

$$\frac{dC}{dt} = 3(10) - 4(\tfrac{1}{50})10 = 30 - 0.8 = 29.2$$

Thus the total cost is increasing at the rate of about $29.20 per hour after 50 units have been produced.

3.6.4   Example

The revenue obtained by selling x yards of carpet is given by $R = 0.03x^2 + 2x$ and the rate of sales is constant at 100 yards per day. At what time and sales level will the revenue be increasing at the rate of $1,000 per day?

Solution:

This time we are given $\frac{dR}{dt}$ and we want to find the corresponding x and t. To make use of the given information, we must differentiate the expression for R and get one for $\frac{dR}{dt}$.

$$\frac{dR}{dt} = 0.06x\frac{dx}{dt} + 2\frac{dx}{dt}$$

Substituting $\frac{dR}{dt} = 1,000$ and $\frac{dx}{dt} = 100$, we get

$$1,000 = 0.06x(100) + 2(100) = 6x + 200$$
$$800 = 6x \qquad \text{so } x = 133\tfrac{1}{3}$$

Thus the revenue will be increasing at $1,000 per day after $133\tfrac{1}{3}$ rugs have been produced, or about one-third of the way through the second day $(t = \tfrac{4}{3})$.

3.6.5   Example in Biology

Suppose that the population P of some predator species depends roughly on the population of its favorite prey according to the formula $P =$

$0.001x + 0.0001x^2$, where x is the number of prey inhabiting a given region. If the number of prey is increasing at the rate of 300 per year at a time when there are 1,000 of the prey in the region, how fast is the population of predators increasing with respect to time t?

Solution:

Differentiating the formula for P with respect to t, we get

$$\frac{dP}{dt} = 0.001 \frac{dx}{dt} + 0.0002x \frac{dx}{dt}$$
$$= 0.001(300) + 0.0002(1,000)(300) = 0.3 + 60 = 60.3$$

Thus the predators will be increasing at the rate of about 60 per year. (In natural situations the number of the prey would *also* depend on the number of the predators, so the mathematics might be more complicated than this problem indicates. Differential equations, the subject of Section 4.4, are useful in analyzing such situations.)

3.6.6   Example in Biology or Pre-Med

The velocity of the flow of blood in an artery is given by Poiseuille's law:

$$V = k(R^2 - r^2)$$

where k is a constant involving length, pressure, and viscosity; R is the radius of the artery; and r is the (variable) distance of a particular point from the axis of the artery. (See Example 1.4.9.) Assuming that $k = 1,100$ and $R^2 = 0.04$, find a formula for the acceleration of blood flow as a function of the rate of change of r with respect to time t.

Solution:

Substituting the given values into the formula, we have

$$V = 1,100(0.04 - r^2) = 44 - 1,100r^2$$

The acceleration a is $\frac{dV}{dt}$. Hence we must think of both V and r as functions of t. Differentiating implicitly, we obtain

$$\frac{dV}{dt} = -2,200r \frac{dr}{dt}$$

SUMMARY

Velocity is the derivative of distance with respect to time. Acceleration is the derivative of velocity with respect to time. This section began by using these facts to solve some elementary problems in physics.

Sometimes a function in business, biology, or another field states a relationship between two variables both of which vary in time. Then we can differentiate the entire expression with respect to time, and use known information about a given rate of change to find another rate of change.

EXERCISES   3.6. A

1. A falling body descends $s = 16t^2$ feet in t seconds.
    (a) What will be its velocity in 10 seconds?
    (b) If it is dropped from a plane 6,400 feet above the earth, how fast will it be traveling when it hits the earth?
2. If a bullet is fired straight up with a velocity of 320 ft/sec, it will travel $s = 320t - 16t^2$ feet in t seconds.
    (a) When will it reach its maximum height?
    (b) What will its maximum height be?
3. Suppose that the price p of a commodity is given by $p = -\frac{1}{2}q^2 + q + 1{,}000$ when q items are sold. What is the rate of change in the price $\left(\text{that is, } \dfrac{dp}{dt}\right)$ if $q = 20$ and the number of items is changing at the rate of 10 per day (that is, if $\dfrac{dq}{dt} = 10$)?
4. If $R = pq$ and other facts are as in problem 3, find the rate of change in the revenue. $\left(Hint: \text{Find } \dfrac{dR}{dt} \text{ using the product rule for differentiation.}\right)$
5. If the cost of producing q items is $C = \$(50 - q + q^2)$, the cost is changing at \$10 per day, and $q = 20$, find the rate of change of q $\left(\text{that is, } \dfrac{dq}{dt}\right)$.

EXERCISES   3.6. B

1. A falling body descends $s = 16t^2$ feet in t seconds.
    (a) What will be its velocity after 20 seconds?
    (b) If it is dropped from a plane 3,600 feet above the earth, how fast will it be traveling when it hits the earth?
2. If a bullet is fired straight up with a velocity of 640 ft/sec, it will travel $s = 640t - 16t^2$ feet in t seconds.
    (a) When will it reach its maximum height?
    (b) What will its maximum height be?
3. Suppose that the price p of a commodity is given by $p = -q^2 + 2q + 500$ when q items are sold. What is the rate of change in the price $\left(\text{that is, } \dfrac{dp}{dt}\right)$ if $q = 20$ and the number of items is changing at the rate of 10 per day (that is, if $\dfrac{dq}{dt} = 10$)?
4. If $R = pq$ and other facts are as in problem 3, find the rate of change in the revenue. $(Hint: \text{Find } \dfrac{dR}{dt} \text{ using the product rule for differentiation.})$
5. If the cost of producing q items is $C = \$(100 - q + q^2)$, the cost is changing at \$20 per day, and $q = 10$, find the rate of change of q $\left(\text{that is, } \dfrac{dq}{dt}\right)$.

EXERCISES   3.6. C

1. The position of an object thrown horizontally with an initial horizontal velocity of v ft/sec is given by

$$y = -16 \frac{x^2}{v^2}$$

Find the rate of change in y with respect to time t (that is, vertical speed) if $v = 4$, $x = 10$, and $\frac{dx}{dt} = v$.

ANSWERS AND SOME SOLUTIONS   3.6. A

1. (a) 320 ft/sec because $v = 32t$. (b) First we solve $6,400 = 16t^2$ to find that it will hit the earth at $t = \sqrt{400} = 20$ seconds after it is dropped. Therefore, it will be traveling at $v = 32t = 32(20) = 640$ ft/sec when it hits.

2. (a) $t = 10$ and (b) $s = 1,600$. This is because $\frac{ds}{dt} = 320 - 32t = 0$ has a solution of $t = 10$ and $s(10) = 320(10) - 16(10)^2 = 1,600$.

3. $\frac{dp}{dt} = -190$ 5. $\frac{dq}{dt} = \frac{10}{39}$

4. $\frac{dR}{dt} = 4,400$

ANSWERS   3.6. C

1. -80 ft/sec

3.7   Rate of Proportional Change and Asset Growth

This section studies two interesting applications of the logarithmic and exponential functions. The first of these, the rate of proportional change (or growth rate), deals with the derivative, y'/y, of the logarithm of a function, $y = f(x)$; this turns out to be a useful measure of "growth." The second application is also concerned with growth. Specifically, it deals with the growth of the value of an asset, such as a collection of fine wines, whose value changes with time. The goal is to find the most advantageous time to sell the asset.

Rate of Proportional Change (Growth Rates)

Consider a person whose height is increasing at a rate of 3 inches per year. Does this represent a rapid rate of growth or not? The answer depends largely on the person's present size. For a baby 30 inches tall such a growth is a 10 percent increase in total height. But for a young adult 60 inches tall an increase of 3 inches represents only a 5 percent increase in height. The important measure is often the ratio between the rate of change and present size (that is, y'/y). We call this ratio the "rate of proportional change" or "growth rate."

3.7.1   Definition

If $y = f(x)$, then

$$\frac{d}{dx}[\ln (f(x))] = \frac{f'(x)}{f(x)}$$

is commonly called the rate of proportional change in y.

If $y = f(t)$, where t is units of time, $f'(t)/f(t)$ is called the growth rate of the function f.

3.7.2   Example

$A(t) = Pe^{rt}$ (the usual continuous compound interest formula); find the growth rate of A.

Solution:

$$\frac{A'(t)}{A(t)} = \frac{Pre^{rt}}{Pe^{rt}} = r \qquad \text{the interest rate}$$

Note that the growth rate of any human or animal population described by a function of the form $P(t) = e^{rt}$ is also r by the reasoning of the above example. (See also problem 3 in Exercises 3.4.A and 3.4.B.)

3.7.3   Example

The cost of a continuous operation is given by $f(t) = 10,000 + t$ at time t. Find the growth rate when $t = 2$.

Solution:

$$\frac{f'(t)}{f(t)} = \frac{1}{10,000 + t}$$

so at $t = 2$ it is $1/10,002$. Notice that in this case, unlike the previous example, the growth rate depends on t and changes with time.

3.7.4   Example

Find the growth rate of the function $f(t) = e^{t^3 - t^2}$ when $t = 1$.

Solution:

$$\frac{f'(t)}{f(t)} = \frac{(3t^2 - 2t)f(t)}{f(t)} = 3t^2 - 2t$$

Hence, when $t = 1$, the growth rate is $3 - 2 = 1$.

3.7.5   Example

Find the minimum growth rate of the function in Example 3.7.4.

Solution:

The growth rate, by the calculations above, is $g(t) = 3t^2 - 2t$. Thus $g'(t) = 6t - 2$ and the only critical point is at $t = \frac{1}{3}$. Since $g''(t) = 6 > 0$, the function has a minimum at $\frac{1}{3}$ by the second derivative test. Hence the minimum growth rate is $3(\frac{1}{3})^2 - 2(\frac{1}{3}) = -\frac{1}{3} \approx -33$ percent.

3.7.6   Example

If $V = P(1 + r)^t$, find the growth rate of V, where r and P are constants and t is a continuous variable.

Solution:

Since we have an exponential function whose base is not e, we use logarithmic differentiation (see pages 173–174):

$$\ln (V) = \ln [P(1 + r)^t] \qquad \text{[taking the logarithm of } V = P(1 + r)^t]$$
$$= \ln (P) + \ln [(1 + r)^t] \qquad \text{[using Rule 3.4.2(a)]}$$
$$= \ln (P) + t[\ln (1 + r)] \qquad \text{[using Rule 3.4.2(b)]}$$

Now we differentiate both sides of $\ln (V) = \ln (P) + t \ln (1 + r)$ with respect to t and immediately get an expression for the growth rate:

$$\frac{V'}{V} = \ln (1 + r)$$

3.7.7   Example in Geology, Paleontology, and Archaeology

Radioactive decay of various elements is often used to date ancient rocks, fossils, and artifacts. A radioactive element will decay with a constant (negative) growth rate that can be measured. For example, the amount A of carbon 14 remaining after t years, if A_0 is the original quantity, is given by the formula

$$A = A_0 e^{-\alpha t} \qquad \text{where } \alpha = \frac{\ln 2}{5,600}$$

(a) Find the growth (decomposition) rate of carbon.
(b) How long will it take for one-half of the carbon to decompose at this rate? (That is, find the half-life of carbon 14.)

Solution:

(a) $A' = \dfrac{dA}{dt} = -\alpha A_0 e^{-\alpha t}$, so the

$$\text{growth rate} = \frac{A'}{A} = \frac{-\alpha A_0 e^{-\alpha t}}{A_0 e^{-\alpha t}} = -\alpha = -\frac{\ln 2}{5,600}$$

(b) If we begin with an amount A_0 of carbon, the amount left after one-half the carbon has decayed is $A = \frac{1}{2}A_0$. We substitute this quantity into the general formula and then solve for t:

$$\tfrac{1}{2}A_0 = A_0 e^{-\alpha t}$$

Dividing both sides by A_0, $\frac{1}{2} = e^{-\alpha t}$.
Taking the log of both sides, $\ln\left(\frac{1}{2}\right) = \ln\left(e^{-\alpha t}\right) = -\alpha t$.*
But $\ln\left(\frac{1}{2}\right) = \ln\left(2^{-1}\right) = (-1)\ln(2)$ and $\alpha = \ln(2)/5{,}600$, so

$$-\ln(2) = -\frac{\ln(2)}{5{,}600}\,t \qquad \text{(substituting the above line into the * line)}$$

which implies that $1 = (1/5{,}600)t$ or $5{,}600 = t$. Thus we have found that it will take 5,600 years for half the carbon 14 to decay.

Asset Growth

We now consider a model of the growth of an asset whose value V changes in time. That is, $V = V(t)$ is a function of time. Therefore, the growth rate of the value will be given by V'/V. Recall that money invested at a fixed rate of 100r percent, compounded continuously, has a growth rate of r. (See Example 3.7.2.) If the bank's rate is guaranteed, we might ask if it would be better to liquidate (sell) a given asset and put the proceeds in the bank. This would be wise if the growth rate of V is less than the growth rate r offered by the bank.

If the growth rate of the asset is initially higher than r but is decreasing, the optimal selling time is the time t_0 such that $V'(t_0)/V(t_0) = r$ because after time t_0 the bank interest gives a higher return than does the original asset. However, before time t_0 the original asset grows at a faster rate than that offered by the bank.

Figure 3.7–1

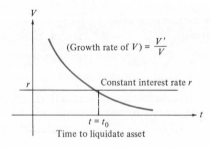

3.7.8   **Example**

Suppose that Ali Baba's Jewelers owns some diamonds whose value at time t years is $V = 100{,}000e^{\sqrt{t}}$. If the investment rate is 5 percent, find the optimal selling time.

Solution:

We can find the growth rate of the diamonds by first calculating

$$V'(t) = (100{,}000e^{\sqrt{t}}) \left(\frac{1}{2} \frac{1}{\sqrt{t}} \right)$$

Thus

$$\text{growth rate of } V = \frac{V'}{V} = \frac{1}{2} \frac{1}{\sqrt{t}}$$

(We notice that as t increases, the growth rate decreases.) Setting the growth rate equal to 5 percent, we have

$$\frac{1}{2} \frac{1}{\sqrt{t}} = 0.05$$

Solving the last equation gives us $t = 100$. Hence, the optimal strategy is to sell the diamonds after letting them increase in value for 100 years. Then invest the proceeds at 5 percent, assuming that this is still the prevailing interest rate.

3.7.9   Example

Find the growth rate of the diamonds in the previous example at times (a) $t = 81$ and (b) $t = 121$ years.

Solution:

(a) $\dfrac{V'(81)}{V(81)} = \dfrac{1}{2} \dfrac{1}{\sqrt{81}} = \dfrac{1}{18} = 0.0555 \ \ldots$

(b) $\dfrac{V'(121)}{V(121)} = \dfrac{1}{2} \dfrac{1}{\sqrt{121}} = \dfrac{1}{22} = 0.0454545 \ \ldots$

Observe that before 100 years has passed, the growth rate is higher than 0.05, and after 100 years has passed, the growth rate is lower than the 0.05 rate offered by the bank.

SUMMARY

The rate of proportional change (often called growth rate) and asset growth were the applications of calculus introduced in this section. The rate of proportional change of a function y is y'/y, the derivative of $\ln (y)$. If the value of an asset, such as wines, antiques, or jewels, is known to vary according to a certain function $V(t)$, we can find the best time to sell the asset by setting the growth rate equal to the prevailing interest rate. In the next section we will discuss still another application of derivatives, an application that will be useful in Chapter 4, where we begin our study of the other major part of calculus — integral calculus.

EXERCISES   3.7. A

1. Find the growth rate of the functions
    (a) $f(t) = t^2$ (d) $f(t) = e^{t^2-t}$
    (b) $f(t) = t^3$ (e) $f(t) = e^{3t-t^3}$
    (c) $f(t) = t^{3/2}$ (f) $f(t) = e^{4t-5}$
    Evaluate these growth rates at time $t = 1$.

2. Following an advertising campaign, the sales often first increase and then decrease. If the sales following a campaign are predicted to be $S(t) = -15t^2 + 50t + 100$ $(t \leqslant 3)$, find the growth rate at $t = 0$, at $t = 1$, at $t = 2$, and at $t = 3$.

3. Diamond Jim and Co. possesses a gem the value of which increases in time according to the formula $V = 1,000e^{0.3t^{1/2}}$. If the interest rate is 5 percent, find the best selling time.

4. Ye Select Wines is storing a batch of wines whose value at time t is $V = 500e^{0.27t^{2/3}}$. Assuming an interest rate of 6 percent, when should Ye Select Wines sell this batch?

5. Suppose that a collection of antiques has value $V = 6,000e^{0.28t^{1/4}}$. If the available interest rate is 7 percent, when should the antiques be sold?

EXERCISES   3.7. B

1. Find the growth rate of the functions
    (a) $f(t) = t$ (d) $f(t) = e^{t-t^2}$
    (b) $f(t) = t^{-1}$ (e) $f(t) = \ln (2t)$
    (c) $f(t) = e^{-t}$
    Evaluate these at time $t = 1$.

2. Following an advertising campaign the sales often first increase and then decrease. If the sales following a campaign are predicted to be $S(t) = -12t^2 + 48t + 120$ $(t \leqslant 3)$, find the growth rate at $t = 0$, at $t = 1$, at $t = 2$, and at $t = 3$.

3. Beautiful Jewels, Inc., has a diamond whose value is increasing in time according to the formula $V = 10,000e^{0.4t^{1/2}}$. If the interest rate is 5 percent, find the best selling time.

4. Suppose that a batch of wines has value $V = 1,000e^{0.16t^{3/4}}$ at time t. When should the batch be sold if the interest rate is 6 percent?

5. Suppose that a collection of antiques has value $V = 8,000e^{0.1t^{2/5}}$. If the available interest rate is 4 percent, when should the antiques be sold?

EXERCISES   3.7. C

1. Grecco Bacchas and Sons sell fine wines. The value of the wines increases over time according to the formula $V = 1,000(1.5)^{\sqrt{t}}$. Find the optimum time to sell the wines if the interest rate is 7 percent.

2. The value of the antiques owned by Ye Olde Curiosity Shoppe varies according to the formula $V = 2,500\sqrt{t}$. If the interest rate is 6 percent, find the best time to sell these antiques and invest the profits.

3. In the case of discontinuous functions, the growth rate of $f(n)$ is given by $(\Delta f/\Delta n)/f(n)$. Show that if $f(n) = P(1 + r)^n$, the growth rate of $f(n)$ is r.

4. A certain population is given at time t by $P(t) = t/(1 + t)$. Find the growth rate of this population.

ANSWERS   3.7. A

1. (a) $2/t$ (b) $3/t$ (c) $3/2t$ (d) $2t - 1$ (e) $3 - 3t^2$ (f) 4.
    When $t = 1$ the answers are (a) 2 (b) 3 (c) $\frac{3}{2}$ (d) 1 (e) 0 (f) 4
2. $g(0) = \frac{1}{2}$; $g(1) = \frac{20}{135}$; $g(2) = -\frac{1}{14}$; $g(3) = -\frac{8}{23}$
3. $t = 9$ 5. $t = 1$
4. $t = 27$

ANSWERS   3.7. C

1. 8.4 years
2. 8.3 years
3. We assume $\Delta n = 1$. Thus $[\Delta f/(\Delta n)]/f(n) = [(P(1 + r)^{n+\Delta n} - P(1 + r)^n)/(\Delta n)]/$
    $[P(1 + r)^n] = [P(1 + r)^n((1 + r)^1 - 1)]/[P(1 + r)^n] = (1 + r) - 1 = r$
4. $\dfrac{1}{t(1 + t)}$

3.8   Differentials

With this section we conclude our presentation of differential calculus. We do so with a discussion of some ideas of Gottfried Wilhelm von Leibniz (1646–1716), who, in addition to Sir Isaac Newton (1642–1727), invented calculus. It was Leibniz who introduced the $\dfrac{dy}{dx}$ symbol for derivatives, and the symbol for "integrals," which we shall use when we turn to integral calculus in the next section. Leibniz used dy and dx separately (apart from the apparent fraction $\dfrac{dy}{dx}$, which is really not a fraction); we shall now carefully explore how this can be done.

3.8.1   Discussion

Suppose that $x = x_1$ and $y = f(x)$ are given and that Δx is some increment from x_1. When we define the derivative $f'(x) = \dfrac{dy}{dx}$, we consider the fraction $\dfrac{\Delta y}{\Delta x}$ for smaller and smaller Δx ($\neq 0$). When Δx is "very small," we say that

$$\frac{\Delta y}{\Delta x} \approx \frac{dy}{dx} \quad \text{or} \quad \Delta y \approx \frac{dy}{dx} \Delta x$$

The geometry of this, expressed in Figure 3.8–1, should now be familiar. The slope of the tangent line to $y = f(x)$ at the point A is precisely $\dfrac{dy}{dx}$ and is approximately the slope of the line between A and D, which is $\dfrac{\Delta y}{\Delta x}$.

Figure 3.8–1

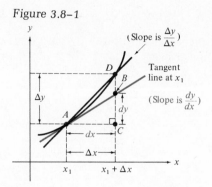

But now for a trick. We call Δx by a new name — dx. If we consider the right triangle ABC in Figure 3.8–1, we see that the slope of the tangent line to the graph at the point A is $\overline{BC}/\overline{AC} = f'(x_1)$.

Since $dx = \Delta x$ is the denominator of the fraction $\overline{BC}/\overline{AC}$, it makes sense to call the numerator dy. It is, in fact, customary to do so, by defining dy properly. (Notice that dy is an approximation to Δy, and an increasingly good approximation as $\Delta x \to 0$. We use this later.)

3.8.2   Definition

Let $y = f(x)$ be a function with a continuous derivative. Then

1. $dx = \Delta x$ is called the <u>differential of x</u>.
2. $dy = f'(x_1)\, dx$ is called the <u>differential of y at x_1</u>. In other words,

$$dy = \frac{dy}{dx}\, dx$$

3.8.3   Example

Find the differential of $y = 3x^2 + 2x - 1$ at $x = 1$ when $\Delta x = 0.1$. Compare it with Δy.

Solution:

First we calculate $f'(x) = \dfrac{dy}{dx} = 6x + 2$. Therefore,

$$f'(1) = 6(1) + 2 = 8$$

Thus

$$dy = \frac{dy}{dx}\, dx = f'(1)\, dx = 8(0.1) = 0.8 \qquad [dx = \Delta x = 0.1]$$

To compare this with Δy, we compute

$$\Delta y = f(x + \Delta x) - f(x) = f(1.1) - f(1) = 4.83 - 4 = 0.83$$

Hence dy and Δy differ by 0.03.

Notice that dy depends both on x and dx. Thus we say that it is a "function of two variables"—something we shall consider in more detail in Chapter 6. It suffices here to observe that such functions are evaluated by substituting in the values of both the independent variables, as in Example 3.8.3.

The rules for obtaining differentials are analogous to those for derivatives and can be derived from them.

3.8.4   Rule

Let u and v be continuously differentiable functions of x and let c and n denote constants. Then

(a) $d(c) = 0$.
(b) $d(x^n) = n(x^{n-1})\,dx$.
(c) $d(u \pm v) = du \pm dv$.
(d) $d(u \cdot v) = v\,du + u\,dv$.
(e) $d(u/v) = (v\,du - u\,dv)/v^2$ if $v \neq 0$.
(f) $d(u^n) = n(u^{n-1})\,du$.

Each part of this rule is illustrated in the corresponding part of the next example. [For example, Rule 3.8.4(a) corresponds to Example 3.8.5(a).]

3.8.5   Example

Let $u = x^2 - 2$ and $v = 3x^3 + x - 1$. Find dy if
(a) $y = 4$
(b) $y = x^8$
(c) $y = u + v$
(d) $y = u \cdot v$
(e) $y = u/v$
(f) $y = u^8$

Solution:
(a) $dy = 0$
(b) $dy = 8x^7\,dx$
(c) $dy = (2x - 0)\,dx + (9x^2 + 1 - 0)\,dx = (2x + 9x^2 + 1)\,dx$
(d) $dy = (3x^3 + x - 1)(2x\,dx) + (x^2 - 2)(9x^2 + 1)\,dx$
$\quad\quad = (15x^4 - 15x^2 - 2x - 2)\,dx$

(e) $dy = \dfrac{(3x^3 + x - 1)\,d(x^2 - 2) - (x^2 - 2)\,d(3x^3 + x - 1)}{(3x^3 + x - 1)^2}$

$\quad\quad = \dfrac{[(3x^3 + x - 1)(2x) - (x^2 - 2)(9x^2 + 1)]\,dx}{(3x^3 + x - 1)^2}$

(f) $dy = 8(x^2 - 2)^7 d(x^2 - 2) = 8(x^2 - 2)^7 2x\,dx = 16x(x^2 - 2)^7\,dx$

Differentials and Approximations

We may use the fact that $\Delta y \approx dy$ to aid in certain computations. For example, the square-root function $f(x) = \sqrt{x}$ is one whose values we often approximate. If we wish to find the square root of 17, it is the same as finding $f(17)$. We know that $f(16) = \sqrt{16} = 4$, and in Example 3.8.6 we use this fact to find the value of the function at 17, since 17 differs from 16 by an increment $dx = 1$. We wish to find the corresponding increment in the square-root function; we add this to $\sqrt{16}$ to get $\sqrt{17}$ since $\Delta y = f(x + \Delta x) - f(x)$ implies that

$$f(x + \Delta x) = f(x) + \Delta y \approx f(x) + dy$$

so

$$\sqrt{17} = f(16 + 1) = f(16) + \Delta y = 4 + \Delta y \approx 4 + dy$$

In the next two examples we use dy to approximate Δy.

3.8.6   **Example**

Approximate the square root of 17.

Solution:

We let $y = \sqrt{x}$, $x = 16$, and $dx = 1$. Then

$$\frac{dy}{dx} = \frac{1}{2\sqrt{x}} = \frac{1}{2\sqrt{16}} = \frac{1}{8} \quad \text{and} \quad dy = \frac{dy}{dx} dx = \frac{1}{8}(1) = 0.125$$

It follows that

$$\sqrt{17} = y + \Delta y \approx y + dy = 4 + 0.125 = 4.125$$

Note that $(4.125)^2 = 17\frac{1}{64}$, so the approximation is close.

3.8.7   **Example**

Approximate ln (1.2) using differentials.

Solution:

We let $x = 1$ since ln (1) $= 0$. Then $dx = 0.2$. Setting $y = f(x) = \ln(x)$, we have $\frac{dy}{dx} = \frac{1}{x}$. Therefore,

$$dy = \frac{dy}{dx} dx = \frac{1}{x} dx = \frac{1}{1}(0.2) \qquad \text{since } x = 1 \text{ and } dx = 0.2 \text{ above}$$

$$\ln (1.2) = f(x + dx) \approx f(x) + dy = \ln (1) + 0.2 = 0 + 0.2 = 0.2$$

[Table IV gives that ln (1.2) $= 0.1823$, so we are correct to the nearest tenth.]

Approximations and Relative Error

Differentials can be used to estimate the error in the dependent variable if we have some idea of the possible error in our estimation of the independent variable. For example, if the revenue R is a function of the number x of items sold, and dx is our possible error in estimating the number sold, then dR is the possible error in estimating the revenue.

$\dfrac{dx}{x}$ is called the <u>relative error in x</u>. $\dfrac{dy}{y}$ is called the <u>relative error in y</u>. Relative error times 100 is called the <u>percent error</u>.

3.8.8   **Example**

Suppose that the cost of producing x items is given by $C(x) = 4x - 0.002x^2 + 1{,}200$. If we expect to produce 200 items with a 5 percent margin of error,

(a) What is the estimated cost of producing the 200 items?
(b) What is the margin of error in this estimation?
(c) What is the percentage margin of error?

Solution:

(a) $C(200) = 4(200) - 0.002(200)^2 + 1{,}200 = 800 - 80 + 1{,}200 = 1{,}920$
(b) $dx = 0.05(200) = 10$, so

$$dC = \frac{dC}{dx}\,dx = (4 - 0.004x)\,dx = [4 - 0.004(200)]10$$
$$= (4 - 0.8)10 = 3.2(10) = 32$$

(c) $\dfrac{dC}{C} = \dfrac{32}{1{,}920} \approx 0.01666$, so the margin of error in the estimate of the cost is about 1.7 percent.

3.8.9   **Example**

Suppose that the revenue is given by the formula $R(x) = \sqrt{x}$, where x is the number of gizmos sold. An experienced manager is able to estimate a production level of 10,000 gizmos per day with a 1 percent margin of error. Estimate the maximum error and percent error in revenue.

Solution:

Since his production estimate has at most a 1 percent error, $\Delta x = dx \leqslant 0.01(10{,}000) = 100$, so

$$dR = \frac{dR}{dx}\,dx = \tfrac{1}{2}x^{-1/2}\,dx \leqslant \frac{100}{2\sqrt{10{,}000}} = \frac{100}{200} = 0.5$$

It follows that $\dfrac{dR}{R} \leqslant \dfrac{0.5}{100} = 0.005$, so the margin of error in the estimate of the revenue is about 0.5 percent.

SUMMARY

In elementary school we defined a fraction a/b after first knowing what its numerator and denominator were. In this section we turn these concepts around. We have a quantity, $\dfrac{dy}{dx}$, which looks like a fraction but is not, and we have a sensible definition for its denominator. We *define* the "numerator," $dy = \left(\dfrac{dy}{dx}\right)(dx)$, in such a way as to make the expression look arithmetically true!

   This turns out to be a useful definition, because if Δx is sufficiently small, dy so defined closely approximates $\Delta y = f(x + \Delta x) - f(x)$. But dy is much easier to calculate than Δy. We used differentials, thus defined, to approximate the values of mathematical functions and to estimate errors in business models.

EXERCISES   3.8. A

1. Find dy when
   (a) $y = x^2 - 4x + 1$ (e) $y = e^{-x^2}$
   (b) $y = 2x^3 + 6x^6 - 4x^{-1} + \sqrt[3]{x}$ (f) $y = \ln(x^2)$
   (c) $y = (3x + 1)^5$ (g) $y = (x + 1)/(x - 1)$
   (d) $y = (5x^3 - 2x + 1)^{-4/3}$

2. Suppose that the predicted profit when x items are produced is given by $P = 10x - 500 - 0.002x^2$ and the estimated sales is 2,000 items with a 5 percent chance of error.
   (a) What is the predicted profit?
   (b) What is the error in the predicted profit?
   (c) What is the percentage error in the predicted profit?

3. Suppose that the predicted cost when x items are produced is given by $C = 5x + 500 - 0.01x^2$ and the estimated production is 200 items with a 5 percent margin of error.
   (a) What is the error in the predicted cost?
   (b) What is the predicted cost?
   (c) What is the percentage error in the predicted cost?

4. If the cost of producing x widgies is $C = 1,000 + \sqrt{x} + x$ and we estimate the number of widgies to be 400 with a possible 2 percent error:
   (a) What is the estimated cost?
   (b) What is the error in the estimated cost?
   (c) What is the percentage error in the estimated cost?

5. Estimate the square root of 27 using differentials and $\sqrt{25} = 5$.

6. Estimate the cube root of 30 using differentials and $\sqrt[3]{27} = 3$.

7. Estimate $\ln(3)$ using differentials and $\ln(2.72) \approx 1$.

HC 8. Refer to problem 2. Compare your error estimation in part (b) to that obtained by computing the true potential error, $P(2,000) - P(1,900)$.

EXERCISES   3.8. B

1. Find dy when
    (a) $y = x^2 + 5x + 1$
    (b) $y = 3x^2 + x^7 - x^{-2} + \sqrt{x}$
    (c) $y = (2x - 1)^4$
    (d) $y = (4x^2 - 2x + 1)^{1/2}$
    (e) $y = e^{3x^2}$
    (f) $y = \ln (x^3)$

2. Suppose that the predicted profit when x items are produced is given by $P = 5x - 250 - 0.001x^2$ and the estimated sales is 1,000 items with a 5 percent chance of error.
    (a) What is the predicted profit?
    (b) What is the error in the predicted profit?
    (c) What is the percentage error in the predicted profit?

3. Suppose that the predicted cost when x items are produced is given by $C = 10x + 1000 - 0.02x^2$ and the estimated production is 200 items with a 5 percent chance of error.
    (a) What is the error in the predicted cost?
    (b) What is the predicted cost?
    (c) What is the percentage error in the predicted cost?

4. If the cost of producing x items is $C = 1,000 - \sqrt{x} + x$ and we estimate the number of items to be 400 with a possible 2 percent error:
    (a) What is the estimated cost?
    (b) What is the error in the estimated cost?
    (c) What is the percentage error in the estimated cost?

5. Estimate the square root of 50 using differentials and $\sqrt{49} = 7$.

6. Estimate the cube root of 65 using differentials and $\sqrt[3]{64} = 4$.

7. Estimate $\ln (3)$ using differentials and $\ln (2.72) \approx 1$.

HC 8. Refer to problem 2. Compare your error estimation in part (b) to that obtained by computing the true potential error, $P(1,000) - P(950)$.

EXERCISES   3.8. C

1. The volume of a spherically shaped fruit is given by $V = \frac{4}{3}\pi r^3$, where r is the radius of the fruit (the distance from the surface to the center). What effect would changing the radius by 0.1 inch have on the volume when the volume is $\frac{4}{3}\pi$ cubic inches?

2. A formula derived by Fechner early in the history of mathematical psychology gave a relation between stimulus s and response R as

$$dR = \alpha \frac{ds}{s}$$

where α is constant. Try to express the meaning of this formula in words.

ANSWERS   3.8. A

1. (a) $dy = (2x - 4)\, dx$ (b) $dy = (6x^2 + 36x^5 + 4x^{-2} + \frac{1}{3}x^{-2/3})\, dx$
    (c) $dy = 15(3x + 1)^4\, dx$ (d) $(-\frac{4}{3}(5x^3 - 2x + 1)^{-7/3}(15x^2 - 2))\, dx = dy$
    (e) $dy = -2xe^{-x^2}\, dx$ (f) $dy = (2/x)\, dx$ (g) $dy = -2/(x - 1)^2\, dx$

2. (a) $P(2,000) = \$11,500$ (b) $dP = 200$ (c) 1.7 percent

3. (a) $dC = 10$ (b) $C(200) = \$1,100$ (c) 0.9 percent

4. (a) $C(400) = 1420$ (b) $dC = 8.2$ (c) 0.6 percent

5.   5.2   7.   1.10
6.   3.11   8. $\Delta P = 220$

ANSWERS   3.8. C

1.   Increases V by $dV = 1.3$ cubic inches.
2.   A small change in the response is proportional to "proportional change in stimulus."

SAMPLE TEST   **Chapter 3**

Do the first *two problems* and any eight of the following ten. Each problem counts 10 points.
1. Find y' if
   (a) $y = (3x^6 - 9x^3)^5$ (b) $y = e^{5x}$
2. Sketch the graphs of the following functions. Three points per graph is enough, but be sure that you indicate the general shape of each curve.
   (a) $y = e^x$ (b) $y = \ln (x)$
3. Find the present value of \$100 due 5 years from now if it is invested at 6 percent
   (a) compounded annually   (c) compounded continuously
   (b) compounded quarterly
4. Suppose that a country has a birth rate of 0.040 and a death rate of 0.010. How long will it take the population to double?
5. Find the marginal cost if the cost function is given by
   (a) $C = \ln (x^3 + 6x)$ (b) $C = x^2 \ln (x)$
6. Find the equation of the line tangent to the hyperbola $3x^2 - y^2 = 3$ at the point $(2, 3)$.
7. Find $\dfrac{dx}{dy}$ when
   (a) $y = \sqrt{2x^4 - 5x^2}$ (b) $y = e^{4x-x^4}$
8. Find the growth rate when
   (a) $y = \sqrt{3x^2 + 5x}$ (b) $y = e^{4x^2+3x}$
9. Suppose that a bullet is shot into the air and travels a distance $s = 96t - 16t^2$ in t seconds. At what time will it be at its maximum height?
10. Find y' when
    (a) $y = 3^{6x}$ (b) $y = \log_3 (x^2)$
11. Suppose that a collection of antiques is growing in value according to the formula $V = 10,000e^{0.18t^{2/3}}$ If the investment rate is 6 percent, when is the best time to sell the antiques?
12. Suppose that the cost of producing x items is given by $C = 3x + 150$. If it is estimated that 150 items will be produced and there is a 6 percent margin of error in this estimation
    (a) What is the expected cost of producing the 150 items?
    (b) What is the expected error in this estimated cost?
    (c) What is the expected percentage error in the cost?

The answers are at the back of the book.

Chapter 4

INTEGRATION: PART I

4.1 Antiderivatives

We now study reversing the process of differentiation. You have often reversed mathematical processes before; the reverse of addition is subtraction, and the reverse of multiplication is division. There are many reasons why one might want to reverse differentiation. We might want to retrieve a revenue function from a known marginal revenue function. Or we might want to discover the cost function when the corresponding marginal cost model has been determined. The following definition is useful in doing this.

4.1.1 Definition

$F(x)$ is an antiderivative of $f(x)$ if $F'(x) = f(x)$. The symbol

$$\int f(x)\, dx$$

represents all antiderivatives of $f(x)$.

4.1.2 Example

Suppose we know that if x units are sold, the additional revenue received from selling an additional unit is $200 - 0.2x$. In other words, the marginal revenue is given by the derivative

$$R'(x) = \frac{dR}{dx} = 200 - 0.2x$$

where $R(x)$ denotes the (unknown) revenue function. To show that we wish to reverse the differentiation, we shall tentatively write

$$R(x) = \int (200 - 0.2x)\, dx$$

indicating that we wish to find a function $R(x)$ whose derivative with respect to x is $(200 - 0.2x)$. The sign $\int$ reverses the process of differentiation.

We can guess at an antiderivative for $(200 - 0.2x)$ by considering 200 and $-0.2x$ separately. We know that if $F(x) = 200x$ then $F'(x) = 200$

and

if $G(x) = -0.1x^2$ then $G'(x) = -0.2x$

so it follows that one antiderivative of $(200 - 0.2x)$ is $(200x - 0.1x^2)$. Others, however, include

$$(200x - 0.1x^2 + 1) \quad \text{and} \quad (200x - 0.1x^2 + 3)$$

and

$$(200x - 0.1x^2 + 100)$$

You notice that in each of these cases we have added a different constant to the first answer. But when we differentiate each new answer, we still obtain the original $(200 - 0.2x)$. Similarly,

$$200x - 0.1x^2 + k$$

is an antiderivative of $(200 - 0.2x)$, where k is *any* constant (that is, any number).

4.1.3   Basic Rule

If $F'(x) = G'(x)$, then $F(x) = G(x) + k$, where k is a constant.

4.1.4   Rule

If $F'(x) = f(x)$, then $\int f(x) \, dx = F(x) + k$ is an expression for all antiderivatives of f, where k denotes an arbitrary constant [that is, $\int F'(x) \, dx = F(x) + k$].

Rule 4.1.4 is an immediate consequence of Rule 4.1.3. $\int f(x) \, dx$ is called the indefinite integral of $f(x)$, where k is the constant of integration. (In other books it is more common to call this constant C, but we shall use k because C appears so often in this text to denote "cost.") The word "indefinite" reflects the fact that $F(x) + k$ indicates no definite (particular) function, but rather a whole class of functions.

If the value of the antiderivative $F(x)$ is known for one value of x, however, the constant of integration is no longer arbitrary. Indeed, in this case it is uniquely determined, as the following examples show.

4.1.5   Example

Find the revenue function $R(x)$ when the marginal revenue function $R'(x) = 200 - 0.2x$ and a production of zero units yields zero revenue.

Solution:

The antiderivatives of $200 - 0.2x$ all have the form

$$R(x) = \int (200 - 0.2x) \, dx = 200x - 0.1x^2 + k$$

But $R(0) = 0$ implies that

$$0 = 200(0) - 0.1(0)^2 + k = k \qquad \text{so } k = 0$$

Thus the unique answer is

$$R(x) = 200x - 0.1x^2$$

4.1.6   Rule

To determine the constant k of integration in an antiderivative $\int f(x)\,dx = F(x) + k$, where it is given that the particular antiderivative takes the value V at some point $x = a$:

1. Write $F(a) + k = V$.
2. Solve for k.

The equation $F(a) + k = V$ in Rule 4.1.6 is called the <u>boundary con-dition</u>. In Example 4.1.5 the boundary condition is $R(0) = 0$. In Example 4.1.7 the boundary condition is $C(10) = 2{,}500$.

4.1.7   Example

Suppose that the marginal cost of production is $C'(x) = 200 - 0.2x$ and the cost of producing 10 units is \$2,500.
(a) Find the cost function.
(b) Find the cost of producing 20 units.

Solution:
(a) As before, we know that $C(x) = \int (200 - 0.2x)\,dx = 200x - 0.1x^2 + k$, where k is to be determined. This time the boundary condition $C(10) = 2{,}500$ yields

$$2{,}500 = 200(10) - 0.1(10)^2 + k = 2{,}000 - 10 + k$$

so

$$510 = k$$

and this particular cost function is

$$C(x) = 200x - 0.1x^2 + 510$$

(b) $C(20) = 200(20) - 0.1(20)^2 + 510 = 4{,}000 - 40 + 510 = 4{,}470$.

Some formulas of differentiation give rise to corresponding formulas of antidifferentiation. Others, such as the product formula, have no direct match in antidifferentiation, although they are useful in advanced tech-niques (see Section 4.3). We list here some antidifferentiation formulas that are easy consequences of differentiation rules. You should check them by differentiation.

4.1.8   Rule

(a) $\int x^n \, dx = \dfrac{1}{n+1} x^{n+1} + k, \, n \neq -1$

(b) $\int a \, dx = ax + k$, where a is any constant (number)

(c) $\int e^x \, dx = e^x + k$

(d) $\int \dfrac{1}{x} \, dx = \ln |x| + k$

(e) $\int af(x) \, dx = a \int f(x) \, dx$, where a is any constant

(f) $\int (f(x) + g(x)) \, dx = \int f(x) \, dx + \int g(x) \, dx$

4.1.9   Example

Find the antiderivatives of the following functions.
(a) $f(x) = 5$
(b) $f(x) = x^5$
(c) $f(x) = x^{1/5}$
(d) $f(x) = 4x$

Solution:

(a) $\int 5 \, dx = 5x + k$ [by Rule 4.1.8(b)]

(b) $\int x^5 \, dx = \frac{1}{6}x^6 + k$ [by Rule 4.1.8(a)]

(c) $\int x^{1/5} \, dx = \frac{5}{6}x^{6/5} + k$ [by Rule 4.1.8(a)]

(d) $\int 4x \, dx = 2x^2 + k$ [by Rule 4.1.8(a) and (e)]

4.1.10   Example

Find the antiderivatives of the following functions.
(a) $f(x) = 5x^2$
(b) $f(x) = x^{-1} + x + x^2 + e^x$
(c) $f(x) = (x + 1)(x + 2)$

Solution:

(a) $\int 5x^2 \, dx = \frac{5}{3}x^3 + k$ [using Rule 4.1.8(a) and (e)]

(b) $\int (x^{-1} + x + x^2 + e^x) \, dx = \ln |x| + \frac{1}{2}x^2 + \frac{1}{3}x^3 + e^x + k$ [using   Rule 4.1.8(a), (c), (d), and (f)]

(c) $\int (x + 1)(x + 2)\ dx = \int (x^2 + 3x + 2)\ dx = \frac{1}{3}x^3 + \frac{3}{2}x^2 + 2x + k$

[using Rule 4.1.8(a), (b), (e), and (f)]

4.1.11   Example from Sociology or Biology

Suppose that the population (of a city or a bacteria culture) is growing at the rate

$$P'(t) = 10{,}000e^{0.5t}$$

and at time $t = 0$, $P(0) = 100{,}000$. Find $P(t)$.

Solution:

Since $P'(t) = 10{,}000e^{0.5t}$, we have

$$P(t) = \int P'(t)\ dt = \int 10{,}000e^{0.5t}\ dt = 20{,}000e^{0.5t} + k$$

(as you can check by differentiating). When we use the boundary condition $P(0) = 100{,}000$ to solve for k, we have

$$P(0) = 100{,}000 = 20{,}000e^{0.5(0)} + k = 20{,}000 + k$$

so

$$80{,}000 = k \quad \text{and} \quad P(t) = 20{,}000e^{0.5t} + 80{,}000$$

4.1.12   Example in Biology or Pre-Med

Fick's law, named after Adolf Fick (1829–1901), states that the diffusion, $c'(t)$ [in other words, the rate of change of the concentration, $c(t)$], across the wall of a cell membrane is

$$c'(t) = \frac{bA}{V}\ [c_0 - c(t)]$$

where A is the area of the cell membrane, V is the volume of the cell, c_0 is the concentration of the solute *outside* the cell, $c(t)$ is the concentration of the solute inside the cell at time t, and b is a constant determined by the structure and thickness of the membrane. If c_1 denotes the concentration of the solute *inside* the cell at time $t = 0$, then it is possible to use a technique, separation of variables, explained in Section 4.4, to find that the concentration of the solute at time t inside the cell membrane is given by

$$c(t) = (c_1 - c_0)e^{-bAt/V} + c_0$$

You can see now that this is correct by differentiating this expression and then substituting $c(t)$ and $c'(t)$ into Fick's law; it checks. You can also check that $c(0) = c_1$.

SUMMARY

We began the study of integral calculus in this section by introducing antiderivatives. If $F(x)$ is a function such that $F'(x) = f(x)$, then $F(x)$ is called an "antiderivative" of $f(x)$. $\int f(x)\, dx$ is the indefinite integral of $f(x)$ and denotes all antiderivatives of $f(x)$. Examples illustrated how finding antiderivatives is useful in retrieving the total cost function from the marginal cost function, or the population function from a function that describes the rate of change in the population. In many of these examples we used boundary conditions to evaluate the constants of integration. In Sections 4.2 and 4.3 we show techniques for calculating the antiderivatives of more complicated functions. These antiderivatives will be called "indefinite integrals" since they include an indefinite constant of integration.

EXERCISES 4.1. A

1. Check all parts of Rule 4.1.8 by differentiation.
2. Evaluate the following:

(a) $\int x^3\, dx$ (e) $\int (x-2)(x+3)\, dx$

(b) $\int x^{1/3}\, dx$ (f) $\int (x+1)^5\, dx$

(c) $\int (3x + \frac{1}{2}x^2 + 1)\, dx$ (g) $\int 3e^x\, dx$

(d) $\int \left(x^{-2} + \dfrac{1}{x^3} + \sqrt{x} \right) dx$ (h) $\int (4/x)\, dx$

3. Find the specific antiderivative of the first four functions in problem 2 if $F(1) = 2$ is the given boundary condition.
4. If the marginal revenue $R'(x)$ of a certain commodity is given by the formula $10x - 2$ and we know that $R(5) = 130$, find
   (a) $R(x)$ (b) $R(10)$ (c) $R(6)$
5. If the marginal cost $C'(x) = 3x - \sqrt{x}$ and $C(1) = 10$, find
   (a) $C(x)$ (b) $C(4)$ (c) $C(9)$

EXERCISES 4.1. B

1. Evaluate the following:

(a) $\int x^4\, dx$ (e) $\int (x+1)(x+1)\, dx$

(b) $\int x^{1/2}\, dx$ (f) $\int (x+4)^4\, dx$

(c) $\int (2x + x^2 + 2)\, dx$ (g) $\int 4e^x\, dx$

(d) $\int (x^{-3} + 1/x^2 + \sqrt[3]{x})\, dx$ (h) $\int (3/x)\, dx$

2. Find the specific antiderivative of the first four functions in problem 1 if $F(1) = 3$ is the given boundary condition.
3. Check all parts of Rule 4.1.8 by differentiation.
4. If the marginal revenue $R'(x)$ of a certain commodity is given by the formula $5x - 1$ and we know that $R(5) = 65$, find
    (a) R(x)   (b) R(10)   (c) R(6)
5. If the marginal cost is $C'(x) = 6x - 2\sqrt{x}$ and $C(1) = 20$, find
    (a) C(x)   (b) C(4)   (c) C(9)

EXERCISES 4.1. C

1. If the rate of change of the marginal cost is given by $x + 1$, show that the total cost must be given by

$$C(x) = \frac{x^3}{6} + \frac{x^2}{2} + kx + b$$

where k and b are arbitrary constants.

2. Suppose that the population $P(t)$ of a colony of muskrats appears to be growing at a rate $P'(t) = 100t - 3t^2$, and that at time $t = 0$, $P(0) = 10$. Find the population function.

ANSWERS 4.1. A

2. (a) $\frac{1}{4}x^4 + k$ (b) $\frac{3}{4}x^{4/3} + k$ (c) $\frac{3}{2}x^2 + \frac{1}{6}x^3 + x + k$ (d) $-x^{-1} - \frac{1}{2}x^{-2} + \frac{2}{3}x^{3/2} + k$
    (e) $\int (x^2 + x - 6)\,dx = x^3/3 + x^2/2 - 6x + k$ (f) $\frac{1}{6}(x + 1)^6 + k$ (g) $3e^x + k$
    (h) $4 \ln |x| + k$
3. (a) $F(x) = \frac{1}{4}x^4 + \frac{7}{4}$ (b) $F(x) = \frac{3}{4}x^{4/3} + \frac{9}{4}$ (c) $F(x) = \frac{3}{2}x^2 + \frac{1}{6}x^3 + x - \frac{2}{3}$
    (d) $F(x) = -x^{-1} - x^{-2}/2 + \frac{2}{3}x^{3/2} + \frac{17}{6}$
4. (a) $R(x) = 5x^2 - 2x + 15$ (b) $R(10) = 495$ (c) $R(6) = 183$
5. (a) $C(x) = \frac{3}{2}x^2 - \frac{2}{3}x^{3/2} + \frac{55}{6}$ (b) $C(4) = 27\frac{5}{6}$ (c) $C(9) = 112\frac{2}{3}$

ANSWERS 4.1. C

1. Differentiate $C(x)$ twice.
2. $P(t) = 50t^2 - t^3 + 10$

4.2   Finding Antiderivatives by Substitution

There are many useful functions whose antiderivatives cannot be found merely by applying the six rules in Rule 4.1.8. Other techniques for finding antiderivatives will be examined in this section and the next. In the present section we shall discuss the method of *substitution*, which reverses the chain rule for differentiation. The chain rule, you may remember, says that

$$\frac{dy}{dx} = \frac{dy}{du} \cdot \frac{du}{dx}$$

Another way of writing this is

$$d(F(u(x))) = F'(u(x))u'(x)\, dx$$

Hence, with $F' = f$, we have

4.2.1   Rule

$$\int f(u(x)) \cdot u'(x)\, dx = \int f(u)\, du, \text{ where we let } u = u(x).$$

4.2.2   Example

Evaluate $\displaystyle\int (x^2 + 1)^5 2x\, dx.$

Solution:

To use Rule 4.2.1 effectively, we first have to guess what to use for u. Often it helps to try whatever is in parentheses. Thus in this case we try

$$u = x^2 + 1$$

Differentiating, we get

$$du = 2x\, dx$$

Now we use these two expressions:

$$\int (x^2 + 1)^5 2x\, dx = \int u^5\, du \qquad \text{(by substituting } u = x^2 + 1 \text{ and } du = 2x\, dx\text{)}$$

$$= \tfrac{1}{6}u^6 + k \qquad \text{(by integrating)}$$

$$= \tfrac{1}{6}(x^2 + 1)^6 + k \qquad \text{(by substituting } u = x^2 + 1 \text{ again)}$$

We urge you to differentiate $\tfrac{1}{6}(x^2 + 1)^6 + k$ and the answers to subsequent examples to check the integration.

The technique for using the substitution method may be summarized as follows.

1. Select a piece of the integrand (that which is integrated) and call it $u(x)$.
2. Differentiate $u(x)$, obtaining $du = u'(x)\, dx$.
3. Write the integral in terms of u and du.
4. Integrate.
5. Substitute for u in terms of x.

You may wonder about the roles of dx and du. They are symbols that enable us to undo the chain rule in a mechanical way. They may not seem necessary when writing the integral, since they merely accompany the integral sign to indicate that integration must be performed. But without them we have to use a lot more creative thinking when finding antiderivatives, so most people find it faster to use them.

4.2.3   Example

Evaluate $\int e^{x^2} 2x \, dx$.

Solution:

Let $u = x^2$. Then $du = 2x \, dx$. When we use Rule 4.2.1, we get

$$\int e^{x^2} 2x \, dx = \int e^u \, du = e^u + k = e^{x^2} + k$$

We see from these examples that the use of Rule 4.2.1 depends upon (1) proper choice of u and (2) recognition of the differential du. Both steps are essential and they are often interdependent. Sometimes, even though du is not obviously present after the choice of u has been made, the problem can be manipulated into a usable form by multiplying and dividing the integrand by a constant [Rule 4.1.8(e)].

4.2.4   Example

Find

(a) $\int (x^2 + 1)^5 x \, dx$

(b) $\int e^{2x} \, dx$

(c) $\int \dfrac{1}{2x} \, dx$

Solution:

(a) Let $u = x^2 + 1$, so $du = 2x \, dx$, as before. Thus we must write

$$\int (x^2 + 1)^5 x \, dx = \tfrac{1}{2} \int (x^2 + 1)^5 2x \, dx = \tfrac{1}{2} \int u^5 \, du = \tfrac{1}{12} u^6 + k$$
$$= \tfrac{1}{12} (x^2 + 1)^6 + k$$

(b) Let $u = 2x$. Then $du = 2 \, dx$, so we write

$$\int e^{2x} \, dx = \tfrac{1}{2} \int e^{2x} 2 \, dx = \tfrac{1}{2} \int e^u \, du = \tfrac{1}{2} e^u + k = \tfrac{1}{2} e^{2x} + k$$

(c) $\displaystyle \int \frac{1}{2x} \, dx = \tfrac{1}{2} \int \frac{1}{x} \, dx = \tfrac{1}{2} \ln |x| + k$

(If we can avoid substitution, we do.)

The cleverest of tricks, however, sometimes fail, as the next example shows.

4.2.5   Example

Find $\int (x^2 + 1)^5 \, dx$.

Solution:

If we try this time to let $u = x^2 + 1$, we still have $du = 2x \, dx$, but there is no x in the equation except for that in $u(x)$ and dx. Since x is not a constant (that is, a number), we cannot introduce it as we introduced 2 in Example 4.2.4 without mixing the expressions in x with those in u, which is no help. So there is no alternative to using the long approach to this problem; we must expand the expression $(x^2 + 1)^5$ as a product:

$$\int (x^2 + 1)^5 \, dx = \int (x^{10} + 5x^8 + 10x^6 + 10x^4 + 5x^2 + 1) \, dx$$
$$= \tfrac{1}{11}x^{11} + \tfrac{5}{9}x^9 + \tfrac{10}{7}x^7 + 2x^5 + \tfrac{5}{3}x^3 + x + k$$

Skill in making a judicious choice for u takes practice. The next examples will illustrate the technique to help you get a feel for it. But the substitution method can be mastered only by actually doing the exercises.

4.2.6   Example

Find the antiderivatives of the following functions.
(a) $2(x^3 + 3)^{1/4}x^2$

(b) $\dfrac{x}{\sqrt{x^2 + 1}}$

Solution:

(a) Let $u = x^3 + 3$; then $du = 3x^2 \, dx$

$$\int 2(x^3 + 3)^{1/4}x^2 \, dx = \tfrac{2}{3} \int 3(x^3 + 3)^{1/4} x^2 \, dx = \tfrac{2}{3} \int u^{1/4} \, du$$
$$= \tfrac{2}{3} \cdot \tfrac{4}{5}u^{5/4} + k = \tfrac{8}{15}(x^3 + 3)^{5/4} + k$$

(b) Let $u = x^2 + 1$, so $du = 2x \, dx$

$$\int \frac{x \, dx}{\sqrt{x^2 + 1}} = \tfrac{1}{2} \int \frac{2x \, dx}{\sqrt{x^2 + 1}} = \tfrac{1}{2} \int u^{-1/2} \, du = \tfrac{1}{2} \cdot \tfrac{2}{1}u^{+1/2} + k$$
$$= \sqrt{x^2 + 1} + k$$

4.2.7   Example

Find the antiderivatives of the following functions.
(a) e^{-3x}

(b) $\dfrac{5e^x}{\sqrt{1-e^x}}$

Solution:

(a) Let $u = -3x$; then $du = -3\,dx$

$$\int e^{-3x}\,dx = -\tfrac{1}{3}\int e^{-3x}(-3)\,dx = -\tfrac{1}{3}\int e^u\,du = -\tfrac{1}{3}e^u + k$$
$$= -\tfrac{1}{3}e^{-3x} + k$$

(b) Let $u = 1 - e^x$; it follows that $du = -e^x\,dx$

$$\int \frac{5e^x\,dx}{\sqrt{1-e^x}} = (-5)\int \frac{-e^x\,dx}{\sqrt{1-e^x}} = -5\int u^{-1/2}\,du = -5(\tfrac{2}{1})u^{1/2} + k$$
$$= -10\sqrt{1-e^x} + k$$

4.2.8   Example

Suppose that the rate of change of the value of a certain bond is $e^{0.05t}$ and its value at time $t = 0$ is \$108.
(a) Write an expression for the value of the bond at time t.
(b) What is it worth after 10 years?

Solution:

(a) Letting $y(t)$ be the value of the bond at time t, we know that $y'(t) = e^{0.05t}$. To get an expression for y, we must therefore take the antiderivative of $e^{0.05t}$. We let $u = 0.05t$, so $du = 0.05\,dt$

$$\int e^{0.05t}\,dt = \frac{1}{0.05}\int 0.05e^{0.05t}\,dt = \frac{1}{0.05}\int e^u\,du = \frac{1}{0.05}e^{0.05t} + k$$

Since $1/0.05 = 20$, the indefinite integral is $y(t) = 20e^{0.05t} + k$. When we use the boundary condition $y(0) = 108$, we get

$$108 = 20e^0 + k$$
$$108 = 20 + k$$
$$88 = k$$

So the value of the bond is given by

$$y(t) = 20e^{0.05t} + 88$$

(b) $y(10) = 20e^{0.05(10)} + 88 \approx 20(1.65) + 88 = 121$.   Therefore, after 10 years the bond will be worth \$121.

SUMMARY

In this section we found antiderivatives using substitution, the method analogous to the chain rule for differentiation. A summary of how to use this rule was given on page 206, and examples of its use were given throughout the section.

EXERCISES 4.2. A

Evaluate the following:

1. $\int 3x^2(x^3 + 1)^5 \, dx$

2. $\int e^{4x} \, dx$

3. $\int 4x^2(x^3 + 1)^5 \, dx$

4. $\int xe^{x^2} \, dx$

5. $\int 4x^{-1} \, dx$

6. $\int 4e^{-x} \, dx$

7. $\int \dfrac{3x^2 \, dx}{\sqrt{x^3 - 1}}$

8. $\int \dfrac{2e^x \, dx}{\sqrt{e^x - 1}}$

9. $\int (3x^2 + x)(x^3 + \tfrac{1}{2}x^2 + 1)^5 \, dx$

10. $\int \dfrac{(3x^2 + x) \, dx}{\sqrt{x^3 + \tfrac{1}{2}x^2 + 1}}$

11. $\int \dfrac{\ln (x)dx}{x}$

12. If the marginal value of a stock is $e^{0.08t}$ at time t and its value when $t = 0$ is $132, write an expression for the value of the stock at time t. What is the value at time $t = 10$?

EXERCISES 4.2. B

Evaluate the following:

1. $\int 4x^3(x^4 + 1)^3 \, dx$

2. $\int e^{-4x} \, dx$

3. $\int x(x^2 + 7)^4 \, dx$

4. $\int x^2 e^{x^3} \, dx$

5. $\int 5x^{-1} \, dx$

6. $\int 5e^{-x/2} \, dx$

7. $\int \dfrac{4x^3}{\sqrt[3]{x^4 - 1}} \, dx$

8. $\int \dfrac{e^{2x}}{\sqrt{e^{2x} + 1}} \, dx$

9. $\int (3x^2 + 2)(x^3 + 2x - 3)^4 \, dx$

10. $\int \dfrac{3x^2 + 2}{\sqrt{x^3 + 2x - 3}} \, dx$

11. $\int \dfrac{1}{x \ln (x)} \, dx$

12. If the marginal value of a stock at time t is $e^{0.05t}$ and its value at time $t = 0$ is $80:
   (a) Write an expression for the value of the stock at time t.
   (b) What is the value at time $t = 10$?

EXERCISES **4.2. C**

Find the integrals of the following functions.

1. $\dfrac{\ln (x^3)}{x}$

2. $xe^{x^2}\sqrt{e^{x^2}+1}$

3. $\dfrac{1+e^x}{x+e^x}\ln (x+e^x)$

ANSWERS **4.2. A**

1. $\frac{1}{6}(x^3+1)^6 + k$
2. $\frac{1}{4}e^{4x} + k$
3. $\frac{4}{18}(x^3+1)^6 + k$
4. $\frac{1}{2}e^{x^2} + k$
5. $4(\ln |x|) + k$
6. $-4e^{-x} + k$

7. $2\sqrt{x^3-1} + k$
8. $4\sqrt{e^x - 1} + k$
9. $\frac{1}{6}(x^3 + \frac{1}{2}x^2 + 1)^6 + k$
10. $2\sqrt{x^3 + \frac{1}{2}x^2 + 1} + k$
11. $\frac{1}{2}[\ln (x)^2 + k]$
12. $y(t) = 12.5e^{0.08t} + 119.5; V(10) \approx \$147$

ANSWERS **4.2. C**

1. $\frac{1}{6}[\ln (x^3)]^2 + k$
2. $\frac{1}{3}(e^{x^2}+1)^{3/2} + k$

3. $\frac{1}{2}[\ln (x+e^x)]^2 + k$

SAMPLE QUIZ   **Sections 3.8, 4.1, and 4.2 (and, Possibly, 3.6 and 3.7)**

Do the first four problems and any two of the last five.

1. (10 pts) If $y = 3x^2 - 2e^{-3x}$, then $dy =$

2. (20 pts) $\displaystyle\int \left(x^7 + \sqrt{x} + \frac{1}{x}\right) dx =$

3. (10 pts) $\displaystyle\int e^{7x}\, dx =$

4. (20 pts) $\displaystyle\int x^2(5x^3 + 7)^9\, dx =$

Do any *two* of the following five problems (20 pts each):

5. If the marginal profit is given by $P'(x) = 3 - 0.06x$ and $P(5) = 4.25$:
    (a) What is the profit function?
    (b) What is the profit when six items are produced?

6. Suppose that the cost of producing x gizmos is $C(x) = 8x - 0.02x^2 + 200$ and we expect to produce 150 gizmos with an 8 percent margin of error.
    (a) How much do we expect these gizmos to cost?
    (b) What is the estimated error in this cost?
    (c) What is the percentage error in the cost estimate?

7. Suppose that an arrow is shot into the air with an initial velocity of 160 ft/sec and its height from the ground is given by the formula $s = 160t - 16t^2$ after t seconds.
    (a) What is its velocity after 3 seconds?
    (b) At what time does it reach its maximum height?
    (c) What is its maximum height?

8. Suppose that a precious jewel has a value at time t of $V = 40,000e^{0.2t^{3/4}}$ and

the interest rate is 7.5 percent. When should the jewel be sold and the proceeds invested?

9.   Approximate the cube root of 29 using differentials.

The answers are at the back of the book.

4.3   Antiderivative Tables and Integration by Parts

The list of rules given in Section 4.1 combined with the substitution
method is not sufficient to find the antiderivatives of some elementary
functions of the type encountered in practical applications. Extensive
lists of antiderivatives, also called integral tables, have been developed
by mathematicians. Table V is such a table but in a shortened form; some
integral tables fill entire books. It takes some skill to use these tables — it
is not just a matter of copying the right antiderivative. Often a substitution is needed to modify the integrand. The following very short list will
be used in the examples of this section:

1. $\int \dfrac{du}{a^2 - u^2} = \dfrac{1}{2a} \ln \left| \dfrac{a + u}{a - u} \right| + k$

2. $\int \dfrac{du}{u^2 - a^2} = \dfrac{1}{2a} \ln \left| \dfrac{u - a}{u + a} \right| + k$

3. $\int u e^u \, du = u e^u - e^u + k$

4. $\int u^2 e^u \, du = e^u (u^2 - 2u + 2) + k$

4.3.1   Example

Find the antiderivatives of $f(x) = 1/(x^2 - 4)$.

Solution:
The function has the form of the integrand in 2 in the list with $u = x$ and
$a = 2$. Hence

$$\int \frac{dx}{x^2 - 4} = \int \frac{dx}{x^2 - (2)^2} = \frac{1}{2(2)} \ln \left| \frac{x - 2}{x + 2} \right| + k$$

4.3.2   Example

Evaluate $\int \dfrac{dx}{4 - 9x^2}$.

Solution:
The integrand resembles that of 1 in the list with $u = 3x$ and $a = 2$. Thus
$du = 3 \, dx$ and we have

$$\int \frac{dx}{4-9x^2} = \frac{1}{3}\int \frac{3\ dx}{(2)^2-(3x)^2} = \frac{1}{3}\int \frac{du}{2^2-u^2}$$

$$= \frac{1}{3}\cdot\frac{1}{2(2)}\cdot\ln\left|\frac{2+3x}{2-3x}\right| + k$$

Examples 4.3.1 and 4.3.2 suggest how much care is needed in selecting the proper form for your integrand in the integral table; many forms look alike at first glance.

4.3.3   Example

Find $\int 4xe^{2x}\ dx$.

Solution:

For this we use 3 in the list with $u = 2x$. Then $du = 2\ dx$ and we have

$$\int 4xe^{2x}\ dx = \int 2xe^{2x}2\ dx = \int ue^u du = ue^u - e^u + k$$

$$= 2xe^{2x} - e^{2x} + k$$

4.3.4   Example

Find the antiderivatives of $3x^2e^{3x}$.

Solution:

This time we use 4 in the list with $u = 3x$. Then $du = 3\ dx$ and $u^2 = 9x^2$.

$$\int 3x^2e^{3x}\ dx = \tfrac{1}{9}\int 9x^2e^{3x}3\ dx = \tfrac{1}{9}\int u^2e^u du$$

$$= \tfrac{1}{9}e^u(u^2 - 2u + 2) + k = \tfrac{1}{9}e^{3x}(9x^2 - 6x + 2) + k$$

Integration by Parts

When integral tables are not available, one can sometimes use a powerful technique known as *integration by parts*. It is based on the product rule for derivatives:

$$\frac{d}{dx}f(x)g(x) = f(x)g'(x) + f'(x)g(x)$$

Taking the antiderivatives of both sides of this equation, we obtain

$$f(x)g(x) = \int f(x)g'(x)\ dx + \int f'(x)g(x)\ dx$$

From this we get the following rule.

4.3.5   Rule

If $f(x)$ and $g(x)$ are differentiable functions, then

$$\int f(x)g'(x)\ dx = f(x)g(x) - \int f'(x)g(x)\ dx$$

This can also be written $\int u\ dv = uv - \int v\ du.$

It is *not* obvious how this rule can be used to find antiderivatives. We shall use an example to show its possibilities.

4.3.6   Example

Find the antiderivatives of $f(x) = xe^x$.

Solution:

Let $f(x) = x$ and $g'(x) = e^x$. Then
$$f'(x) = 1 \quad \text{and} \quad g(x) = e^x$$

$$\int xe^x\ dx = xe^x - \int 1(e^x)\ dx = xe^x - e^x + k$$

(See 3 in the list at the beginning of the section.)

The preceding example shows that $f(x) = x$ and $g(x) = e^x$ are good choices, but it does not show how we got them. It is often difficult to know how to select $f(x)$ and $g(x)$, since an arbitrary choice may get you nowhere; the resulting integral may be as difficult to solve as the original one. It takes considerable practice to use integration by parts effectively. But a good rule of thumb is to factor the integrand into $f(x)$ and $g'(x)$ such that

1. It is easy to determine $g(x)$ by integrating $g'(x)$.
2. $f'(x)$ is simpler in some sense than $f(x)$.
3. $\int f'(x)g(x)\ dx$ can be determined.

4.3.7   Example

Find $\int x^2 e^x\ dx$.

Solution:

Let $f(x) = x^2$ and $g'(x) = e^x$. Then
$$f'(x) = 2x \quad \text{and} \quad g(x) = e^x$$

Hence $\int x^2 e^x \, dx = x^2 e^x - \int 2x e^x \, dx$. When we apply Exercise 4.3.6 to $\int 2x e^x \, dx$, we get

$$\int x^2 e^x \, dx = x^2 e^x - 2x e^x + 2e^x + k$$

Compare this with 4 in the list.

4.3.8   Example

Find the antiderivatives of ln (x).

Solution:

We first notice that $\int \ln\,(x)\,dx = \int [\ln\,(x) \cdot 1]\,dx$. Set $f(x) = \ln x$ and $g'(x) = 1$. We have

$$f'(x) = \frac{1}{x} \qquad \text{and} \qquad g(x) = x$$

$$\int \ln\,(x)\,dx = x \ln |x| - \int x \frac{1}{x}\,dx = x \ln |x| - \int 1\,dx$$
$$= x \ln |x| - x + k$$

4.3.9   Example

Find $\int x \ln\,(x)\,dx$.

Solution:

Let $f(x) = \ln\,(x)$ and $g'(x) = x$, so

$$f'(x) = \frac{1}{x} \qquad \text{and} \qquad g(x) = \tfrac{1}{2}x^2$$

Hence

$$\int x \ln\,(x)\,dx = \frac{x^2}{2} \ln |x| - \int \frac{x^2}{2} \frac{1}{x}\,dx = \frac{x^2}{2} \ln |x| - \int \frac{x}{2}\,dx$$
$$= \frac{x^2}{2} \ln |x| - \frac{x^2}{4} + k$$

SUMMARY

This section examined finding antiderivatives using tables and integration by parts. A short table is given at the back of the book; many tables are much longer. Integration by parts is an adaptation of the multiplication rule for differentiation to antidifferentiation.

EXERCISES   4.3. A

Evaluate the following, using integration by parts.

1. $\int xe^{2x}\ dx$

4. $\int \ln{(2x)}\ dx$

2. $\int x^2 e^{2x}\ dx$

5. $\int x \ln{(x)}\ dx$

3. $\int x^3 e^x\ dx$

6. $\int [\ln{(x)}]^2\ dx$

Evaluate the following, using any method, including Table V.

7. $\int \sqrt{x^2 + 1}\ dx$

12. $\int (2x) \ln{(2x)}\ dx$

8. $\int \sqrt{9x^2 - 4}\ dx$

13. $\int 3xe^{-3x}\ dx$

9. $\int (1/\sqrt{9x^2 - 4})\ dx$

14. $\int \ln{(3x)}\ dx$

10. $\int (1/\sqrt{16 + 25x^2})\ dx$

15. $\int 2x[\ln{(x^2)}]^2\ dx$

11. $\int (2x/\sqrt{x^2 - 9})\ dx$

EXERCISES   4.3. B

Evaluate the following, using integration by parts.

1. $\int 3xe^{x/2}\ dx$

4. $\int 2x \ln{(x^2)}\ dx$

2. $\int x^2 e^{-2x}\ dx$

5. $\int 2x \ln{(2x)}\ dx$

3. $\int x^3 e^{x^2}\ dx$

6. $\int [\ln{(x)}]^2\ dx$

Evaluate the following, using any method, including Table V.

7. $\int \sqrt{4x^2 + 9}\ dx$

12. $\int \ln{(3x)}\ dx$

8. $\int \sqrt{x^2 - 4}\ dx$ •

13. $\int 4x^2 e^{-4x^3}\ dx$

9. $\int (1/\sqrt{9x^2 + 4})\ dx$

14. $\int 4x^2 \ln{(2x)}\ dx$

10. $\int (1/\sqrt{16 + 49x^2})\ dx$

15. $\int \dfrac{\ln{(2x)}}{2x}\ dx$

11. $\int (4x/\sqrt{16x^2 - 9})\ dx$

ANSWERS **4.3. A**

1. $\frac{1}{2}xe^{2x} - \frac{1}{4}e^{2x} + k$
2. $\frac{1}{2}x^2e^{2x} - \frac{1}{2}xe^{2x} + \frac{1}{4}e^{2x} + k$
3. $x^3e^x - 3x^2e^x + 6xe^x - 6e^x + k$
4. $x \ln (2x) - x + k$
5. $\frac{1}{2}x^2 \ln (x) - \frac{1}{4}x^2 + k$
6. $x[\ln (x)]^2 - 2x \ln (x) + 2x + k$
7. $\frac{1}{2}[(x\sqrt{x^2 + 1}) + \ln |x + \sqrt{x^2 + 1} |] + k$
8. $\frac{1}{2}(x\sqrt{9x^2 - 4}) - \frac{2}{3} \ln |3x + \sqrt{9x^2 - 4} | + k$
9. $\frac{1}{3}(\ln |3x + \sqrt{9x^2 - 4})|) + k$
10. $\frac{1}{5}(\ln |5x + \sqrt{25x^2 + 16})|) + k$
11. $2\sqrt{x^2 - 9} + k$
12. $x^2 \ln (2x) - \frac{1}{2}x^2 + k$
13. $-xe^{-3x} - \frac{1}{3}e^{-3x} + k$
14. $x \ln (3x) - x + k$
15. $x^2(\ln (x^2))^2 - 2x^2 \ln (x^2) + 2x^2 + k$

4.4   Differential Equations

Many mathematical models of the real world involve explicit statements about rates of change. Specifically, it might be possible to devise a marginal cost function, a radioactive decay function, or a function describing the rate of change of a population. For example, we might observe that the marginal cost is given by

$$\frac{dC}{dx} = 200 - 0.2x$$

This is an example of a differential equation.

4.4.1   Definition

A differential equation is any equation that contains differentials or derivatives. A solution of a differential equation is a function which does not involve derivatives and which, when substituted into the given equation, makes the equation true.

If the derivatives in the equation indicate marginal cost, then a solution is the corresponding cost function. If the derivatives indicate marginal profit, the solution function is the profit function, and so forth.

Section 4.1 introduced the basic method of solving differential equations, integration. Recall, however, that a particular solution is determined only when boundary conditions are given. If these are not specified, one is left with indefinite constants of integration, and hence a general solution.

4.4.2   Example

(a) Find the general solution of the differential equation $\dfrac{dB}{dx} = x^2 - 3x + 1$.

(b) Find the particular solution for the boundary condition $B(0) = 10$.

Solution:

(a) Integrating $\int dB = \int (x^2 - 3x + 1)\, dx$, we have

$$B(x) = \int dB = \int (x^2 - 3x + 1)\, dx = \frac{x^3}{3} - \tfrac{3}{2}x^2 + x + k$$

where k is a constant.

(b) $10 = B(0) = 0 - 0 + 0 + k$, so $10 = k$. Hence

$$B(x) = \frac{x^3}{3} - \tfrac{3}{2}x^2 + x + 10$$

is the unique solution for the boundary condition $B(0) = 10$.

Sometimes a differential equation that seems to have the dependent and independent variables mixed up can, through algebraic manipulation, be put into the form

$$f(x)\, dx = g(y)\, dy$$

Such a differential equation is called separable because the x's (on the left side of the equal sign) are separated from the y's (on the right side of the equal sign). These are among the easiest to solve, because they can be solved by separating the variables and then integrating both sides of the equation.

4.4.3   Rule

If f and g are continuous functions, solutions of $f(x)\, dx = g(y)\, dy$ are given by $\int f(x)\, dx = \int g(y)\, dy + k$.

4.4.4   Example

Show that the following differential equations are separable, and then solve them.

(a) $y\, dy + x\, dx = 0$

(b) $\dfrac{dy}{dx} = x^2 y^3$

(c) $y' = xe^y$

Solution:

(a) Write $y\, dy = -x\, dx$, and integrate to get $\int y\, dy = -\int x\, dx + k$.

Thus $y^2/2 = -(x^2/2) + k$ gives solutions implicitly to the original equation. Explicitly, the solution is $y = \pm\sqrt{-x^2 + 2k}$.

(b) Separating variables, we get

$$\frac{dy}{y^3} = x^2\, dx$$

So, integrating both sides, we have

$$\frac{y^{-2}}{-2} = \frac{x^3}{3} + k$$

which defines the solutions implicitly.

(c) Remembering that $y' = \dfrac{dy}{dx}$, write $e^{-y}\, dy = x\, dx$ and integrate to get

$$-e^{-y} = \frac{x^2}{2} + k$$

Explicitly, the solution is

$$y = -\ln\left|-\frac{x^2}{2} - k\right|$$

4.4.5   Rule

The technique for solving separable differential equations is to

1. Write them in differential notation.
2. Separate the variables on either side of the equal sign.
3. Integrate both sides of the resulting equation.

Solving differential equations is both an important and a complicated subject, and whole courses are devoted to it. Here we have illustrated only the simplest techniques, and those with no mathematical proofs. But even these elementary techniques are sufficient for some important applications.

4.4.6   Example

The marginal cost of producing x items of a commodity is given by the differential equation $C' = x^2 - x$. If the fixed cost is $1,000, find the cost of producing 10 items.

Solution:

To find the cost function, we solve the differential equation $dC = (x^2 - x)\, dx$, getting $C(x) = (x^3/3) - (x^2/2) + k$. $C(0) = 1,000$ implies that $k = 1,000$, so

$$C(x) = \frac{x^3}{3} - \frac{x^2}{2} + 1{,}000 \quad \text{and} \quad C(10) = \frac{1{,}000}{3} - \frac{100}{2} + 1{,}000$$

$$= \frac{7{,}700}{6}$$

We have done problems like this one before. The next has some new aspects.

4.4.7   Example

Suppose that the *growth rate* (or rate of proportional change in value) of a certain stock is a constant, r, and its initial value is P. In other words, its value at time $t = 0$ is P. Remember that "growth rate" was defined to be y'/y in Section 3.7, where y is the value of the stock.

(a) Find an equation that expresses y as a function of t for a given growth rate r and a given principal P.

(b) If the growth rate is 0.7 and the initial value is $1,000$, find the value of the stock after 3 years.

Solution:

(a) Since the growth rate is r, we have the differential equation

$$\frac{y'}{y} = r$$

Writing this in differential notation, we have

$$\frac{\frac{dy}{dt}}{y} = r$$

Separating the variables by putting the y's on the left and the t's on the right, this becomes

$$\frac{dy}{y} = r \, dt$$

Integrating, we get

$$\ln y = rt + k \qquad \text{(note that } y > 0\text{)}$$

Exponentiating both sides and remembering the relationship between the logarithmic and exponential functions, we have

$$y = e^{\ln y} = e^{rt+k} = e^{rt}e^{k}$$

so

$$y = e^{rt}e^{k}$$

is the general solution of this equation. Using the boundary condition

$y(0) = P$, this becomes

$$P = e^{r0}e^k = e^0 e^k = 1 \cdot e^k = e^k$$

Putting $P = e^k$ back into the original general solution, we have

$$y = e^{rt}P \quad \text{or} \quad y = Pe^{rt}$$

You will remember having seen this final solution in Section 3.3; it is the expression for the value of an investment when the interest is compounded continuously. What we have proved in this section is that any investment that has a constant growth rate must indeed be in this form.
(b) Setting $r = 0.7$, $P = 1{,}000$, and $t = 3$ in the equation $y = Pe^{rt}$, we get

$$y(3) = 1{,}000e^{0.7(3)} = 1{,}000e^{2.1} \approx \$8{,}166$$

4.4.8   Example

Suppose that the elasticity of demand of a certain type of price–demand equation is constantly 1—that is, $E = 1$. Find the general form of such equations. Remember that the elasticity of demand was defined in Section 2.8 to be

$$E = -\frac{p}{q\left(\dfrac{dp}{dq}\right)}$$

Solution:

In the equation we separate the variables by taking the p's to the left and the q's to the right:

$$1\frac{dp}{p} = E\frac{dp}{p} = -\frac{dq}{q}$$

Integrating both sides, we get

$$\ln p = -\ln q + k$$

Exponentiating both sides, we have

$$e^{(\ln p)} = e^{-\ln q + k} = e^{\ln q^{-1}}e^k$$

which is

$$p = q^{(-1)}e^k$$

Since k is a constant, so is e^k; we can set $a = e^k$, where a is also a constant. Thus we conclude that if $E = 1$,

$$p = q^{(-1)}a = \frac{a}{q}$$

for some constant a. Thus $p = a/q$ is the form of these price-demand equations.

4.4.9   Example in Paleontology

The phenomenon of radioactive decay is commonly used to date fossil remains. This method uses the fact that the rate of decay $\dfrac{dA}{dt}$ of an element will be proportional to the amount A of the element that exists at the time t.

In other words, there is a constant, α (which depends on the element), such that

$$\frac{dA}{dt} = -\alpha A$$

(a) Determine a formula for the amount A remaining after time t if the initial quantity of the element is A_0.
(b) How long will it take for half the element to decay? This length of time is called the *half-life* of the element.
(c) Radioactive carbon, ^{14}C, is often used to date fossils. In living organisms ^{12}C and ^{14}C are continually being made in fixed proportions, but in dead organisms, ^{14}C decays with a half-life of 5,600 years. If a fossil insect is found in metamorphic (volcanic) rock which has only $\frac{1}{32}$ the usual amount of ^{14}C, how old is the fossil?
(d) If a fossil is found that has only $\frac{1}{20}$ the usual amount of radioactive carbon, how old is the fossil?

Solution:

(a) Separating variables, the differential equation becomes

$$\frac{dA}{A} = -\alpha \, dt$$

Integrating both sides, we have

$$\ln A = -\alpha t + k$$

Exponentiating both sides of this equation, we get

$$e^{\ln A} = A = e^{-\alpha t + k} = e^{-\alpha t} e^k = e^{-\alpha t} K$$

where $K = e^k$ is another constant. Rewriting this, we have

$$A = K e^{-\alpha t}$$

To evaluate K, we use the fact that the initial quantity of the element is A_0. Speaking mathematically, this means that when $t = 0$, $A = A_0$. Thus we can write

$$A_0 = K e^{-\alpha \cdot 0} = K$$

so the desired formula is

$$A = A_0 e^{-\alpha t}$$

(b) This question asks at which time t will the amount A be equal to $\frac{1}{2}A_0$. Thus we solve the following equation for t:

$$\tfrac{1}{2}A_0 = A_0 e^{-\alpha t}$$

Since $A_0 \neq 0$, this equation is the same as

$$\tfrac{1}{2} = e^{-\alpha t}$$

To extract the t, we take the logarithm of both sides.

$$\ln \tfrac{1}{2} = \ln \left(e^{-\alpha t}\right) = -\alpha t$$

so

$$t = \frac{\ln \tfrac{1}{2}}{-\alpha} = \frac{-\ln 2}{-\alpha} = \frac{\ln 2}{\alpha}$$

Thus the half-life of any element can be calculated by dividing the corresponding proportionality constant, α, into $\ln 2$.

(c) This problem is relatively easy if we recognize that $32 = 2^5$. [If not, we must use the technique of part (d) below.] Then we can see that one-half the original quantity will be left after 5,600 years, one-half of that (that is, $\frac{1}{4}$) after $2 \cdot 5,600$ years, one-half of that ($\frac{1}{8}$) after $3 \cdot 5,600$ years, one-half of that ($\frac{1}{16}$) after $4 \cdot 5,600$ years, and one-half of that ($\frac{1}{32}$ of the original quantity) after $5 \cdot 5,600$ years. Thus we estimate that the fossil is about $5 \cdot 5,600 = 28,000$ years old.

Figure 4.4–1

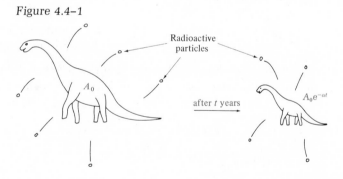

Radioactive particles

A_0

after t years

$A_0 e^{-\alpha t}$

(d) Since 20 is not an integer power of 2, we use a more mechanical solution for this problem. We want the t that satisfies the equation

$$\tfrac{1}{20}A_0 = A_0 e^{-\alpha t}$$

Since $A_0 \neq 0$, $\frac{1}{20} = e^{-\alpha t}$; taking the logarithm of each side, we have $\ln \left(\frac{1}{20}\right) = \ln \left(e^{-\alpha t}\right) = -\alpha t$, so [using $\ln \left(\frac{1}{20}\right) = -\ln 20$]

$$-\ln (20) = -\alpha t \qquad \text{and} \qquad \frac{\ln (20)}{\alpha} = t$$

From part (b) we know that $5{,}600 = \dfrac{\ln 2}{\alpha}$, so $\alpha = \dfrac{\ln 2}{5{,}600}$. Thus

$$t = \ln (20) \Big/ \frac{\ln 2}{5{,}600} = \frac{5{,}600 \cdot \ln (20)}{\ln (2)} \approx \frac{5{,}600 \cdot 3}{0.69}$$

$$\approx 24{,}348$$

4.4.10   Example in Biology or Pre-Med

It is often reasonable to assume that the number of people that will come down with a certain illness is directly proportional to both the number that currently are infected with the illness and the number that are not yet infected but are susceptible to it. Speaking mathematically, this means that if x is the number of people susceptible but not yet infected and y is the number that are already infected,

$$\frac{dy}{dt} = Bxy \qquad \text{for a constant } B$$

Suppose that we have a population of n uninfected individuals into which one infected person is introduced at time $t = 0$. Then at all times we have a total population of $n + 1$ people, so

$$x + y = n + 1$$

and thus

$$\frac{dy}{dt} = B(n + 1 - y)y \qquad \begin{array}{l}\text{(substituting } x = n + 1 - y \text{ into the}\\ \text{differential equation above)}\end{array}$$

We shall only sketch the solution of this differential equation here, leaving the details (marked in lowercase letters) for the reader. (Refer, if you like, to the hints in problem 3 of Exercises 4.4.C.) Separating the variables, we get

$$\frac{dy}{(n + 1 - y)y} = B \, dt$$

But

$$B \, dt = \frac{dy}{(n + 1 - y)y} = \frac{1}{n + 1} \left(\frac{1}{n + 1 - y} + \frac{1}{y} \right) dy \qquad \text{(a)}$$

so, after multiplying through by $(n + 1)$ and integrating, we have

$$(n + 1)Bt + k = -\ln (n + 1 - y) + \ln (y) = \ln \left(\frac{y}{n + 1 - y} \right) \qquad \text{(b)}$$

Exponentiating, we have

$$e^{(n+1)Bt+k} = \frac{y}{n + 1 - y}$$

Solving this equation for y, we get

$$y = \frac{n+1}{1 + Ae^{-(n+1)Bt}} \qquad \text{where } A = e^{-k} \qquad \text{(c)}$$

If we make the assumption that at the beginning (time $t = 0$) there was only one infected person in the group [that is, that $y(0) = 1$], we have that

$$1 = \frac{n+1}{1 + Ae^0} = \frac{n+1}{1 + A} \qquad \text{which implies that } A = n$$

Thus the particular solution of the differential equation under the assumption $y(0) = 1$ is

$$y = \frac{n+1}{1 + ne^{-B(n+1)t}}$$

The graph of this equation is the sigmoid or S-curve, which was mentioned in Example 1.2.9 in connection with fruit flies.

Figure 4.4–2

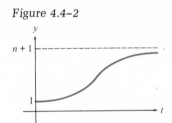

4.4.11   Example in Sociology

A rumor tends to spread in direct proportion to the number of people that have heard it, x, and to the number of people that are uninformed, y. In other words,

$$\frac{dy}{dt} = Bxy \qquad \text{for a constant } B$$

The mathematics of this situation is exactly the same as that in the previous problem.

Figure 4.4–3

$$\frac{dy}{dt} = Bxy$$

SUMMARY

Differential equations often arise when describing how various real-world situations are changing. The solution of a differential equation is a function that makes the equation true. Sometimes we can find this function by separating the variables in the differential equation and then integrating. This section showed how to do this and gave examples of separable differential equations involving growth rate, elasticity of demand, radioactive decay, and spread of contagious diseases or rumors.

EXERCISES 4.4. A

1. Give the general solutions for the following separable differential equations.

    (a) $\dfrac{dy}{dx} = \dfrac{x}{y}$

    (b) $\dfrac{dy}{dx} = \sqrt{yx}$

    (c) $x\,dy = y\,dx$

    (d) $\dfrac{dy}{dx} = e^{-y}$

    (e) $dy = e^{y}\,dx$

    (f) $\dfrac{dy}{dx} = xe^{y}$

2. The marginal cost of producing x items of a commodity is given by the differential equation $C' = 3x^2 - 4x$.
    (a) If the fixed cost is $200, write the cost equation.
    (b) Find the cost of producing 20 items.

3. The marginal revenue obtained by producing x items of a commodity is given by the differential equation $\dfrac{dR}{dx} = 7 - 0.4x$, and the revenue received from one item is $10.
    (a) Find the revenue function.
    (b) What is the revenue obtained from selling four items?

4. The growth rate y'/y of an animal population tends to be constant if the environment does not change. In other words, the rate of growth is proportional to the number alive. ($y' = By$, where B is the "growth rate" or the "proportionality constant.") The human population of the earth in 1930 was 2 billion and in 1960 was 3 billion.
    (a) Letting 1930 be time $t = 0$, write an equation for the population of the earth at time t. (Hint: Look at Examples 4.4.7 and 4.4.9.)
    (b) Use this equation and Tables III and IV to find the projected population in 2000 A.D.
    (c) Find the projected population in 2100 A.D.

EXERCISES 4.4. B

1. Give the general solutions for the following separable differentiable equations.

    (a) $\dfrac{dy}{dx} = xy$

    (b) $\dfrac{dy}{dx} = \sqrt{\dfrac{x}{y}}$

    (c) $x^2\,dy = y\,dx$

    (d) $\dfrac{dy}{dx} = \dfrac{1}{y}e^{-y^2}$

    (e) $dy = ye^{x}\,dx$

    (f) $\dfrac{dy}{dx} = x^2y^2$

2. The marginal cost of producing x items of a commodity is given by the differential equation $C' = 4x^2 - 3x$.
    (a) If the fixed cost is $100, write the cost equation.
    (b) Find the cost of producing 10 items.

3. The marginal revenue obtained by producing x items of a commodity is given by the differential equation $\frac{dR}{dx} = 6 - 0.6x$, and the revenue received from one item is $10.
    (a) Find the revenue function.
    (b) What is the revenue obtained from selling four items?

4. In 1900 the population of the United States was about 75 million and in 1950 it was about 150 million.
    (a) If the growth of the population is proportional to its present size (that is, $\frac{dP}{dt} = BP$ for some constant B), write an expression for the population at time t letting 1900 be time $t = 0$. (*Hint:* Look at Examples 4.4.7 and 4.4.9.)
    (b) What is the projected population in the year 2000 A.D.?
    (c) What is the projected population in the year 2100 A.D.?

EXERCISES 4.4. C

1. Radioactive beryllium is sometimes used to date deep-sea sediment. Beryllium decay satisfies the differential equation $\frac{dy}{dt} = -1.5(10)^{-7}y$. What is the half-life of beryllium?

2. Thorium 230 is sometimes used to date coral and shells. Its decay satisfies the equation $\frac{dy}{dt} = -92(10^{-7})y$. What is the half-life of radioactive thorium?

3. Complete the steps omitted in Example 4.4.10. For (a) multiply out the right side to verify that it equals the left side. For (b) use the substitution $u = n + 1 - y$ to integrate the first term on the right side of the equation. Part (c) involves messy algebra. It helps to define $C = e^{(n+1)Bt+k}$, solve for y, and then substitute the long-winded expression for C back into the final expression.

4. The growth of an unchecked population tends to be proportional to the size of the population, as we saw in problems 4 of Exercises 4.4.A and 4.4.B. (This is true of either human or animal populations.) If, however, there is a maximum number M that the environment can sustain, then the growth is proportional to both the size of the existing population, x, and the difference between the maximum size and the present size, $M - x$. Thus the growth satisfies the differential equation

    $$\frac{dx}{dt} = Bx(M - x) \qquad \text{for a constant } B$$

    Solve this differential equation. (*Hint:* Pattern your solution on Example 4.4.10.) For a graph of a typical resulting solution, see Example 1.2.9.

5. The growth rate (or rate of proportional change) of a certain stock is given by $0.02t$ and its initial value is $1,000. Find the time t it takes to double its value.

6. The growth rate of a certain corporation is given by $\frac{1}{2}\sqrt{t}$ and its initial net worth was $250,000. (Remember that growth rate is y'/y.)
   (a) Write an equation for the net worth of the company at time t.
   (b) Find the time that it takes the corporation to become worth $1,000,000.

7. The elasticity of demand of a certain type of price-demand equations is given by $10q$. Find the general form of the equation. (See Example 4.4.8.)

8. The half-life of the potassium isotope ^{42}K, used as a biological tracer, is 12.5 hours.
   (a) Write a differential equation giving the rate of change of A with respect to t.
   (b) Write an equation giving the amount A remaining from an initial amount A_0 of ^{42}K after t hours.
   (c) How long will it be before only $\frac{1}{4}A_0$ of the potassium remains? (Refer to Example 4.4.9.)

9. Fechner's law of stimulus–response states that the rate of change of an organism's response R to a stimulus S is inversely proportional to the stimulus.
   (a) Express Fechner's law as a differential equation.
   (b) If the detection threshold is given as S_0, that is, if $R(S_0) = 0$, write the functional relationship between R and S. (Note: You should compare the formula you obtain with the very similar formula $L = 10[\log_{10}(I) - \log_{10}(I_0)]$ that was given for loudness in Example 3.4.6.)

10. The Brentano–Stevens law of stimulus–response can be written as the differential equation $\dfrac{dR}{dS} = k \cdot R/S$, where k is a positive constant and R and S are response and stimulus, respectively. If the detection stimulus is given as S_0, that is, $R(S_0) = 0$, find the functional relationship between R and S. (Hint: You will discover that we must have $S_0 = 0$. Furthermore, this relationship is often called the power law of stimulus–response.)

ANSWERS 4.4. A

1. (a) $y^2 = x^2 + 2k$ or $y = \pm\sqrt{x^2 + 2k}$ (b) $2\sqrt{y} = \frac{2}{3}x^{3/2} + k$ or $y = (x^{3/2}/3 + k/2)^2$
   (c) $\ln(y) = \ln(x) + k$ or $y = xe^k$ (d) $e^y = x + k$ or $y = \ln(x + k)$
   (e) $-e^{-y} = x + k$ or $y = -\ln|-x - k|$
   (f) $-e^{-y} = x^2/2 + k$ or $y = -\ln|-x^2/2 - k|$
2. (a) $C = x^3 - 2x^2 + 200$ (b) $7,400
3. (a) $R(x) = 7x - 0.2x^2 + 3.2$ (b) $R(4) = 28.00
4. (a) $y = 2(10^9)e^{0.0135t}$ (b) $y(70) \approx 5$ billion (c) $y(170) \approx 20$ billion

ANSWERS 4.4. C

1. 4.6 million years
2. 75,000 years
4. $y = M/(1 + Ae^{-MBt})$ for a constant of integration $A = e^k$
5. $t = 10\sqrt{\ln 2}$
6. (a) $y = 250{,}000e^{1/3t^{3/2}}$ (b) $t = (6 \ln 2)^{2/3}$
7. $p = e^{(q^{-1}/10)+k}$
8. (a) $\dfrac{dA}{dt} = -\alpha A$, where α is determined in part (b)

(b) $A = A_0 e^{-\alpha t}$, where $\alpha = \dfrac{\ln{(2)}}{12.5} \approx 0.06$ (c) 25 hours

9. (a) $\dfrac{dR}{dS} = \dfrac{k}{S}$ (b) $R = k[\ln{(S)} - \ln{(S_0)}]$

10. $R = C \cdot S^k$, where C is a positive constant.

4.5   The Definite Integral as Total Change

We began Chapter 2 by saying that differential calculus is the mathematical study of change. Then we showed how the derivative of a function describes the instantaneous change of the function. In business terminology, the marginal cost (revenue, profit, etc.) describes how the cost (revenue, profit, etc.) is changing at each point.

There are three types of change we might consider: total change, average change, and marginal change. Speaking more precisely, as a function goes from $F(a)$ to $F(b)$ (for example, as the quantity sold rises from a units to b units), we might want to investigate:

1. $F(b) - F(a)$: The total change tells the gross change over the whole interval and depends greatly on which a and b are chosen.
2. $[F(b) - F(a)]/(b - a)$: The average change gives a rough estimate of what is happening per unit in the interval. Notice that although it gives us the change "per unit," it does not treat the unit intervals individually.
3. $F'(x)$: The marginal change describes how much change tends to take place at each point in the interval. We can use it to discover the approximate change in F as x makes a small change going from any x_1 to $x_1 + \Delta x$. By Section 3.8, we know that the change in $F(x)$ over this (small) interval will be about $F'(x) \, \Delta x$.

It is clear that "average change" can be easily obtained from "total change" by dividing. We obtain marginal change, $F'(a)$, from average change by letting b approach a (where $b = a + \Delta a$ and Δa approaches 0).

Figure 4.5–1

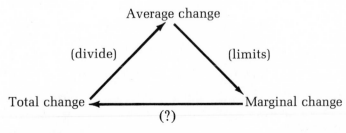

The answer to the question mark in Figure 4.5–1 gives us what may be the most significant relationship in all of calculus. The next few sections explore in detail the relationship between total change and marginal change.

4.5.1   Example

Suppose that marginal revenue is given by $R'(x) = 3$. Find the total change in revenue: $R(5) - R(2)$.

Solution:

The revenue function will be given by

$$R(x) = 3x + k$$

for a constant k. Hence

$$\begin{aligned} R(5) - R(2) &= (3 \cdot 5 + k) - (3 \cdot 2 + k) \\ &= 15 + k - 6 - k \\ &= 9 \end{aligned}$$

Notice that the k's cancel out. This suggests that in order to pass from marginal change to total change we need only antidifferentiate without concern for the arbitrary k's. For instance, in Example 4.5.1, we shall write

$$R(5) - R(2) = \int_2^5 3 \, dx$$

This motivates the following definition, which is valid for continuous functions $f(x)$.

4.5.2   Definition

If $F(x)$ is *any* antiderivative of $f(x)$,

$$\int_a^b f(x) \, dx = F(b) - F(a) \quad \left(\text{This difference is often written } F(x) \Big|_a^b \right)$$

is called the definite integral of $f(x)$ from a to b.

In other words, $\int_a^b f(x) \, dx$ is the total change of *any* antiderivative of $f(x)$ as x changes from a to b.

You may wonder if the expression $\int_a^b f(x) \, dx$ is, as predicted, uniquely defined by the above definition. Suppose that F and G are two antiderivatives of f. Will we have $F(b) - F(a) = G(b) - G(a)$? In other words, will $\int_a^b f(x) \, dx$ have the same value according to Definition 4.5.2 no matter which antiderivative of f is chosen?

It is not hard to see that the answer to this question is yes. This is because any two antiderivatives F and G of a function f must differ by a constant k. In other words, for some constant k, $F(x) = G(x) + k$ for all x. (See Appendix III.) Therefore,

$$F(b) - F(a) = [G(b) + k] - [G(a) + k] = G(b) - G(a)$$

We have proved

4.5.3   Rule

If $F(x)$ and $G(x)$ are any two antiderivatives of f, then

$$F(b) - F(a) = G(b) - G(a) = \int_a^b f(x)\,dx$$

Thus $\int_a^b f(x)\,dx$ is uniquely defined—just find *any* antiderivative of f, evaluate it at the points a and b, and subtract. This is why $\int_a^b f(x)\,dx$ is called the "definite" integral; it has one definite value.

Figure 4.5–2 may give you a geometric idea of the reason $G(b) - G(a) = F(b) - F(a)$ if $G' = F'$. Remember that if two curves have the same derivative everywhere, they are "parallel" in some sense.

Figure 4.5–2

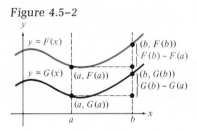

4.5.4   Example

Evaluate the following definite integrals.

(a) $\displaystyle\int_1^2 x\,dx$

(b) $\displaystyle\int_0^1 e^x\,dx$

(c) $\displaystyle\int_3^{-1} y^2\,dy$

Solutions:

(a) $\displaystyle\int_1^2 x\,dx = \tfrac{1}{2}x^2\Big|_1^2 = \tfrac{1}{2}\cdot 2^2 - \tfrac{1}{2}\cdot 1^2 = \tfrac{3}{2}$

(b) $\int_0^1 e^x \, dx = e^x \Big|_0^1 = e^1 - e^0 = e - 1$

(c) $\int_3^{-1} y^2 \, dy = \dfrac{y^3}{3} \Big|_3^{-1} = -\tfrac{1}{3} - 9 = -9\tfrac{1}{3}$

4.5.5   Example

Find the total change in cost if production rises from $x = 20$ to $x = 40$ and the marginal cost of producing the xth item is $4 - 0.06x$.

Solution:

Let the total cost be called $C(x)$. Then we know that $C'(x) = 4 - 0.06x$, so the change in cost is given by

$$C(40) - C(20) = \int_{20}^{40} (4 - 0.06x) \, dx$$

We cannot determine the cost function $C(x)$, but we don't need it to solve this problem. We merely compute

$$\int_{20}^{40} (4 - 0.06x) \, dx = (4x - 0.03x^2) \Big|_{20}^{40} = (160 - 48) - (80 - 12)$$
$$= 112 - 68 = 44$$

and discover that the cost increases 44 as x goes from 20 to 40.

4.5.6   Example

As the number produced changes from 2 to 4 in Example 4.5.5, find the total change in *profit* if the marginal revenue is given by $R'(x) = 20$.

Solution:

Your first impulse may be to find the profit function: $P = R - C$. We do not have enough information to do this and, again, it is not necessary to be so specific. Instead, we recall that the marginal profit equals the marginal revenue minus the marginal cost. Thus in this case

$$P' = 20 - (4 - 0.06x)$$

So the total change in P as x goes from 2 to 4 is

$$\int_2^4 [20 - (4 - 0.06x)] \, dx = (16x + 0.03x^2) \Big|_2^4 = (64 + 0.48)$$
$$- (32 + 0.12) = 64.48 - 32.12$$

Thus $P(4) - P(2) = 32.36$; the profit increases approximately 32 units.

4.5.7   Example

The rate of change of labor hours needed to assemble the xth unit of a product has been found to be $x - \frac{3}{2}\sqrt{x}$ by considering the assembly of 49 units of the product. What is the total additional number of hours needed to assemble another 51 units?

Solution:

We seek

$$\int_{49}^{100} (x - \tfrac{3}{2}x^{1/2})\, dx = \tfrac{1}{2}x^2 - x^{3/2}\Big|_{49}^{100}$$

$$= [\tfrac{1}{2}100^2 - (\sqrt{100})^3] - [\tfrac{1}{2}49^2 - (\sqrt{49})^3]$$

$$= (5{,}000 - 1{,}000) - (1{,}200.5 - 343)$$

$$= 4{,}000 - 857.5 = 3{,}142.5$$

Hence we need an additional 3,142.5 hours of labor.

We now state some rules about the total change of a function in terms of a definite integral.

4.5.8   Rule

(a) $\displaystyle\int_a^a f(x)\, dx = 0$

(b) $\displaystyle\int_a^b f(x)\, dx = -\int_b^a f(x)\, dx$

(c) $\displaystyle\int_a^b (f(x) \pm g(x))\, dx = \int_a^b f(x)\, dx \pm \int_a^b g(x)\, dx$

(d) $\displaystyle\int_a^b f(x)\, dx + \int_b^c f(x)\, dx = \int_a^c f(x)\, dx$

We can justify these rules as follows [where $F'(x)=f(x)$ and $G'(x)=g(x)$].

(a) $\displaystyle\int_a^a f(x)\, dx = F(a) - F(a) = 0$

(b) $\displaystyle\int_a^b f(x)\, dx = F(b) - F(a) = -[F(a) - F(b)] = -\int_b^a f(x)\, dx$

(c) $\displaystyle\int_a^b (f(x) + g(x))\, dx = [F(b) + G(b)] - [F(a) + G(a)]$

$$= F(b) - F(a) + G(b) - G(a)$$

$$= \int_a^b f(x)\, dx + \int_a^b g(x)\, dx$$

(d) $\displaystyle\int_a^b f(x)\,dx + \int_b^c f(x)\,dx = F(b) - F(a) + F(c) - F(b)$

$$= F(c) - F(a) = \int_a^c f(x)\,dx$$

4.5.9   Example in Archaeology

(See also Example 4.4.9) We can use the concept of "total change" as explained in this section to determine the age of archaeological artifacts. Radioactive carbon, ^{14}C, decays at the rate

$$\frac{dA}{dt} = -A_0\alpha e^{-\alpha t}$$

where A_0 was the original amount of ^{14}C and $\alpha = \ln(2)/5{,}600$. When the organism is alive, it replaces the ^{14}C continuously, but after it dies the decay is steady and the amount of ^{14}C remaining in an artifact enables us to calculate, using a definite integral, how long it has been changing (decaying).

(a) An archaeological expedition to the Yucatan Peninsula has unearthed some Indian artifacts. The scientists guess that they are 5,600 years old. What would have been the total change in the ^{14}C content of these artifacts during that time?

(b) If it is later determined that only about one-fourth of the original ^{14}C has decayed, how old are the artifacts?

Solution:

(a) To find the total change, we integrate the decay rate (the "marginal change") from 0 to 5,600:

$$\int_0^{5{,}600} -A_0\alpha e^{-\alpha t}\,dt = A_0 e^{-\alpha t}\Big|_0^{5{,}600} = A_0 e^{-\ln(2)} - A_0 = A_0(\tfrac{1}{2}) - A_0$$

$$= -\frac{A_0}{2}$$

In other words, in 5,600 years, half the original quantity has disappeared; the minus sign indicates that the total change has been negative, a decrease. Since 5,600 years is the time required for half the original ^{14}C to decay, 5,600 is the *half-life* of ^{14}C.

(b) Let x be the number of years that the artifact has been decaying. Then

$$-\tfrac{1}{4}A_0 = \int_0^x -A_0\alpha e^{-\alpha t}\,dt = A_0 e^{-\alpha t}\Big|_0^x = A_0 e^{-\alpha x} - A_0$$

so

$$\tfrac{3}{4}A_0 = A_0 e^{-\alpha x} \qquad \text{and} \qquad \tfrac{3}{4} = e^{-\alpha x}$$

Taking the logarithm of both sides, we have

$$\ln(\tfrac{3}{4}) = \ln(e^{-\alpha x}) = -\alpha x$$

It follows that

$$x = \frac{-\ln\left(\frac{3}{4}\right)}{\alpha} = (\ln 4 - \ln 3)\,\frac{5{,}600}{\ln 2} \approx 2{,}324$$

4.5.10   Example in Biology

Suppose the rate of change of bacteria in a culture at time t minutes is given by $f'(t) = 10e^{0.05t}$. What is the total change in the size of the culture from time $t = 60$ to time $t = 80$ minutes?

Solution:

The total change can be written as a definite integral of the rate of change:

$$\int_{60}^{80} 10e^{0.05t}\,dt = 10\,\frac{1}{0.05}\,e^{0.05t}\Big|_{60}^{80} = 200(e^{0.05(80)} - e^{0.05(60)})$$

$$\approx 200(55 - 20) = 200 \cdot 35 = 7{,}000$$

Thus the culture grows a total of 7,000 bacteria between time $t = 60$ and time $t = 80$.

SUMMARY

The definite integral was defined in this section. If $F'(x) = f(x)$, then $\int_a^b f(x)\,dx$ is defined to be $F(b) - F(a)$. This leads to the natural interpretation of the definite integral as total change; if $f(x)$ is the marginal cost, for example, then $F(b) - F(a)$ is the total change in the cost as the quantity changes from a to b.

EXERCISES   4.5. A

1.   Evaluate the following definite integrals.

(a) $\displaystyle\int_2^3 (x^2 - x)\,dx$

(f) $\displaystyle\int_1^2 x\sqrt{x^2 + 1}\,dx$

(b) $\displaystyle\int_3^2 (x^2 - x)\,dx$

(g) $\displaystyle\int_0^1 e^x\,dx$

(c) $\displaystyle\int_{-1}^4 (x^3 + 1)\,dx$

(h) $\displaystyle\int_{-1}^0 xe^{x^2}\,dx$

(d) $\displaystyle\int_0^1 \sqrt{x + 1}\,dx$

(i) $\displaystyle\int_1^e \frac{\ln(x)}{x}\,dx$

(e) $\displaystyle\int_1^2 (x^{-1} + x^{-2})\,dx$

2.   Find the total change in the cost if the production level changes from $x = 3$ to $x = 5$ items and the marginal cost of producing x items is $C'(x) = 10 - 0.4x$.

3.   Find the total change in the revenue if the marginal revenue is $R'(x) = 12 - 0.6x$ and the amount sold changes from $x = 3$ to $x = 5$ items.

4. (a) Using the facts of problems 2 and 3, find the total change in the profit as x changes from 3 to 5.

    (b) For what quantity is the profit a maximum?

5. Suppose that the rate of change of labor hours needed to assemble the xth unit of a product has been determined to be $f(x) = 1{,}000x^{-1/2}$ by observing the assembly of the first 100 units. What is the additional labor needed to assemble another 100 units? (That is, what is the total change as $x = 100$ changes to $x = 200$?)

6. The marginal price of a commodity is given by $p' = -2q + 10$. Find the total change in price per unit as $q = 5$ goes to $q = 10$.

7. Using problem 1(a), illustrate Rule 4.5.8 by
    (a) reversing the limits of integration
    (b) adding an intermediate point between the given limits
    (c) breaking up the polynomial

EXERCISES 4.5. B

1. Evaluate the following definite integrals.

    (a) $\displaystyle\int_1^2 (x - x^2)\, dx$ (f) $\displaystyle\int_2^1 (3x\sqrt{x^2 + 1})\, dx$

    (b) $\displaystyle\int_2^1 (x - x^2)\, dx$ (g) $\displaystyle\int_{-1}^0 e^x\, dx$

    (c) $\displaystyle\int_0^4 (x^2 + 2)\, dx$ (h) $\displaystyle\int_0^1 x^2 e^{x^3}\, dx$

    (d) $\displaystyle\int_1^0 \sqrt[3]{x + 1}\, dx$ (i) $\displaystyle\int_1^e \frac{\ln (2x)}{x}\, dx$

    (e) $\displaystyle\int_1^2 (x^{-1} + x^{-3})\, dx$

2. Find the total change in the cost if the production level changes from $x = 3$ items to $x = 5$ items and the marginal cost of producing x items is $C'(x) = 5 - 0.04x$.

3. Find the total change in the revenue if the marginal revenue is $R'(x) = 7 - 0.06x$ and the amount sold changes from $x = 3$ to $x = 5$.

4. (a) Using the facts of problems 2 and 3, find the total change in profit as x changes from 3 to 5.

    (b) Find the quantity for which the profit is a maximum.

5. Suppose that the rate of change of labor hours needed to assemble the xth unit of a product has been determined to be $f(x) = 100x^{-1/3}$ by observing the assembly of the first 1,000 units. What is the additional labor needed to assemble another 1,000 units? (That is, what is the total change as $x = 1{,}000$ changes to $x = 2{,}000$?)

6. Suppose that the marginal price $\left(\dfrac{dp}{dq}\right)$ of a quantity is given by $p' = 0.2$. Find the total change in price per unit as $q = 90$ rises to $q = 100$.

7. Using problem 1(a), illustrate Rule 4.5.8 by
    (a) reversing the limits of integration
    (b) adding an intermediate point between the given limits
    (c) breaking up the polynomial

EXERCISES   4.5. C

1. The rate of change in a population of rabbits was observed to be given by $P'(t) = 4t$ if unchecked by predators, disease, or famine. Find the total change in this population from month $t = 4$ to month $t = 5$.

2. In a biological experiment, one grain of radioactive potassium ^{42}K was injected into the bloodstream of a guinea pig. If $\dfrac{dA}{dt} = -\alpha e^{-\alpha t}$, where $\alpha =$ ln (2)/12.5 gives the rate of change of the potassium with time, what is the total change in potassium between the tenth and twentieth hours?

ANSWERS   4.5. A

1. (a) $3\frac{5}{6}$ (b) $-3\frac{5}{6}$ (c) $68\frac{3}{4}$ (d) $\frac{2}{3}(2\sqrt{2} - 1)$ (e) $\frac{1}{2} + \ln 2$ (f) $\frac{1}{3}(5\sqrt{5} - 2\sqrt{2})$
    (g) $e - 1$ (h) $\frac{1}{2}(1 - e)$ (i) $\frac{1}{2}$ [Solution: Let $u = \ln (x)$; $du = 1/x\ dx$;
    $\displaystyle\int_{1}^{e} \frac{\ln (x)}{x}\ dx = \int u\ du = \frac{1}{2}[\ln (x)]^2\Big|_{1}^{e} = \frac{1}{2}[\ln (e)^2 = \ln (1)^2] = \frac{1}{2}(1 - 0) = \frac{1}{2}.$]

2. \$16.80   5. $20,000(\sqrt{2} - 1)$
3. \$19.20   6. -25
4. (a) \$2.40 (b) $x = 10$

ANSWERS   4.5. C

1. 18 rabbits
2. $e^{-20\alpha} - e^{-10\alpha} \approx e^{-1.1} - e^{-0.6} \approx -0.24$ grain

4.6 The Definite Integral as Area

The definite integral can be used not only to calculate total change of a function but also to find the area of figures with curved boundaries. This geometrical application of the definite integral probably was the original motivation of integral calculus (the study of antiderivatives), and it turns out to be useful in many applied fields of study. The relationship between definite integrals and areas is remarkably simple; the area "under" the curve $f(x)$ if $f(x) \geq 0$ is exactly $\displaystyle\int_{a}^{b} f(x)\ dx$.

4.6.1   Illustration

Figure 4.6–1

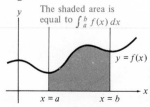

The shaded area is equal to $\int_{a}^{b} f(x)\ dx$

The shaded area in Figure 4.6–1 is equal to $\int_a^b f(x)\,dx$.

4.6.2   Rule

If $f(x) \geqslant 0$ is a continuous function, then the area bounded by the curve $y = f(x)$, the x-axis, and the lines $x = a$ and $x = b$ where $a < b$ is given by $\int_a^b f(x)\,dx$.

We shall examine why this rule is true in the case where $f(x)$ is increasing. Let $A(x_1)$ be the gray area in Figure 4.6–2. When x_1 changes the "little bit" Δx, we shall write an expression for the change in A, ΔA (the area in color of Figure 4.6–2). Then

$$f(x_1)\,\Delta x \leqslant \Delta A \leqslant f(x_1 + \Delta x)\,\Delta x$$

where the area of the region in color is seen to be between the areas of the rectangles with base Δx and heights $f(x_1)$ and $f(x_1 + \Delta x)$, respectively.

Figure 4.6–2

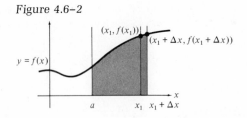

Dividing the preceding inequality by the positive number Δx, we get

$$f(x_1) \leqslant \frac{\Delta A}{\Delta x} \leqslant f(x_1 + \Delta x)$$

and taking the limit of these as $\Delta x \to 0$, we have

$$f(x_1) \leqslant A'(x_1) \leqslant f(x_1) \qquad \text{so } f(x_1) = A'(x_1)$$

We can conclude that A is an antiderivative of f.

By Basic Rule 4.1.3 it remains to find the constant of integration in order to have $A(x)$ completely determined. Since

$$A(a) = 0 = \int_a^a f(x)\,dx$$

$A(x_1)$ and $\int_a^{x_1} f(x)\,dx$ are two antiderivatives of f, which take the same value at $x_1 = a$. Letting $x_1 = b$, we have shown that the area under the curve, $A(b)$, equals $\int_a^b f(x)\,dx$.

4.6.3   **Example**

Find the area bounded by the curve $y = f(x)$, the x-axis, and the lines
$x = 0$ and $x = 1$ if
(a) $f(x) = x$
(b) $f(x) = x^2$

Solution:

(a) $A = \int_0^1 x \, dx = \tfrac{1}{2}x^2 \Big|_0^1 = \tfrac{1}{2} - \tfrac{0}{2} = \tfrac{1}{2}$

This answer can be checked by elementary geometry, since the boundary lines are straight; the area of a triangle is one-half the base times
the height.

Figure 4.6–3

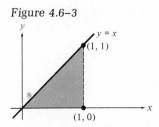

(b) $A = \int_0^1 x^2 \, dx = \tfrac{1}{3}x^3 \Big|_0^1 = \tfrac{1}{3} - \tfrac{0}{3} = \tfrac{1}{3}$

This answer cannot be checked by elementary geometry since the
boundary was curved. The use of calculus was essential. It is reassur-
ing to note, however, that the area under the parabola is less than
one-half the unit square. (See problem 3, Exercises 4.6.C.)

Figure 4.6–4

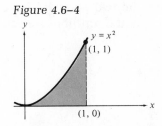

The following rule is somewhat more general than Rule 4.6.2, but it
can be proved using the same considerations.

4.6.4   **Rule**

If $f(x)$ and $g(x)$ are continuous functions with $f(x) \geqslant g(x)$ and if $a < b$,
then the area bounded by the curves $y = f(x)$ and $y = g(x)$, $x = a$, and $x = b$

is given by

$$\int_a^b [f(x) - g(x)] \, dx$$

Figure 4.6–5

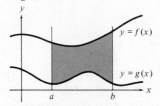

Rule 4.6.4 is an immediate consequence of Rules 4.6.2 and 4.5.8, part (c). Notice that Rule 4.6.2 is a specific case of Rule 4.6.4.

4.6.5   **Example**

Find the area bounded by $f(x) = -x^2$ and $g(x) = x^2 - 2$.

Solution:

The first step in solving such a problem is to sketch the functions. Even a very rough sketch should convince you that these two parabolas intersect in two points and therefore enclose a region whose area we can calculate. The points of intersection occur when the two functions are equal, so to find them we set the functions equal and solve for x:

$$-x^2 = x^2 - 2$$
$$0 = 2x^2 - 2 = 2(x^2 - 1) = 2(x + 1)(x - 1)$$

so $x = -1$ and $x = 1$ give the two points of intersection. Looking at the graph to see that $y = -x^2$ is the top line, we can now apply Rule 4.6.4.

Figure 4.6–6

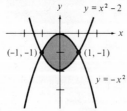

$$\int_{-1}^{1} [-x^2 - (x^2 - 2)] \, dx = \int_{-1}^{1} (-2x^2 + 2) \, dx = (-\tfrac{2}{3}x^3 + 2x)\Big|_{-1}^{1}$$
$$= (-\tfrac{2}{3} + 2) - (\tfrac{2}{3} - 2) = \tfrac{8}{3}$$

4.6.6   Rule

(a) If $f(x) \geqslant 0$ and $a < b$, then $\int_b^a f(x)\,dx$ is the negative of the area bounded by $y = f(x)$, the x-axis, $x = a$, and $x = b$. [See Rule 4.5.8(b).]

(b) If $f(x) \leqslant 0$ and $a < b$, then $\int_a^b f(x)\,dx$ is the negative of the area bounded by $y = f(x)$, the x-axis, $x = a$, and $x = b$. [See Example 4.6.7. This is another special case of Rule 4.6.4, where $f(x) = 0$.]

4.6.7   Example

Find the area from $x = 0$ to $x = 1$ bounded by the x-axis and $y = x^2 - 1$.

Solution:

A sketch shows that $x^2 - 1 \leqslant 0$ in the interval of interest, so the area is the negative of its definite integral.

Figure 4.6–7

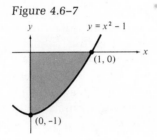

$$A = -\int_0^1 (x^2 - 1)\,dx = -\left(\frac{x^3}{3} - x\right)\Big|_0^1 = -[(\tfrac{1}{3} - 1) - 0] = \tfrac{2}{3}$$

4.6.8   Example

Find the area from $x = 0$ to $x = 1$ bounded by the x-axis and $y = x - \tfrac{1}{2}$.

Solution:

In this problem a good sketch is indispensable; it tells which boundary is "on top" in different sections of the region. If we merely integrate the whole function as one quantity, the two sections cancel each other and we get zero:

$$\int_0^1 (x - \tfrac{1}{2})\,dx = \left(\frac{x^2}{2} - \tfrac{1}{2}x\right)\Big|_0^1 = \frac{1^2}{2} - \tfrac{1}{2} = 0$$

Figure 4.6-8

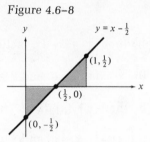

But if we separate the two parts, we get the total shaded area:

$$A = -\int_0^{1/2} \left(x - \frac{1}{2}\right) dx + \int_{1/2}^1 \left(x - \frac{1}{2}\right) dx = -\left(\frac{x^2}{2} - \frac{1}{2}x\right)\Big|_0^{1/2}$$

$$+ \left(\frac{x^2}{2} - \tfrac{1}{2}x\right)\Big|_{1/2}^1$$

$$= -\left[\frac{(\frac{1}{2})^2}{2} - \frac{1}{4}\right] + \left[\left(\frac{1}{2} - \frac{1}{2}\right) - \left(\frac{(\frac{1}{2})^2}{2} - \frac{1}{4}\right)\right]$$

$$= -\left(\frac{1}{8} - \frac{1}{4}\right) - \left(\frac{1}{8} - \frac{1}{4}\right) = \frac{1}{4}$$

The selling price of a commodity depends upon supply and demand. Let $S(q)$ and $D(q)$ [called $p(q)$ earlier] denote the supply and demand functions, respectively, for a given commodity. Then $S(q)$ must be increasing and $D(q)$ must be decreasing. Therefore, they intersect at precisely one point, (q^*, p^*), called the equilibrium point. The equilibrium point tells the market price p^* for which the quantity demanded equals the quantity supplied. It is written (q^*, p^*), where q^* is this quantity and $p^* = D(q^*) = S(q^*)$.

Consumer surplus is a useful economic concept defined as

$$\int_0^{q^*} D(q)\, dq - q^* p^* = \int_0^{q^*} D(q)\, dq - q^* D(q^*)$$

Geometrically, it is the area above the line $p = p^* = D(q^*)$ and below $p = D(q)$. (See Figure 4.6-9.) Notice that the term $q^* D(q^*)$ of this expression is precisely the total revenue when q^* items are bought at price $p^* = D(q^*)$. At the end of Section 4.7 we shall show that the first term (the integral) is approximately the total revenue when q^* items are bought and the qth item (for $q \le q^*$) is bought at the price $D(q)$.

Similarly, the producer surplus is defined to be

$$q^* p^* - \int_0^{q^*} S(q)\, dq = q^* S(q^*) - \int_0^{q^*} S(q)\, dq$$

the area under $p = p^* = S(q^*)$ and above $p = S(q)$.

In economics courses you may study more about the uses of producer and consumer surplus. Now we consider how they can be computed using integrals.

Figure 4.6-9

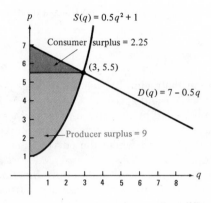

4.6.9 Example

Find the value of consumer surplus and producer surplus if the demand function is $D(q) = 7 - 0.5q$ and the supply function is $S(q) = 0.5q^2 + 1$.

Solution:
The first step is to find the equilibrium point—that q^* for which $S(q^*) = D(q^*)$. To do this we set the supply function equal to the demand function and solve for q.

$$0.5q^2 + 1 = 7 - 0.5q \qquad \text{so } 0.5q^2 + 0.5q - 6 = 0$$

or

$$0 = q^2 + q - 12 = (q - 3)(q + 4) \qquad \text{so } q = 3 \text{ or } q = -4$$

Since negative values for q make no sense in this context, we must have $q = 3$. Since $q^* = 3$, $p^* = D(q^*) = S(q^*) = 5.5$.

$$\text{Consumer surplus} = \int_0^3 (7 - 0.5q)\, dq - 3(5.5)$$
$$= 18.75 - 16.5 = 2.25$$

$$\text{Producer surplus} = 3(5.5) - \int_0^3 (0.5q^2 + 1)\, dq = 16.5 - 7.5 = 9$$

SUMMARY

In this section we interpreted the definite integral as an area and showed how we can calculate the area of regions with curved boundaries using integral calculus. We used this interpretation of the definite integral to

compute consumer surplus and producer surplus, two important economic concepts.

EXERCISES   4.6. A

1.   Find the area in the regions between the x-axis, the lines $x = -5$ and $x = +5$,
and the curve
(a) $y = x + 10$ (d) $y = 2x - 1$
(b) $y = x^2$ (e) $y = x^3$
(c) $y = x^2 + 1$
2.   Find the area in the region between the curves $f(x)$ and $g(x)$ from $x = 0$ to
$x = 5$ if $f(x) = x^2 + 1$ and $g(x) = x$.
3.   Find the area bounded by the curves $y = x$ and $y = x^2$.
4.   Find the area in the region bounded by the curves $f(x) = x^2 + 3x + 1$ and
$g(x) = x + 4$. (*Hint:* Find where the curves intersect and determine which
curve is on top.)
5.   Graph the following demand and supply functions. Find the equilibrium
point and shade the consumer and producer surplus regions. Find the value
of consumer and producer surplus. $S(q) = q + 1$; $D(q) = 10 - q$; $q \geqslant 0$.

EXERCISES   4.6. B

1.   Find the area in the regions between the x-axis, the lines $x = -2$ and $x = +2$,
and the curve
(a) $y = x + 10$ (d) $y = 2x - 1$
(b) $y = x^2$ (e) $y = x^3$
(c) $y = x^2 + 1$
2.   Find the area in the region between the curves $f(x)$ and $g(x)$ from $x = 0$ to
$x = 5$ if $f(x) = x^2$ and $g(x) = x - 1$.
3.   Find the area bounded by the curves $y = x^3$ and $y = x$ from $x = 0$ to $x = 1$.
4.   Find the area in the region bounded by the curves $f(x) = x^2 + x$ and $g(x) =
x + 1$.
5.   Graph the following demand and supply functions. Find the equilibrium
point and shade the consumer and producer surplus regions. Find the value
of consumer and producer surplus. $S(q) = q^2 + 1$; $D(q) = 7 - q$; $q \geqslant 0$.

EXERCISES   4.6. C

1.   Find the area bounded by the curves $y = x^2 + 2x$ and $y = -x^2 + 2x + 8$.
2.   Find the area bounded by the curves $y = x$ and $y = x^3$.
3.   Prove that if $y = px^2$ is any parabola with vertex at the origin and concave
upward, the area under the parabola between $x = 0$ and $x = b$ is one-third
the area of the enveloping rectangle. In other words, the shaded area in the
figure is always $\frac{1}{3} \cdot b \cdot pb^2 = pb^3/3$.

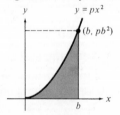

4. (a) The <u>average value of a continuous function</u> $y = f(x)$ over an interval $a \leqslant x \leqslant b$ can be pictured as that value $f(x_0)$ attained in the interval such that the area under the curve is equal to the area $(b - a) \cdot f(x_0)$ of the rectangle in the accompanying figure. Show that the average value over the interval $a \leqslant x \leqslant b$ of $f(x)$ is $[1/(b - a)] \int_a^b f(x)\, dx$.

    (b) If a certain population of rabbits grows according to the formula $P(t) = 10 + 6t^2 - 2t$ over a period of 5 years, find the average number of rabbits over this period of time starting at $t = 0$.

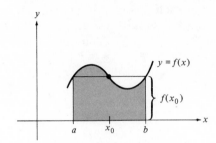

ANSWERS   4.6. A

1. (a) 100 (b) $83\frac{1}{3}$ (c) $93\frac{1}{3}$ (d) $50\frac{1}{2}$ (e) $312\frac{1}{2}$
2. $\frac{205}{6}$
3. $\frac{1}{6}$
4. $10\frac{2}{3}$
5. Equilibrium point: $q^* = \frac{9}{2}$, $p^* = \frac{11}{2}$; consumer surplus = producer surplus = $10\frac{1}{8}$

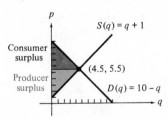

ANSWERS   4.6. C

1. 64/3
2. $\frac{1}{2}$
4. (a) $(b - a)f(x_0) = \int_b^a f(x)\, dx$ (b) $[1/(5 - 0)] \int_0^5 (10 + 6t^2 - 2t)\, dt = 55$

4.7   The Definite Integral as the Limit of a Sum

We now turn to yet another approach to the definite integral, one that has widespread applications in the financial world and elsewhere. The integral can be used to add up a large number (actually an infinite number) of numbers.

4.7.1   Illustration

To begin this discussion, we consider how to approximate the area under the curve in Figure 4.7–1 with a large number of thin rectangles. Assume that $f(x) > 0$ between a and b. Divide the interval from a to b into n equal subintervals so that each has length $\Delta x = (b - a)/n$. We let x_k denote the left endpoint of the kth subinterval and let the kth rectangle have height $f(x_k)$; each base is $\Delta x = (b - a)/n$. The sum of the n rectangles so described

Figure 4.7–1

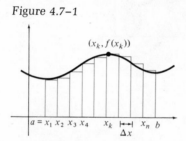

is "near" the area given by the integral $\int_a^b f(x)\,dx$. To approximate the area under the curve, we can sum the rectangles. Using the "sigma" notation, which is customary for large sums, we have

$$A \approx f(x_1)\,\Delta x + f(x_2)\,\Delta x + f(x_3)\,\Delta x + \cdots + f(x_n)\,\Delta x = \sum_{k=1}^{n} f(x_k)\,\Delta x$$

4.7.2   Example

Approximate the area A between the curve $y = 3x^2$ and the x-axis from $x = -5$ to $x = 5$ using
(a) $n = 5$ (see Figure 4.7–2)
(b) $n = 10$

Solution:

(a) If $n = 5$, we have that $\Delta x = [5 - (-5)]/5 = 2$, so each subinterval is of length 2.

$$\begin{aligned}
A &\approx 2f(-5) + 2f(-3) + 2f(-1) + 2f(1) + 2f(3) \\
&= 6(-5)^2 + 6(-3)^2 + 6(-1)^2 + 6(1)^2 + 6(3)^2 \\
&= 270
\end{aligned}$$

Figure 4.7–2

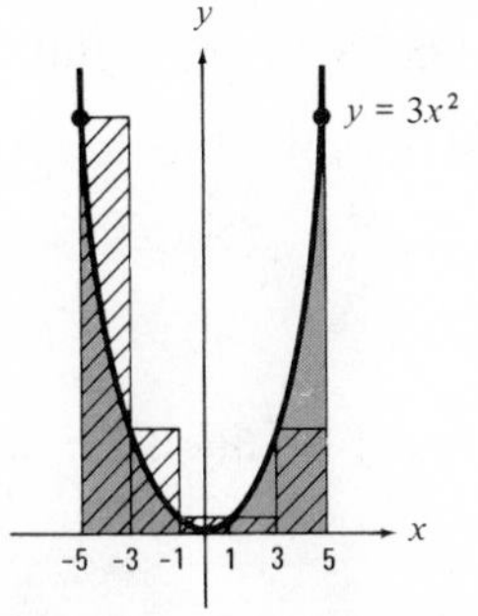

(b) If $n = 10$, we have that $\Delta x = [5 - (-5)]/10$, so this time each sub-interval has length 1.

$$A \approx f(-5) \cdot 1 + f(-4) \cdot 1 + f(-3) \cdot 1 + \cdots + f(3) \cdot 1 + f(4) \cdot 1$$
$$= 3(-5)^2 + 3(-4)^2 + 3(-3)^2 + \cdots + 3(3)^2 + 3(4)^2 = 255$$

As a comparison, we can note that

$$A = \int_{-5}^{5} 3x^2 \, dx = x^3 \Big|_{-5}^{5} = 250$$

4.7.3   Example

Approximate the area between the curve $y = x + 1$ and the x-axis from $x = 0$ to $x = 12$ using
(a) $n = 4$
(b) $n = 24$

Solution:
(a) $\Delta x = (12 - 0)/4 = 3$, so

$$A \approx (0 + 1) \cdot 3 + (3 + 1) \cdot 3 + (6 + 1) \cdot 3 + (9 + 1) \cdot 3 = 66$$

(b) $\Delta x = (12 - 0)/24 = \frac{1}{2}$, so

$$A \approx (0 + 1)\tfrac{1}{2} + (\tfrac{1}{2} + 1)\tfrac{1}{2} + (1 + 1)\tfrac{1}{2} + (\tfrac{3}{2} + 1)\tfrac{1}{2} + \cdots$$
$$+ (\tfrac{23}{2} + 1)\tfrac{1}{2} = 81$$

As a comparison, we calculate

$$\int_{0}^{12} (x + 1) \, dx = \left(\frac{x^2}{2} + x\right)\Big|_{0}^{12} = 84$$

4.7.4   Example

Apply the above summation process to the function $f(x) = 3x^2 - 2$ from $x = -5$ to $x = 5$ using

(a) $n = 5$

(b) $n = 10$. [Notice that $\int_{-5}^{5} (3x^2 - 2)\, dx = 230$ is *not* the area between the curve and the x-axis, merely the total change of the function $x^3 - 2x$ as one moves from -5 to 5.]

Solution:

(a) $\sum\limits_{k=1}^{5} f(x_k)\, \Delta x = [3(-5)^2 - 2] \cdot 2 + [3(-3)^2 - 2] \cdot 2 + \cdots$

$$+ (3 \cdot 3^2 - 2) \cdot 2 = 250$$

(b) $\sum\limits_{k=1}^{10} f(x_k)\, \Delta x = [3(-5)^2 - 2] \cdot 1 + \cdots + (3 \cdot 4^2 - 2) \cdot 1 = 235$

Figure 4.7–3

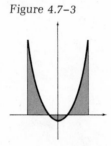

In this case the sum is an approximation to "total change" of the antiderivative of f. The approximation gets better (by which we mean the value of the sum gets closer to the integral) as n increases.

Example 4.7.4 suggests, correctly, that the idea underlying approximation of areas by rectangles can be applied to total change — any definite integral — as well. This notion is so important that it is called the "fundamental theorem of calculus."

4.7.5   Fundamental Theorem of Calculus

Let $f(x)$ be continuous in the interval $a \leqslant x \leqslant b$. Let this interval be divided into n subintervals of length $(b - a)/n = \Delta x_n$. Let x_k be any value of the variable x in the kth subinterval. Then the limit of the sum

$$\sum\limits_{k=1}^{n} f(x_k) \frac{b - a}{n} = \sum\limits_{k=1}^{n} f(x_k)\, \Delta x_n$$

as n increases without bound equals the definite integral

$$\int_{a}^{b} f(x)\, dx = F(b) - F(a)$$

where $F(x)$ is an antiderivative of $f(x)$.

[Incidentally, $\sum_{k=1}^{n} f(x_k)\, \Delta x_n$ is called a <u>Riemann sum</u>.] This theorem has many uses. One is to calculate the value of a definite integral of a function whose antiderivative is unknown.

4.7.6   Example

The marginal revenue function for producing x units of a product is $\$100[x/(x + 1)]$. Approximate the total change in the revenue if we increase production from a level of 2 to a level of 6 units.

Solution:

Using $n = 4$ subintervals, we obtain the approximation

$$\sum_{k=1}^{4} f(x_k)\, \Delta x_4 = f(2) \cdot 1 + f(3) \cdot 1 + f(4) \cdot 1 + f(5) \cdot 1$$

$$= 100 \cdot \tfrac{2}{3} + 100 \cdot \tfrac{3}{4} + 100 \cdot \tfrac{4}{5} + 100 \cdot \tfrac{5}{6} = 305$$

It happens that $F(x) = 100(x - \ln |x + 1|)$ is an antiderivative of $f(x) = 100[x/(x + 1)]$. You can check this by differentiation. Hence

$$\int_{2}^{6} 100\, \frac{x}{x + 1}\, dx = 100(x - \ln |x + 1|) \Big|_{2}^{6}$$

$$= 100 \cdot [(6 - \ln 7) - (2 - \ln 3)] \approx 315$$

The error by using the approximation is, therefore, less than 3.5 percent, although n is relatively small.

The antiderivative in Example 4.7.6 may not have occurred to you, in which case it would have been easier to do the summation than the integration. Some important functions do not have an antiderivative in closed form at all. (We devote Section 5.4 to the study of some such functions.) In such cases one *must* approximate the area. Even so, we might want to speak of an integral corresponding to the area, so we now expand the definition of "definite integral" to include these cases.

4.7.7   Definition

Let $f(x)$ be defined and continuous in the interval $a \leqslant x \leqslant b$, and let this interval be divided into n subintervals of length $\Delta x_n = (b - a)/n$. Let x_k be any value of the variable x in the kth subinterval. Then the limit of the sum $\sum_{k=1}^{n} f(x_k)\, \Delta x_n$ as n increases without bound is defined to be the <u>definite integral of $f(x)$ from a to b</u>. As before, it is denoted by $\int_{a}^{b} f(x)\, dx$.

If $f(x)$ has an antiderivative in closed form, this definition is equivalent to the first definition of "definite integral" given in Section 4.5 on page 230 because of the Fundamental Theorem of Calculus. So we have not thrown anything away. But the new definition is useful in more situations, as the following examples show.

4.7.8   Example

A stream of capital is flowing continuously into an investment fund at the rate of p dollars per year. The investments are compounded continuously at a rate of r percent per year. Find the value A_t of the investment at the end of t years.

Solution:

Divide each year into n equal parts and think of investing $\dfrac{p}{n} = \dfrac{pt}{nt}$ at the beginning of each part. Then the kth payment, made at time x_k, will be compounded continuously for $t - x_k$ years and will accrue to the sum of $\dfrac{p}{n} e^{r(t-x_k)}$. Summing up all $n \cdot t$ of these investments, we have a total value of

$$\sum_{k=1}^{nt} \frac{p}{n} e^{r(t-x_k)} = \sum_{k=1}^{nt} p e^{r(t-x_k)} \, \Delta x_n$$

where $\Delta x_n = 1/n$. Letting n increase without bound, this sum approaches the integral

$$A_t = \int_0^t p e^{r(t-x)} \, dx = \frac{p}{r} (e^{rt} - 1)$$

4.7.9   Example

Approximate the amount accruing to an investor if he invests $100 per year for 10 years compounded continuously at 5 percent per year.

Solution:
Using the formula at the end of Example 4.7.8 we have

$$A_{10} = \frac{100}{0.05} [e^{0.05(10)} - 1] = 2{,}000(e^{0.5} - 1) \approx \$1{,}297$$

In a real situation the money is not being pumped into an investment in a uniform manner, so the integral is only an approximation. If the $100 is invested at the beginning of each year, we say that A_{10} represents the amount of an annuity.

4.7.10   Example

Recall that if $p(q)$ [or, as in the previous section, $D(q)$] is a price function, giving the price per unit when q units are demanded, then $q \cdot p(q)$ is the revenue obtained by buying q units at price $p(q)$.

We say that a seller practices <u>perfect price discrimination</u> if he sells each unit or fractional part thereof at the price determined by $p(q)$. In other words, he sells the first unit at price $p(1)$, the second unit at price $p(2)$, and the kth unit at price $p(k)$. Show that the total seller's revenue under perfect price discrimination for q^* units is given by $\int_0^{q*} p(q)\, dq$.

Solution:

If all the units are whole numbers, we have the sum $R = \sum_{n=1}^{q*} p(n)$ as the total revenue. Dividing each unit into k equal parts, we have that

$$R = \sum_{n=1}^{q*k} p\left(\frac{n}{k}\right) \frac{1}{k} = \sum_{n=1}^{q*k} p\left(\frac{n}{k}\right) \Delta x_k \qquad \text{where } \Delta x_k = \frac{1}{k}$$

When k increases without bound, this sum approaches the integral $\int_0^{q*} p(q)\, dq$, so the problem is solved.

Notice that as on page 242, the <u>consumer surplus</u> is the difference between $\int_0^{q*} p(q)\, dq$ and $q^* p(q^*)$.

SUMMARY

In this section we showed that integrals and sums can be used to approximate each other. (Indeed, in many books the definite integral is defined to be the limit of sums as all Δx's get "very small.") We stated without proof the fundamental theorem of calculus, which says that the limit of sums is closely related to the derivative. Understanding the definite integral as the limit of a sum is essential in certain economic derivations, such as producer and consumer surplus and continuous annuities.

EXERCISES   4.7. A

1. Approximate the area between the function $y = 2x - 1$ and the x-axis from $x = 1$ to $x = 7$ using left endpoints of each subinterval and
    (a) $n = 2$ (c) $n = 6$
    (b) $n = 3$ (d) What is the exact area of this region?
2. Approximate the area between the function $y = x/(x^2 + 1)$ and the x-axis from $x = 0$ to $x = 4$ by using
    (a) $n = 2$ (c) What is the correct area of this region?
    (b) $n = 4$

3. Approximate the value of an annuity after 15 years if $60 is invested at the beginning of each year and it is compounded continuously at 3 percent.
4. Find the producer surplus if $S(q) = 50 + (q^2/4)$ and $q^* = 20$.
5. Find the consumer surplus if $D(q) = 150 - 0.3q^2$ and $q^* = 10$.

HC 6. In problem 2 approximate the area by using $n = 8$. Compute your answer to two decimal places. Is the approximation nearer to the correct area, computed in part (c), than those obtained in parts (a) and (b)?

x	0	0.5	1	1.5	2	2.5	3	3.5
$\dfrac{x}{x^2 + 1}$								

EXERCISES 4.7. B

1. Approximate the area between the function $y = 2x + 1$ and the x-axis from $x = 1$ to $x = 7$ using left endpoints of each subinterval and
    (a) $n = 2$ (c) $n = 6$
    (b) $n = 3$ (d) What is the exact area of this region?
2. Approximate the area between the function $y = 2x/(x^2 + 1)$ and the x-axis from $x = 0$ to $x = 4$ by using
    (a) $n = 2$ (c) What is the correct area of this region?
    (b) $n = 4$
3. Approximate the value of an annuity after 10 years if $60 is invested at 2 percent.
4. Find the consumer surplus if $D(q) = 100 - 2q$ and $q^* = 10$.
5. Find the producer surplus if $S(q) = 50 + 2q$ and $q^* = 20$.

HC 6. In problem 2 approximate the area by using $n = 8$. Compute your answer to two decimal places. Is the approximation nearer to the correct area, computed in part (c), than those obtained in parts (a) and (b)?

x	0	0.5	1	1.5	2	2.5	3	3.5
$\dfrac{2x}{x^2 + 1}$								

EXERCISES 4.7. C

1. Approximate the area between the curve $x = y^3$ and the y-axis from $y = 0$ to $y = 6$ using
    (a) $n = 2$ (c) $n = 6$
    (b) $n = 3$ (d) What is the exact area of this region?
2. Approximate the value of an annuity at the end of 20 years if $100 is invested at the beginning of each year and it is compounded continuously at 3 percent.
3. Approximate the present value of a continuous stream of capital if $1,000 per year is invested for 20 years at the rate 4 percent compounded continuously.

4. Psychologists sometimes try to predict the time that it will take a group to perform a given task. Generally speaking, the time required to perform a given task decreases with additional successful attempts at that task. Suppose that it has been found that the time needed for performing the task the xth time is $f(x) = 10 + 200x^{-2}$ minutes. Show that $\int_1^6 (10 + 200x^{-2})\, dx$ is an approximation to the total time needed to perform the task 5 times.

5. The average value of a continuous function $y = f(x)$ over an interval can be obtained using the limit of a Riemann sum: Divide up the interval $a \leq x \leq b$ into n subintervals of length x. Evaluate $f(x)$ at the point x_i in the ith subinterval as $y_i = f(x_i)$. Then take the limit as x goes to zero of the weighted averages

$$\frac{y_1\, \Delta x + y_2\, \Delta x + \cdots + y_n\, \Delta x}{\Delta x + \Delta x + \cdots + \Delta x}$$

(a) Show that this limit $= \dfrac{1}{b-a} \int_a^b f(x)\, dx = \underset{a \leq x \leq b}{\text{average } f(x)}$.

(b) The amount of a pollutant in the air over a period of 10 years was given by the function $P(t) = 100 - 10t$ tons from $t = 0$ to $t = 10$. Find the average amount of pollution in the air over this 10-year period.

ANSWERS 4.7. A

1. (a) 24   (b) 30   (c) 36   (d) 42
2. (a) 0.8   (b) 1.2   (c) ln (17)/2 ≈ 1.42
3. $1,137

4. $1,333⅓
5. $200
6. 1.34; yes

ANSWERS 4.7. C

1. (a) 81   (b) 144   (c) 225   (d) 324
2. $2,740
3. $\dfrac{1,000}{0.04} \left[e^{0.04(20)} - 1 \right] \approx 30{,}638$

4. $\int_1^6 (10 + 200x^{-2})\, dx = 216\frac{2}{3} \approx (10 + 200 \cdot 1^{-2}) \cdot 1 + (10 + 200 \cdot 2^{-2}) \cdot 1 +$
    $(10 + 200 \cdot 3^{-2}) \cdot 1 + (10 + 200 \cdot 4^{-2}) \cdot 1 + (10 + 200 \cdot 5^{-2}) \cdot 1$

5. (a) $\lim \dfrac{y_1\, \Delta x + y_2\, \Delta x + \cdots + y_n\, \Delta x}{\Delta x + \Delta x + \cdots + \Delta x}$

    $= \dfrac{\lim (y_1\, \Delta x + y_2\, \Delta x + \cdots + y_n\, \Delta x)}{b - a} = \dfrac{1}{b-a} \int_a^b f(x)\, dx$

    (b) $\dfrac{1}{10 - 0} \int_0^{10} (100 - 10t)\, dt = 50$ tons

SAMPLE QUIZ   ## Sections 4.4, 4.5, 4.6, and 4.7

Do either part (a) or part (b) of each of the following four problems. Each problem [part (a) or (b)] has a value of 25 points.

1. (a) Solve the following differential equations:

$$\frac{dy}{dx} = x^3 y^4 \qquad \frac{y'}{y} = 0.05$$

(b) Suppose that the growth rate y'/y of a stock is 5 percent and its value at time $t = 0$ is \$200. Write an equation for its value at time t. What will be its value after 20 years?

2. (a) Find the total change in the cost as the production goes from $x = 50$ to $x = 100$ if $C'(x) = 2 - 0.01x$.

(b) $\displaystyle\int_{-1}^{2} x e^{x^2}\, dx =$

3. (a) Find the area under the curve $y = x^2(x^3 + 2)^3$ (that is, between this curve and the x-axis) between the lines $x = 0$ and $x = 1$. (Do not try to graph this function.)

(b) Find the area between the curves $f(x) = -x^2$ and $g(x) = x - 6$. Graph these functions first.

4. (a) If $S(q) = 2q + 1$ and $D(q) = 9 - q^2$ are the supply and demand functions, respectively, find the equilibrium point and the consumer surplus.

(b) Approximate the area between the function $y = x^2$ and the x-axis between the lines $x = 1$ and $x = 2$ using $n = 4$ intervals and left endpoints.

The answers are at the back of the book.

SAMPLE TEST   # Chapter 4

1. (15 pts) Suppose that the marginal revenue is given by $R'(x) = 2x - \frac{2}{3}$ and the revenue obtained from selling one item is 10. [$R(1) = 10$.] Write an equation for the revenue. What is $R(4)$?

2. (a) (15 pts) $\displaystyle\int \left(x^4 + x^{1/3} + \frac{3}{x} \right) dx = $ _____.

(b) (10 pts) $\displaystyle\int e^{-3x}\, dx = $ _____.

(c) (10 pts) $\displaystyle\int (x^2 + 7)^4 x\, dx = $ _____.

(d) (10 pts) $\displaystyle\int x e^{-6x}\, dx = $ _____.

3. (10 pts) Suppose that the growth rate of a stock is 8 percent. If its value at time $t = 0$ is \$300, write an equation for its value at time t. What will be its value after 12.5 years?

4. (15 pts) Find the area between the curves $y = x^2$ and $y = 2x + 3$. Graph the functions first.

5. (15 pts) Approximate the area under the curve $x^2 + 1$ between $x = 0$ and $x = 4$ using left endpoints and $n = 4$ subintervals. Then find the exact area.

The answers are at the back of the book.

Chapter 5

INTEGRATION: PART II

5.1   Improper Integrals

It is often useful to be able to find the area between a curve and the x-axis for a region whose "base" is infinitely long. This is true in statistics and in mathematical models having to do with projections into the future. It may seem that such an area would surely be infinite. In some cases this is correct, but in others we can take a limit and get a finite answer. (You may want to review Section 1.5 now, since the ideas there are basic to this section.) Consider the following example.

5.1.1   Example

Find the area between the curve $y = e^{-x}$ and the x-axis to the right of the y-axis.

Figure 5.1–1

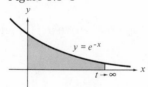

Solution:

Although the function is above the x-axis for all x, the curve approaches the x-axis ever more closely as x increases in size. (We sometimes say that the x-axis is an "asymptote" to the curve $y = e^{-x}$.) If we were to consider the area only to the left of a point $x = t$, the integral would be

$$\int_0^t e^{-x}\, dx = -e^{-x}\Big|_0^t = -e^{-t} - (-e^{-0}) = -\frac{1}{e^t} + 1$$

Letting t increase without bound (letting $t \to \infty$), we have

$$\lim_{t \to \infty} \int_0^t e^{-x}\, dx = \lim_{t \to \infty} \left(1 - \frac{1}{e^t}\right) = 1 - 0 = 1$$

Therefore, we write $\int_0^\infty e^{-x}\, dx = 1$. The area is equal to 1!

5.1.2   Definition

An integral of the form

$$\int_a^\infty f(x)\, dx = \lim_{t \to \infty} \int_a^t f(x)\, dx$$

or

$$\int_{-\infty}^b f(x)\, dx = \lim_{t \to -\infty} \int_t^b f(x)\, dx$$

or

$$\int_{-\infty}^\infty f(x)\, dx = \int_{-\infty}^c f(x)\, dx + \int_c^\infty f(x)\, dx$$

is called an <u>improper integral</u>. (There are other kinds of improper integrals not mentioned in this book because they are not often used in biology and the social sciences.) If the limit is infinite, we say that the integral is divergent or does not exist. In particular, if either $\int_{-\infty}^c f(x)\, dx$ or $\int_c^\infty f(x)\, dx$ or both are infinite, then $\int_{-\infty}^\infty f(x)\, dx$ is divergent.

5.1.3   Example

Let $f(x) = 1/\sqrt{x}$ and find the area between $f(x)$ and the x-axis to the right of the line $x = 1$.

Solution:

Again the curve is always above the x-axis in the region of interest and again it approaches the x-axis as x increases—but not quickly enough. The area, if it exists, equals

$$\int_1^\infty x^{-1/2}\, dx = \lim_{t \to \infty} \int_1^t x^{-1/2}\, dx = \lim_{t \to \infty} 2x^{1/2}\Big|_1^t = \lim_{t \to \infty} (2\sqrt{t} - 2) = \infty$$

Figure 5.1–2

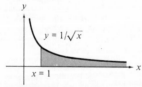

Hence the integral diverges and the area is infinite.

5.1.4   Example

Find the value of the following improper integrals.

(a) $\displaystyle\int_0^\infty e^{-2x}\,dx$

(b) $\displaystyle\int_0^\infty \frac{1}{(x+1)^2}\,dx$

(c) $\displaystyle\int_1^\infty \frac{1}{x^{1/3}}\,dx$

Solution:

(a) $\displaystyle\int_0^\infty e^{-2x}\,dx = \lim_{t\to\infty}\int_0^t e^{-2x}\,dx = \lim_{t\to\infty} -\tfrac{1}{2}e^{-2x}\Big|_0^t = \lim_{t\to\infty} -\tfrac{1}{2}(e^{-2t} - e^0)$

$$= \lim_{t\to\infty} -\tfrac{1}{2}\left(\frac{1}{e^{2t}} - 1\right) = -\tfrac{1}{2}(-1) = \tfrac{1}{2}$$

(b) $\displaystyle\int_0^\infty \frac{dx}{(x+1)^2} = \lim_{t\to\infty}\int_0^t \frac{dx}{(x+1)^2} = \lim_{t\to\infty} -(x+1)^{-1}\Big|_0^t$

$$= \lim_{t\to\infty} -\left(\frac{1}{t+1} - 1\right) = -(-1) = 1$$

(c) $\displaystyle\int_1^\infty \frac{dx}{x^{1/3}} = \lim_{t\to\infty}\int_1^t x^{-1/3}\,dx = \lim_{t\to\infty} \tfrac{3}{2}x^{2/3}\Big|_1^t = \lim_{t\to\infty} (\tfrac{3}{2}t^{2/3} - \tfrac{3}{2}) = \infty$, so this integral diverges.

5.1.5   Example in Ecology

Suppose that a certain amount A_0 of radioactive material is released into the atmosphere each year. If the material decays according to the formula $A = A_0 e^{-kt}$, (a) show that the amount of this material in the air at time t is $\dfrac{A_0}{k}\left(1 - \dfrac{1}{e^{kt}}\right)$. (b) Show that the total accumulation of this material will approach $\dfrac{A_0}{k} = \displaystyle\int_0^\infty A_0 e^{-kx}\,dx$.

Solution:

(a) As in Example 4.7.8, we divide each year into n parts and think of accumulating A_0/n at the beginning of each part. Then the ith addition of radioactive material, made at time x_i, will have a value of $(A_0/n)e^{-k(t-x_i)}$. Letting $\Delta x = 1/n$ and summing over i, the total of these amounts is

$$\sum_{i=1}^{nt} A_0 e^{-k(t-x_i)}\,\Delta x$$

and letting $n \to \infty$, this sum becomes the integral $\int_0^t A_0 e^{-k(t-x)} \, dx =$
$$\frac{A_0}{k} e^{-k(t-x)} \Big|_0^t = \frac{A_0}{k} \left(1 - \frac{1}{e^{kt}} \right).$$

(b) As $t \to \infty$, $\frac{1}{e^{kt}} \to 0$. Hence, $\lim_{t \to \infty} \frac{A_0}{k} \left(1 - \frac{1}{e^{kt}} \right) = \int_0^\infty A_0 e^{-kt} \, dt = \frac{A_0}{k}$.

SUMMARY

Improper integrals involve calculating the area of a region whose boundary is infinite. The area may or may not be infinite, depending on whether certain limits exist as x "goes to infinity." If the area exists, we say the integral "converges." In this section it is useful to know that

$$\lim_{t \to \infty} \frac{1}{t^n} = 0 \qquad \text{if } n > 0$$

$$\lim_{t \to \infty} e^{-kt} = 0 \qquad \text{if } k > 0$$

$$\lim_{t \to \infty} te^{-kt} = 0 \qquad \text{if } k > 0$$

EXERCISES   5.1. A

1. Find the value of the following improper integrals, if they exist.

(a) $\int_3^\infty \frac{dx}{x^3}$

(d) $\int_0^\infty e^{-3x} \, dx$

(b) $\int_0^\infty 2xe^{-x^2} \, dx$

(e) $\int_2^\infty \frac{dx}{(1-x)^{5/3}}$

(c) $\int_1^\infty \frac{dx}{x}$

2. Suppose that one ton of strontium 90, a particularly dangerous kind of nuclear fallout, is released into the atmosphere each year. If the formula for the decay of strontium 90 is given by $A = A_0 e^{-kx}$ with $k \approx 0.024$, find the total limiting accumulation of strontium 90.

3. Find the area between the curve $y = 1/x^2$ and the x-axis to the right of the line whose equation is $x = 3$.

4. Solve for k: $\int_0^\infty 4e^{-kx} \, dx = 1$.

5. Find a value of t such that $\int_0^t e^{-x} \, dx > 0.9$.

EXERCISES   5.1. B

1. Find the value of the following improper integrals, if they exist.

(a) $\int_2^\infty \frac{1}{x^2} \, dx$

(c) $\int_1^\infty \frac{dx}{\sqrt[3]{x}}$

(e) $\int_2^\infty \frac{dx}{(x+3)^{4/3}}$

(b) $\int_0^\infty 3x^2 e^{-x^3} \, dx$

(d) $\int_0^\infty e^{-2x} \, dx$

2. Referring to Exercise 5.1.A, number 2, above, find the accumulation of stron-
    tium 90 after (a) 10 years; (b) 100 years.

3. Find the area between the curve $y = 1/x^4$ and the x-axis to the right of the
    line $x = 4$.

4. Solve for k: $\int_0^\infty ke^{-5x}\,dx = 1$.

5. Find a value of t such that $\int_0^t e^{-x}\,dx$ is greater than 0.95.

EXERCISES   5.1. C

1. The improper integral $\int_0^\infty e^{-x^2}\,dx$ is very important in statistics. Convince

    yourself that this integral exists by comparing it with $\int_0^\infty e^{-x}\,dx$. It can be

    shown that $\int_0^\infty e^{-x^2}\,dx = \tfrac{1}{2}\sqrt{\pi}$. If we want to approximate this integral for

    other limits, we must use sums as in Section 4.7; we shall see more about
    this technique in Section 5.4. Meanwhile, graph $y = e^{-x^2}$ using the tech-
    niques of Chapter 2; this will help you to better understand this important
    function.

2. $\int_{-\infty}^\infty 2xe^{-x^2}\,dx =$

ANSWERS   5.1. A

1. (a) $\frac{1}{18}$ (b) 1 (c) does not exist (d) $\frac{1}{3}$ (e) $-\frac{3}{2}$
2. about $41\frac{2}{3}$ tons
3. $\frac{1}{3}$
4. $k = 4$
5. $t = 2.4$ or anything larger

ANSWERS   5.1. C

1. $e^{-x^2} < e^{-x}$ if $x > 1$ (see answer 5 on page 179)
2. 0 because $\int_0^\infty 2xe^{-x^2}\,dx = 1$ and $\int_{-\infty}^0 2xe^{-x^2}\,dx = -1$

5.2   Integrals as Measures of Probability

Statistics and probability are used to analyze many practical situations,
including business forecasting, customer waiting times, and quality control. This book makes no effort to explain these topics in detail, but the

next three sections will include enough brief definitions so that the student can use his skill in calculus to compute answers to statistical problems that require an application of the definite integral.

5.2.1   Illustration

Probably the best known graph in all of statistics is the "bell-shaped curve," shown in Figure 5.2–1. This curve is often used to describe phenomena that are randomly distributed in nature. The particular curve

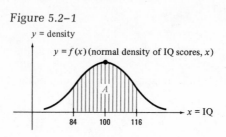

Figure 5.2–1

y = density

$y = f(x)$ (normal density of IQ scores, x)

A

x = IQ

84   100   116

in Figure 5.2–1 describes the distribution of human intelligence as given by IQ scores. However, IQ scores are usually given in whole numbers, whereas the bell-shaped curve is continuous — hence only a rough model. Moreover, the height $f(x)$ indicates density of concentration rather than frequency of each score x. Thus a score of $x = 100$ has the highest density; that is, it is the "most typical."

Even though an "average" or typical IQ (as measured by some tests) is about 100, the likelihood of a randomly selected person having an IQ of exactly 100 is practically nil, as given by our curve. On a continuous scale, there are infinitely many IQ scores that an individual could have very close to, but not exactly, 100.

Even "very small" quantities may add up in a significant way if there are enough of them. Indeed, the probability that a randomly selected individual has an IQ score *between* 84 and 116, for example, is quite large; in fact, it is better than $\frac{6}{10}$. This is reflected in Figure 5.2–1 by the fact that the region A shaded under the curve is more than 60 percent of the area in the (unbounded) region under the entire curve.

It is natural to think of area in this context. We are concerned with adding up many very small quantities, the probabilities of all scores between 84 and 116. We have done such summations in the past with a definite integral (see the Fundamental Theorem of Calculus, page 248), which, with a *positive* function such as the bell-shaped curve, gives the area under the curve. Since the area of A is about two-thirds of the total area under the curve, we say that this represents a probability of about $\frac{2}{3}$.

Summing up, in our example the variable x labels individuals according to their IQ scores. The function $f(x)$ indicates the concentration or density of some label x in a population — but not the likelihood. Likelihood, or probability, corresponds to the *area* under the curve $y = f(x)$. We shall now discuss these ideas more generally.

By a <u>random variable</u> x, we shall mean a measurable quantity that varies continuously over an interval $a \le x \le b$, where either a or b might be infinite. (In some texts x is allowed to take on only whole-number values — but not here.) A random variable x might represent the IQ scores of some group of students, the height of patrons of a clothing store, the interval of arrival times between consecutive customers at a service window, or the deviation of the exact weight of a particular item in a lot from the advertised weight.

The density function $f(x)$ (often called the "probability density function") has two properties that are shared by every mathematical function which can be used to model the concentration of a random variable x.

5.2.2 Definition

A <u>probability density function</u> f is a function such that

1. $f(x) \ge 0$ for all x.
2. $\int_{-\infty}^{\infty} f(x)\, dx = 1.$

Figure 5.2–2

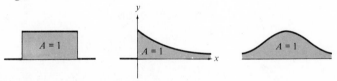

Remember that the integral of a positive function can be interpreted to be the area between that function and the x-axis. Thus the second condition of this definition says that the shaded areas in Figure 5.2–2 are always 1. We are showing here several density functions f, to give the idea that no matter which probability function is chosen, the area under the curve must be 1. Notice that the probability function on the left is discontinuous in two places and the function in the middle is discontinuous at $x = 0$.

The practical significance of the integral being 1 is the assumption that there is probability 1 (certainty) that x must take some value. Every patron of the store must have some height. There must be some interval between the arrival of consecutive customers at a service window; some such intervals may be of length zero, but they exist. And every item that is sold must have some weight. Thus we say that the probability of everything possible happening is 1.

We shall understand the phrase "the probability that x is between a and b" to mean the proportion of a population (or the measurements as given by x) lying between the two values a and b. More precisely, we have:

5.2.3   Definition

The <u>probability that a random variable x falls between two values</u> $a \leqslant b$ [which we abbreviate Prob($a \leqslant x \leqslant b$)] is given by a definite integral,

$$\text{Prob}(a \leqslant x \leqslant b) = \int_a^b f(x) \, dx$$

where $f(x)$ is the probability density function of x.

5.2.4   Example

Suppose that the residents of a particular "new" community are chosen so that the income level of all the families in the community is between $5,000 and $15,000, and the income levels are distributed according to a function $f(x) = kx$ for x in the interval $5,000 \leqslant x \leqslant 15,000$ and $f(x) = 0$ outside this interval. Assume that $f(x)$ is a probability density function.
(a) Find the value of k.
(b) Find the proportion of the population having incomes between $8,000 and $10,000.
(c) Find the probability that a randomly selected family has income over $12,000.

Solution:

We ignore all intervals where $f(x) = 0$.

(a) $\displaystyle\int_{5,000}^{15,000} kx \, dx = 1$ by Definition 5.2.2(2), so

$$1 = \frac{kx^2}{2}\Big|_{5,000}^{15,000} = \frac{k}{2}(225{,}000{,}000 - 25{,}000{,}000) = 100{,}000{,}000k$$

which implies that $k = 1/100{,}000{,}000 = 10^{-8}$.

(b) Prob($8,000 \leqslant x \leqslant 10,000$) $= 10^{-8} \displaystyle\int_{8,000}^{10,000} x \, dx = 0.18$

(c) Prob($x \geqslant 12,000$) $= 10^{-8} \displaystyle\int_{12,000}^{15,000} x \, dx = \dfrac{405}{1,000}$

The probability density function (often called simply the "density function") in Example 5.2.4 was designed to fit a particular situation. But there are classes of standard density functions that appear repeatedly in statistical situations. One common class of density functions is the uniform density function. It is used where the probabilities are equal that x will lie in any two subintervals of $a \leqslant x \leqslant b$ if the two subintervals are of equal length.

5.2.5   Definition

$f(x)$ is a <u>uniform density function</u> if there are two constants $a < b$ such that

$$f(x) = \begin{cases} \dfrac{1}{b-a} & \text{if } a \leqslant x \leqslant b \\ 0 & \text{otherwise} \end{cases}$$

Figure 5.2–3

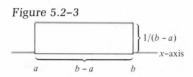

Notice that a uniform density function always has two discontinuities (see Section 1.5), one at $x = a$ and the other at $x = b$. It is zero for "most" values of x, and positive only on a finite interval. Over this interval it takes a constant value that forces its integral to be 1. Since the definite integral can be interpreted to be an area, and the area enclosed is precisely a rectangle, we can calculate it by merely multiplying length times width: $(b-a)[1/(b-a)] = 1$. Thus we have shown that every uniform density function is indeed a probability density function.

5.2.6   Example

Describe a uniform density function that is nonzero only in the interval $7 \leqslant x \leqslant 10$.

Solution:

Since $10 - 7 = 3$, the interval has length 3. Therefore, in order to have the area of the rectangle equal to 1, its height must be $1/(10 - 7) = \frac{1}{3}$. Thus the uniform density function satisfying the given conditions is

$$f(x) = \begin{cases} \frac{1}{3} & \text{if } 7 \leqslant x \leqslant 10 \\ 0 & \text{otherwise} \end{cases}$$

Figure 5.2–4

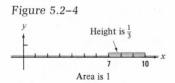

5.2.7   Example

A certain machine completes an operation every 5 minutes. What is the probability that someone arriving randomly will have to wait 2 minutes or more for the operation to end?

Solution:

It is reasonable to assume that the waiting time x is uniformly distributed over $0 \leqslant x \leqslant 5$. The probability that the person will wait more than 2 minutes is, then,

$$\text{Prob}(x \geqslant 2) = \text{Prob}(2 \leqslant x \leqslant 5) = \int_2^5 \frac{1}{5-0}\, dx = \tfrac{1}{5}x\Big|_2^5 = \tfrac{1}{5}(5-2) = \tfrac{3}{5}$$

Figure 5.2–5

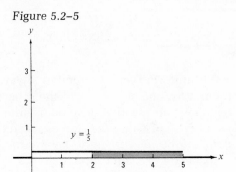

Another class of common density functions that is often used as models for waiting times or for failure rates of machines is the exponential density functions. Next we shall define and illustrate these functions.

5.2.8 **Definition**

An exponential density function $f(x)$ is such that for some fixed $k > 0$,

$$f(x) = \begin{cases} ke^{-kx} & \text{if } x \geqslant 0 \\ 0 & \text{otherwise} \end{cases}$$

Figure 5.2–6

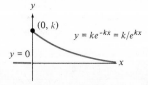

To convince ourselves that every exponential density function is a probability density function, we must check the conditions of Definition 5.2.2. Since $k > 0$ it is clear that the first condition is satisfied because $ke^{-kx} = k/e^{kx} > 0$. To verify the second condition, we note that since the function is zero for $x < 0$, we need to integrate only from 0 to $+\infty$. Letting $u = kx$, we have $du = k\, dx$ and

$$\int_0^\infty ke^{-kx}\, dx = \int_0^\infty e^{-u}\, du = \lim_{t \to \infty} -e^{-kx}\Big|_0^t = \lim_{t \to \infty} -\frac{1}{e^{kx}}\Big|_0^t$$

$$= \lim_{t \to \infty} \left[-\frac{1}{e^{kt}} - \left(-\frac{1}{e^0} \right) \right] = 0 - (-1) = 1$$

Thus every exponential density function satisfies the two conditions required of a probability density function.

We now turn to two typical applications of exponential density functions.

5.2.9   Example

The length x of telephone calls made by the executives of a certain company has been found to be given by the exponential density function $f(x) = \frac{1}{2}e^{-(1/2)x}$. What is the probability that a random telephone call will last over 3 minutes?

Solution:

$$\text{Prob}(x \geqslant 3) = \int_3^\infty \tfrac{1}{2}e^{-(1/2)x}\,dx = \lim_{t\to\infty} -e^{-(1/2)x}\Big|_3^t$$

$$= \lim_{t\to\infty} -(e^{-(1/2)t} - e^{-3/2})$$

$$= e^{-3/2} \approx 0.2231 \qquad \text{(by Table III)}$$

5.2.10   Example

A manufacturer of electronic equipment finds that the minimum time t for which the first breakdown of a certain type of radio occurs has the exponential density function $f(t) = 0.1e^{-0.1t}$. If he guarantees his equipment to last 1 year, what proportion of his customers will be eligible for free service because of failure during the first year?

Solution:

The proportion of disappointed customers is the probability

$$\text{Prob}(t \leqslant 1) = \int_0^1 0.1e^{-0.1t}\,dt = -e^{-0.1t}\Big|_0^1 = -(e^{-0.1} - e^0)$$

$$= 1 - e^{-0.1} \approx 1 - 0.905 = 0.095$$

Another concept in the subject of probability closely related to the probability density function is that of the probability distribution function. Roughly speaking, if $F(x)$ is a probability distribution, then $F(a)$ tells the probability that x will be less than or equal to a. We mention this here, not with the intention of explaining this concept, but so that the reader will be less likely to confuse the density and the distribution in the future. If $f(x)$ is a given density function, then the related distribution is given by

$$F(b) = \int_{-\infty}^b f(x)\,dx$$

(See Table VI for an application of this to the normal density function.) It can be shown that the density function is the derivative of its related distribution function.

SUMMARY

A probability density function is a function which is never negative and for which $\int_{-\infty}^{\infty} f(x)\ dx = 1$. If we know that a density function is of a particular form involving a constant, we can use the fact that the integral is 1 to solve for the constant, as in Example 5.2.4. Given a particular density function, we can find the probability that the random variable will lie between any two values, a and b, by evaluating the integral, $\int_{a}^{b} f(x)\ dx$. We gave examples of this using a uniform density function (Example 5.2.7) and two exponential density functions (Examples 5.2.9 and 5.2.10).

EXERCISES   5.2. A

1. (a) If $f(x) = 0.2e^{-0.2x}$ for $x \geqslant 0$ and $f(x) = 0$ otherwise is a probability density function, find the probability that x will lie between 3 and 5.
    (b) Find the probability that x will be greater than 6.
    (c) Graph this function, indicating the answer to part (a). (*Hint:* See Examples 5.2.9 and 5.2.10.)
2. Suppose that a probability density function is given by $f(x) = \frac{1}{8}$ if x lies between 2 and 10 and $f(x) = 0$ elsewhere.
    (a) What is the probability that x lies between 4 and 7?
    (b) What is the probability that x is greater than 8? (*Hint:* See Example 5.2.7.)
    (c) Graph this function, indicating the answer to part (a).
3. Suppose that a probability density function is described by $f(x) = kx$ for x between 4 and 6, and $f(x) = 0$ otherwise.
    (a) What is k?
    (b) What is the probability that x is larger than 5? (*Hint:* See Example 5.2.4.)
    (c) Graph this function, indicating the answer to part (b).
4. Suppose that you know that a bus runs every half hour, but you don't know when it arrives at your stop. You decide to get there at a random time and take your chances.
    (a) What is the probability that you will have to wait over 15 minutes?
    (b) What is the probability that you will have to wait over 10 minutes?
    (c) What is the probability that you will have to wait over 5 minutes? (*Hint:* Use the uniform density function.)
5. Suppose that you want to catch a subway that runs every 20 minutes and you show up at the station without any knowledge about its schedule. Answer the same three questions as in problem 4.
6. In a certain city the daily consumption of electricity (in megawatt-hours) can be regarded as a random variable that has an exponential density function with $k = 3$. If the capacity of the city's power plant is 3 megawatt-hours,

find the probability that the power supply is adequate on a randomly selected day. (Leave your answer in terms of e; our table does not give you the necessary information to convert to decimal form.)

7. Suppose that a manufacturer of cars finds that the time when his new cars first need a major repair can be described by the exponential density function when $k = 0.07$.

    (a) If he guarantees his cars for 2 years, what proportion can he expect to provide servicing for during the guarantee period?

    (b) If he sells 10,000 of this make, how many can he expect to need major repairs in the first 2 years?

8. Suppose that the number of cars that arrive at the Lincoln Tunnel from the New Jersey side per minute during early afternoon hours on Tuesdays, Wednesdays, and Thursdays is given by the exponential density function with $k = 0.02$.

    (a) What is the probability that in any given minute more than 100 cars will show up at the toll booths?

    (b) What is the probability that 35 or fewer cars will arrive in a given minute?

EXERCISES 5.2. B

1. (a) If $f(x) = 0.3e^{-0.3x}$ for $x \geq 0$ and $f(x) = 0$ otherwise is a probability density function, find the probability that x will lie between 3 and 5.

    (b) Find the probability that x will be greater than 6. (*Hint*: See Examples 5.2.9 and 5.2.10.)

    (c) Graph this function, indicating the answer to part (a).

2. Suppose that a probability density function is given by $f(x) = \frac{1}{6}$ if x lies between 4 and 10 and $f(x) = 0$ elsewhere.

    (a) What is the probability that x lies between 4 and 7?

    (b) What is the probability that x is greater than 8? (*Hint*: See Example 5.2.7.)

    (c) Graph this function, indicating the answer to part (a).

3. Suppose that a probability density function is described by $f(x) = kx$ for x between 5 and 7, and $f(x) = 0$ otherwise.

    (a) What is k?

    (b) What is the probability that x is larger than 6? (*Hint*: See Example 5.2.4.)

    (c) Graph this function, indicating the answer to part (b).

4. Suppose that you know that a train runs every hour, but you don't know when it arrives at your stop. You decide to get there at a random time and take your chances.

    (a) What is the probability you will have to wait over 15 minutes?

    (b) What is the probability you will have to wait over 10 minutes?

    (c) What is the probability you will have to wait over 5 minutes? (*Hint*: Use the uniform density function.)

5. Suppose that you want to catch a bus that runs every 40 minutes and you show up at the station without any knowledge of its schedule. Answer the same three questions as in problem 4.

6. In a certain city the daily consumption of electricity (in megawatt-hours) can be regarded as a random variable having an exponential density function with $k = 4$. If the capacity of the city's power plant is 3 megawatt-hours, find the probability that the power supply is adequate on a randomly selected

day. (Leave your answer in terms of e; our table does not give you the necessary information to convert to decimal form.)

7. Suppose that a manufacturer of cars finds that the time when his new cars first need a major repair can be described by the exponential density function when $k = 0.08$.

(a) If he guarantees his cars for 2 years, what proportion can he expect to provide servicing for during the guarantee period?

(b) If he sells 10,000 of this make, roughly how many will need major repairs in the first 2 years?

8. Suppose that the number of cars that arrive at the Golden Gate Bridge from the Sausalito side per minute during early afternoon hours on Tuesdays, Wednesdays, and Thursdays is given by the exponential density function with $k = 0.06$.

(a) What is the probability that in any given minute more than 100 cars will show up at the toll booths?

(b) What is the probability that 35 or fewer cars will arrive in a given minute?

EXERCISES   5.2. C

The function $f(x) = kxe^{-x}$ for $x \geq 0$ and $f(x) = 0$ otherwise is an example of a "gamma probability density function" for appropriate k.

1. Find the appropriate value of k. (*Hint:* $\lim\limits_{x \to \infty} xe^{-x} = 0$.)

2. Find Prob($x \geq 10$) for the gamma density function above.

3. A team of experimental psychologists have found that the density function $f(x)$ for the time x that it takes a population to learn a certain task is given by

$$f(x) = \frac{3}{216}(8x - x^2) \qquad \text{if } 0 \leq x \leq 8$$
$$f(x) = 0 \qquad\qquad\quad \text{otherwise}$$

What percentage of the population can be expected to learn the task in less than 3 minutes?

ANSWERS   5.2. A

1. (a) 0.18 (b) 0.30
    (c)

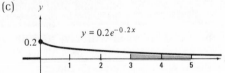

2. (a) $\frac{3}{8}$ (b) $\frac{1}{4}$
    (c)

3. (a) $\frac{1}{10}$ (b) $\frac{11}{20}$
    (c)

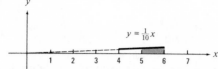

4. (a) $\frac{1}{2}$ (b) $\frac{2}{3}$ (c) $\frac{5}{6}$
5. (a) $\frac{1}{4}$ (b) $\frac{1}{2}$ (c) $\frac{3}{4}$
6. $1 - e^{-9} \approx 0.99988$
7. (a) 13 percent (b) roughly 1,300
8. (a) 0.135 (b) about 0.5

ANSWERS 5.2. C

1. $k = 1$
2. Prob$(x \geqslant 10) = 11e^{-10}$
3. $\frac{81}{216}$

SAMPLE QUIZ   **Sections 5.1 and 5.2**

1. (30 pts) Find the value of the following integrals, if they exist.

    (a) $\displaystyle\int_1^\infty \frac{3}{x^3}\, dx =$

    (b) $\displaystyle\int_3^\infty xe^{-x^2}\, dx =$

    (c) $\displaystyle\int_4^\infty \frac{1}{\sqrt[3]{x}}\, dx =$

2. (10 pts) Find the area under the curve $y = 4e^{-4x}$ to the right of the y-axis.
3. (20 pts) Suppose that a trolley arrives every 4 minutes.
    (a) Sketch the probability density function describing how long you must wait for it if you arrive at a random time. Label your graph carefully.
    (b) Calculate the probability that you will have to wait between 3 and 4 minutes.
4. (20 pts) Suppose that a probability density function is described by $f(x) = kx$ for $4 \leqslant x \leqslant 8$ and $f(x) = 0$ otherwise.
    (a) Sketch this function.
    (b) Find k.
5. (20 pts) Suppose that the length of phone calls of executives of a certain firm is described by the exponential density function for $k = 0.15$.
    (a) What is the probability that a certain phone call will last more than 2 minutes?
    (b) Graph this density function.

The answers are at the back of the book.

5.3   Integrals as Measures of Central Tendency

The word "average" has many uses in everyday language: we aim to please the average customer; we guarantee the average item is up to a certain standard; or we consider the average growth of the economy as significant. More precise definitions are needed in mathematics. There are two common definitions of average which involve integrals; we shall study these two meanings of average in this section.

The Median

5.3.1   Definition

Let x be a random variable with density function $f(x)$. Then the <u>median of x</u> is a value m of x such that

$$\int_{-\infty}^{m} f(x)\,dx = \tfrac{1}{2}$$

This is probably a new way of expressing a definition for you. We do not explicitly express the "m" on one side of an equation (or sentence) and then say what it means on the other side. It is intrinsically involved in the expression; we can say that it is defined "implicitly." We do this because of its nature; calculating the median involves un-

Figure 5.3–1

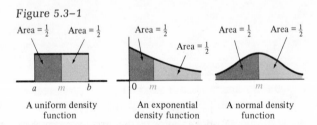

| A uniform density function | An exponential density function | A normal density function |

wrapping the variable m from the middle of a mathematical expression, as we shall see in the examples that follow.

The median m is that value for which half the values of the variable are above m and half of the values are below m. The median is often called the "fiftieth percentile score." (Some density functions may have more than one m satisfying our definition. This situation will never arise in this book, however; we always get a unique answer when solving for m.)

For example, if a class takes a test, the median score for that test is the number m such that half the scores are below m and half are above m. It is an easy measure for a teacher to use in evaluating the general performance of the class. We shall now calculate the median for several of the probability densities in Section 5.2.

5.3.2   Example

(Refer to Example 5.2.4) Suppose that the income levels of the families of a certain community have the population density $f(x) = 10^{-8}x$ for $5{,}000 \leqslant x \leqslant 15{,}000$ and $f(x) = 0$ elsewhere. Find the median income of these families.

Solution:

We seek an m such that $10^{-8} \int_{5{,}000}^{m} x \, dx = \frac{1}{2}$. We can formally integrate to get

$$10^{-8} \frac{x^2}{2}\Big|_{5{,}000}^{m} = 10^{-8} \left(\frac{m^2}{2} - \frac{5{,}000^2}{2} \right) = \frac{1}{2}$$

It follows that

$$m^2 - 5{,}000^2 = 10^8 \qquad \text{so } m^2 = 125{,}000{,}000 \text{ and } m \approx \$11{,}180$$

5.3.3   Example

(See Example 5.2.7) Find the median if the random variable x has the uniform density function $f(x) = \frac{1}{5}$ if $0 \leqslant x \leqslant 5$ and $f(x) = 0$ elsewhere.

Solution:

We seek m such that $\frac{1}{5} \int_{0}^{m} dx = \frac{1}{2}$. Evaluating the integral, we have

$\frac{1}{5}x\Big|_{0}^{m} = m/5 = \frac{1}{2}$, which implies that $m = \frac{5}{2}$.

5.3.4   Example

(See Example 5.2.10) Find the median if the random variable x has the exponential density function defined by $f(x) = (0.1)e^{-0.1x}$ for $x \geqslant 0$ and $f(x) = 0$ for $x < 0$.

Solution:

We seek m such that $\int_{0}^{m} 0.1e^{-0.1x} \, dx = \frac{1}{2}$. Evaluating the definite integral, we get $-(e^{-0.1m} - e^{0}) = \frac{1}{2}$, so $1 - \frac{1}{2} = e^{-0.1m}$, which implies that $m \approx 6.931$.

The Mean

The median is an important kind of "average," but it is not as easy to manipulate mathematically as another average called the "mean." The mean arises from the idea of the arithmetic average of n different scores obtained by adding up all n of the scores and dividing their sum by n. If a score occurs more than once, we can multiply by the number of times it occurs before we add; in other words, we "weight" each score by the frequency of its occurrence.

Now, if x takes on all values in an interval (instead of just a finite number of scores), and if $f(x)$ is a probability density that describes how often each value of x occurs, we can obtain the mean by multiplying x by $f(x)$ and then integrating.

5.3.5   Definition

If x is a random variable with density function $f(x)$, the mean of x is

$$\mu = \int_{-\infty}^{\infty} xf(x) \, dx$$

(μ is pronounced "miew.")

5.3.6   Example

Find the mean income in Example 5.3.2.

Solution:
In that example $f(x) = 10^{-8}x$ when $5{,}000 \leqslant x \leqslant 15{,}000$ and $f(x) = 0$ otherwise. Therefore,

$$\mu = 10^{-8} \int_{5{,}000}^{15{,}000} x \cdot x \, dx = 10^{-8} \left.\frac{x^3}{3}\right|_{5{,}000}^{15{,}000} = 10{,}833\tfrac{1}{3}$$

which is lower than m.

5.3.7   Example

Find the mean in Example 5.3.3.

Solution:
In that example $f(x) = \tfrac{1}{5}$ if $0 \leqslant x \leqslant 5$ and $f(0) = 0$ elsewhere. Thus

$$\mu = \tfrac{1}{5} \int_0^5 x \, dx = \tfrac{1}{5} \left.\frac{x^2}{2}\right|_0^5 = \tfrac{5}{2} \qquad \text{which equals } m$$

5.3.8   Example

Find the mean in Example 5.3.4.

Solution:

In that example $f(x) = 0.1e^{-0.1x}$ for $x \geqslant 0$ and $f(x) = 0$ for $x < 0$. Therefore,

$$\mu = 0.1 \int_0^\infty xe^{-0.1x} \, dx$$

$$= \lim_{t \to \infty} 0.1 \int_0^t xe^{-0.1x} \, dx$$

$$= \lim_{t \to \infty} \left. (-xe^{-0.1x} - 10e^{-0.1x}) \right|_0^t \qquad \text{(see Rule 5.3.9)}$$

$$= \lim_{t \to \infty} (-te^{-0.1t} - 10e^{-0.1t} + 0 + 10e^0) = 10$$

because $e^{-0.1t}$ goes to zero much faster than t goes to ∞. Notice that this time μ is *greater* than m.

In Example 5.3.8 we used the following rule, which will be useful in solving the exercises of this section. You can check it by differentiation. You should not memorize this rule, but you should be aware of it.

5.3.9   **Rule**

$$k \int xe^{-kx} \, dx = -xe^{-kx} - \frac{1}{k} e^{-kx} + C$$

The Variance

In Examples 5.3.6 and 5.3.8 you may have noticed that the mean is more sensitive to extreme values of x than the median; this is especially true if the density function is nonnegative for large values of x. Unlike the median, the mean has a tendency to be "pulled" in the direction of extreme values.

The *variance* measures how the values of x are distributed around the mean—whether they are clustered closely about it or are scattered widely. We give the integral definition of the variance here, but defer its interesting features until the exercises and the next section.

5.3.10   **Definition**

If x is a random variable with density function $f(x)$ and mean μ, the variance is

$$\text{var}(x) = \int_{-\infty}^\infty (x - \mu)^2 f(x) \, dx$$

5.3.11   Example

Find the variance for the uniform density function $f(x) = \frac{1}{5}$ for $0 \leqslant x \leqslant 5$, $f(x) = 0$ otherwise.

Solution:

Using the fact that $\mu = \frac{5}{2}$ (proved in Example 5.3.7), we have

$$\text{var}(x) = \int_0^5 (x - \mu)^2 (\tfrac{1}{5}) \, dx = \tfrac{1}{5} \int_0^5 (x^2 - 2x\mu + \mu^2) \, dx$$

$$= \tfrac{1}{5} \left(\frac{x^3}{3} - \mu x^2 + \mu^2 x \right) \bigg|_0^5 = \tfrac{1}{5} \left(\frac{5^3}{3} - \frac{5^3}{2} + \frac{5^3}{4} \right)$$

$$= \tfrac{1}{5} \left(\frac{5^3}{12} \right) = \tfrac{25}{12}$$

The following expression for the variance is often used because it is simpler than the definition.

5.3.12   Rule

$$\text{var}(x) = \int_{-\infty}^{\infty} x^2 f(x) \, dx - \mu^2$$

Rule 5.3.12 can be derived from Definition 5.3.10 by expanding the square in the definition.

$$\text{var}(x) = \int_{-\infty}^{\infty} x^2 f(x) \, dx - \int_{-\infty}^{\infty} 2x\mu f(x) \, dx + \int_{-\infty}^{\infty} \mu^2 f(x) \, dx$$

$$= \int_{-\infty}^{\infty} x^2 f(x) \, dx - 2\mu \int_{-\infty}^{\infty} x f(x) \, dx + \mu^2 \int_{-\infty}^{\infty} f(x) \, dx$$

$$= \int_{-\infty}^{\infty} x^2 f(x) \, dx - 2\mu(\mu) + \mu^2(1) \qquad \text{(by Definitions 5.2.2}$$
$$\text{and 5.3.5)}$$

so

$$\text{var}(x) = \int_{-\infty}^{\infty} x^2 f(x) \, dx - \mu^2$$

5.3.13   Example

Find the variance for the uniform density function $f(x) = \frac{1}{5}$ for $0 \leqslant x \leqslant 5$, $f(x) = 0$ otherwise, using the expression just derived for the variance.

Solution:

Again using the fact that $\mu = \frac{5}{2}$, we have

$$\text{var}(x) = \int_0^5 \tfrac{1}{5}x^2 \, dx - \tfrac{25}{4} = \tfrac{1}{15}x^3 \Big|_0^5 - \tfrac{25}{4}$$

$$= \tfrac{125}{15} - \tfrac{25}{4} = \tfrac{25}{3} - \tfrac{25}{4} = \tfrac{25}{12}$$

which is the same answer (as we would hope!) as in Exercise 5.3.11.

In the next example we shall use the following formula, which you can check by differentiation.

5.3.14   Rule

$$\int x^2 k e^{-kx} \, dx = -e^{-kx} \left(x^2 + \frac{2x}{k} + \frac{2}{k^2} \right) + C$$

5.3.15   Example

Find the variance for the exponential density function $f(x) = 0.1e^{-0.1x}$ for $x \geqslant 0$ and $f(x) = 0$ for $x < 0$.

Solution:

Using the fact that $\mu = 10$ (proved in Example 5.3.8), we have

$$\text{var}(x) = \int_0^\infty 0.1x^2 e^{-0.1x} \, dx - 100$$

$$= (-x^2 e^{-0.1x} - 20x e^{-0.1x} - 200 e^{-0.1x}) \Big|_0^\infty - 100$$

$$= 200 - 100 = 100$$

The square root of the variance is called the "standard deviation" and is generally denoted by σ (called "sigma"). We shall discuss the standard deviation in more detail in the next section, in our study of the normal distribution.

SUMMARY

The median of a random variable was defined for continuous functions in a manner analogous to that of the fiftieth percentile for discrete functions. The mean of a random variable for continuous functions was defined to parallel that of the arithmetic average for discrete functions. The median and mean of random variables having specific probability densities were computed. The variance, a measure of how far values of a random variable lie from the mean, was defined and an alternative expression was given. The variance of two density functions was calculated.

EXERCISES   5.3. A

1.   Find the median and the mean of the uniform density $f(x) = \tfrac{1}{20}$ for $15 \leqslant x \leqslant 35$ and $f(x) = 0$ elsewhere. Compare them.

2. Find the median and the mean of the uniform density $f(x) = \frac{1}{4}$ for $6 \leqslant x \leqslant 10$ and $f(x) = 0$ elsewhere.
3. Find the median and the mean of the exponential density $f(x) = 2e^{-2x}$ for $x \geqslant 0$ and $f(x) = 0$ elsewhere. Compare them. (*Hint:* Use Rule 5.3.9.)
4. Find the median and the mean of the exponential density $f(x) = 3e^{-3x}$ for $x \geqslant 0$ and $f(x) = 0$ elsewhere.
5. Show that the mean of any uniform density function $f(x) = 1/(b-a)$ for $a \leqslant x \leqslant b$ and $f(x) = 0$ elsewhere is $\mu = (a+b)/2$.
6. Show that the mean of the exponential density function $f(x) = ke^{-kx}$ for $x \geqslant 0$ and $f(x) = 0$ elsewhere is $\mu = 1/k$.

EXERCISES 5.3. B

1. Find the median and the mean of the uniform density $f(x) = \frac{1}{10}$ for $15 \leqslant x \leqslant 25$ and $f(x) = 0$ elsewhere. Compare them.
2. Find the median and the mean of the uniform density $f(x) = \frac{1}{8}$ for $2 \leqslant x \leqslant 10$ and $f(x) = 0$ elsewhere.
3. Find the median and the mean of the exponential density $f(x) = 4e^{-4x}$ for $x \geqslant 0$ and $f(x) = 0$ elsewhere. Compare them. (*Hint:* Use Rule 5.3.9.)
4. Find the median and the mean of the exponential density $f(x) = 0.2e^{-0.2x}$ for $x \geqslant 0$ and $f(x) = 0$ elsewhere.
5. Show that the mean of any uniform density function $f(x) = 1/(b-a)$ for $a \leqslant x \leqslant b$ and $f(x) = 0$ elsewhere is $\mu = (a+b)/2$.
6. Show that the mean of the exponential density function $f(x) = ke^{-kx}$ for $x \geqslant 0$ and $f(x) = 0$ elsewhere is $\mu = 1/k$.

EXERCISES 5.3. C

1. Find the variance of the density function in
   (a) problem 2, Exercises 5.3.A   (c) problem 2, Exercises 5.3.B
   (b) problem 3, Exercises 5.3.A   (d) problem 3, Exercises 5.3.B
2. Show that the variances for the density functions in problems 5 and 6 of Exercises 5.3.B are $(b-a)^2/12$ and $1/k^2$, respectively. (*Hint:* Use Rules 5.3.12 and 5.3.14.)
3. Show that the median m of an exponential density function $f(x) = ke^{-kx}$ if $x \geqslant 0$ and $f(x) = 0$ elsewhere is given by $m = \ln(2)/k$.
4. Prove that for every exponential function the median is less than the mean.

ANSWERS 5.3. A

1. $m = \mu = 25$; they are equal.
2. $m = \mu = 8$
3. $m \approx 0.35$, $\mu = 0.5$; the median is less than the mean.
4. $m \approx 0.23$, $\mu = \frac{1}{3}$
5. $\mu = \int_a^b x \frac{1}{b-a} dx = \frac{1}{b-a} \frac{1}{2}x^2 \Big|_a^b = \frac{1}{b-a} \frac{1}{2}(b^2 - a^2) = \frac{(b-a)(b+a)}{(b-a)2} = \frac{b+a}{2}$

6. $\mu = \displaystyle\int_0^\infty xke^{-kx}\,dx =$ (use the integral table or Rule 5.3.9)

$$= -xe^{-kx} - \frac{1}{k}\,e^{-kx}\Big|_0^\infty = -\left(-\frac{1}{k}\right) = \frac{1}{k}$$

ANSWERS 5.3. C

1. (a) $1\frac{1}{3}$ (b) 0.25 (c) $5\frac{1}{3}$ (d) $\frac{1}{18}$
2. For the uniform density function

$$\text{var}(x) = \int_{-\infty}^\infty x^2 f(x)\,dx - \mu^2 \qquad \text{(Rule 5.3.12)}$$

$$= \int_a^b x^2\,\frac{1}{b-a}\,dx - \frac{(b+a)^2}{4} \qquad \begin{array}{l}\text{(substitution of the}\\ \text{uniform density func-}\\ \text{tion and problem 5,}\\ \text{Exercises 5.3.B)}\end{array}$$

$$= \frac{1}{b-a}\,\tfrac{1}{3}x^3\Big|_a^b - \frac{(b+a)^2}{4} \qquad \text{(integration)}$$

$$= \frac{1}{b-a}\,\tfrac{1}{3}(b^3 - a^3) - \frac{(b+a)^2}{4}$$

$$= \frac{b^2 + ab + a^2}{3} - \frac{b^2 + 2ab + a^2}{4} \qquad \begin{array}{l}\text{(factoring and canceling}\\ \text{in the first term and}\\ \text{multiplication in the}\\ \text{second)}\end{array}$$

$$= \frac{4b^2 + 4ab + 4a^2 - 3b^2 - 6ab - 3a^2}{12} \qquad \begin{array}{l}\text{(finding a common}\\ \text{denominator)}\end{array}$$

$$= \frac{b^2 - 2ab + a^2}{12} = \frac{(b-a)^2}{12}$$

and the first part is proved.

For the exponential density function

$$\text{var}(x) = \int_{-\infty}^\infty x^2 f(x)\,dx - \mu^2 \qquad \text{(Rule 5.3.12)}$$

$$= \int_0^\infty x^2 k e^{-kx}\,dx - \frac{1}{k^2} \qquad \begin{array}{l}\text{(substituting of exponential}\\ \text{function and problem 6, 5.3.B)}\end{array}$$

$$= -e^{-kx}\left(x^2 + \frac{2x}{k} + \frac{2}{k^2}\right)\Big|_0^\infty - \frac{1}{k^2} \qquad \text{(Rule 5.3.14)}$$

$$= \frac{2}{k^2} - \frac{1}{k^2} = \frac{1}{k^2}$$

3. We solve $\displaystyle\int_0^m ke^{-kx}\,dx = \tfrac{1}{2}$ for m by first integrating:

$$-e^{-kx}\Big|_0^m = \tfrac{1}{2}$$

$$-e^{-km} + 1 = \tfrac{1}{2}$$

$$-e^{-km} = \tfrac{1}{2} - 1 = -\tfrac{1}{2}$$

$$e^{-km} = \tfrac{1}{2}$$

$$-km = \ln\,(e^{-km}) = \ln\,(\tfrac{1}{2}) = \ln\,(2^{-1}) = -\ln\,(2)$$

$$m = \frac{\ln\,(2)}{k}$$

4.   Since $\ln\,(2) \approx 0.69 < 1$, we have

$$m = \frac{\ln\,(2)}{k} < \frac{1}{k} = \mu \qquad \text{(by problem 6, 5.3.A and 5.3.B)}$$

5.4   Integrals of the Normal Density Function

Our final application of the definite integral is to the most important of all probability density functions, the normal or Gaussian density. It is given by the formula

$$f(x) = \frac{1}{\sigma\sqrt{2\pi}}\,e^{-(x-\mu)^2/(2\sigma^2)} \qquad -\infty < x < \infty$$

Figure 5.4–1

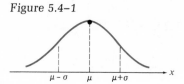

$\mu - \sigma$ μ $\mu + \sigma$ x

It can be shown that μ and σ^2 are the mean and variance, respectively. When σ^2 is the variance, σ is called the standard deviation.

5.4.1   Example

Find the mean and variance of a random variable x with the normal density functions

(a) $\dfrac{1}{\sqrt{2\pi}}\,e^{-x^2/2}$

(b) $\dfrac{1}{4\sqrt{2\pi}}\,e^{-(x-1)^2/32}$

(c) $\dfrac{1}{2\sqrt{2\pi}}\,e^{-(x+1)^2/8}$

Solution:

Solving these is a matter of seeing the particular μ and σ which make them fit the general formula.

(a) $\mu = 0$, $\sigma^2 = \text{var}(x) = 1$

(b) $\mu = 1$, $\sigma^2 = \text{var}(x) = 16$

(c) $\mu = -1$, $\sigma^2 = \text{var}(x) = 4$

To find the probability that x will lie in an interval $a \leqslant x \leqslant b$, the normal density should be integrated (as before):

$$\text{Prob}(a \leqslant x \leqslant b) = \frac{1}{\sigma \sqrt{2\pi}} \int_a^b e^{-(x-\mu)^2/(2\sigma^2)} \, dx$$

The unfortunate truth is, however, that the integrand of this expression does not have an elementary antiderivative. Therefore, to evaluate it, it is necessary to use the approximation methods of Section 4.7, where we took the limits of sums to get a value for an integral. Since this is such an important function and the computations are so tedious, the results have been tabulated in detail by high-speed computers.

Such tables are usually available only for the values $\mu = 0$ and $\sigma = 1$. These values define the standard normal density:

$$f(z) = \frac{1}{\sqrt{2\pi}} e^{-z^2/2} \qquad -\infty < z < \infty$$

We can use the values of the standard normal density to evaluate other normal density functions by using the substitution process described in Section 4.2 as suggested in the following rule.

5.4.2   **Rule**

Use the substitution $z = (x - \mu)/\sigma$ in order to change

$$\frac{1}{\sigma \sqrt{2\pi}} \int_a^b e^{-(x-\mu)^2/(2\sigma^2)} \, dx \quad \text{into} \quad \frac{1}{\sqrt{2\pi}} \int_{(a-\mu)/\sigma}^{(b-\mu)/\sigma} e^{-z^2/2} \, dz$$

Then use the following to calculate specific probabilities:

1. $\text{Prob}(a \leqslant z \leqslant b) = \text{Prob}(z \leqslant b) - \text{Prob}(z \leqslant a)$.
2. $\text{Prob}(0 < a \leqslant z) = 1 - \text{Prob}(z \leqslant a)$.
3. $\text{Prob}(z \leqslant -a < 0) = \text{Prob}(z \geqslant a > 0)$ (since the standard normal density function is symmetric around zero).

5.4.3   **Example**

Find the value of $\int_{-\infty}^1 f(x) \, dx$ if $f(x)$ is a normal density function having

(a) $\mu = 0$, $\sigma = 1$

(b) $\mu = 0$, $\sigma = 2$

(c) $\mu = 1$, $\sigma = 1$
(d) $\mu = 2$, $\sigma = 2$

Solution:

(a) These values define the standard normal density function, so we need merely to look up the value of 1 in Table VI. The answer is 0.8413.

(b) In the next three parts we must use the substitution $z = (x - \mu)/\sigma$ to change from a normal density function to the standard normal density function. As we change the variable in the integrand, we will also change the limits. In all these cases we can note with joy that if $x = -\infty$, so does $z = -\infty$. Thus the lower limit does not change as the variable changes. But the upper limit does.

In part (b), to be specific, we set $z = (x - 0)/2$, so when $x = 1$, $z = \frac{1}{2}$. Thus we have

$$\int_{-\infty}^{1} f(x)\, dx = \frac{1}{\sqrt{2\pi}} \int_{-\infty}^{1/2} e^{-z^2/2}\, dz = \text{Prob}(z \leqslant \tfrac{1}{2}) = 0.6915$$

using Table VI.

(c) This time $z = (x - 1)/1$, so when $x = 1$, $z = 0$. Thus

$$\int_{-\infty}^{1} f(x)\, dx = \frac{1}{\sqrt{2\pi}} \int_{-\infty}^{0} e^{-z^2/2}\, dz = \text{Prob}(z \leqslant 0) = 0.5$$

since the standard normal distribution is symmetric about 0.

(d) Now $z = (x - 2)/2$, so when $x = 1$, $z = -\frac{1}{2}$. Thus

$$\int_{-\infty}^{1} f(x)\, dx = \frac{1}{\sqrt{2\pi}} \int_{-\infty}^{-1/2} e^{-z^2/2}\, dz = \text{Prob}(z \leqslant -\tfrac{1}{2}) = \text{Prob}(z \geqslant \tfrac{1}{2})$$

$$= 1 - \text{Prob}(z \leqslant \tfrac{1}{2}) = 1 - 0.6915 = 0.3085$$

5.4.4   Example

If $f(x)$ is a normal density function with $\mu = 1$ and $\sigma = 2$, find the value of

(a) $\displaystyle\int_{2}^{3} f(x)\, dx$

(b) $\displaystyle\int_{0}^{1} f(x)\, dx$

(c) $\displaystyle\int_{-1}^{0} f(x)\, dx$

Solution:

We let $z = (x - 1)/2$ and make the appropriate substitutions for the limits of integration. Then we get the following integrals of the standard normal density function.

(a) $\int_2^3 f(x)\,dx = \dfrac{1}{\sqrt{2\pi}} \int_{1/2}^{1} e^{-z^2/2}\,dz = \text{Prob}(\tfrac{1}{2} \leqslant z \leqslant 1) = \text{Prob}(z \leqslant 1)$

$- \text{Prob}(z \leqslant \tfrac{1}{2}) = 0.8413 - 0.6915 = 0.1498$

(b) $\int_0^1 f(x)\,dx = \dfrac{1}{\sqrt{2\pi}} \int_{-1/2}^{0} e^{-z^2/2}\,dz = \text{Prob}(-\tfrac{1}{2} \leqslant z \leqslant 0) = \text{Prob}(0 \leqslant z \leqslant \tfrac{1}{2})$

$= \text{Prob}(z \leqslant \tfrac{1}{2}) - \text{Prob}(z \leqslant 0) = 0.6915 - 0.5000 = 0.1915$

(c) $\int_{-1}^0 f(x)\,dx = \dfrac{1}{\sqrt{2\pi}} \int_{-1}^{-1/2} e^{-z^2/2}\,dz = \text{Prob}(-1 \leqslant z \leqslant -\tfrac{1}{2})$

$= \text{Prob}(\tfrac{1}{2} \leqslant z \leqslant 1) = 0.1498 \;[\text{by part (a) above}]$

Explaining why the normal density function is so very important in statistics is beyond the scope of this book. One useful feature, however, is the fact that about 68 percent of a population with a normal density function lies in the interval $\mu - \sigma \leqslant x \leqslant \mu + \sigma$ and about 95 percent lies in the interval $\mu - 2\sigma \leqslant x \leqslant \mu + 2\sigma$.

An important application of the normal density function historically has been to model the astonishing degree of regularity in errors of measurement, and it is to this application that we now turn.

5.4.5 Example

The diameters of a certain manufacturer's ball bearings have a mean of 2 inches with a standard deviation of $\sigma = 0.4$ inch.

(a) What proportion of the ball bearings can the manufacturer expect to have with diameters less than 2.8 inches?

(b) If the producer loses 5 cents for each ball bearing that deviates more than 0.8 inch from 2 inches, and he makes a batch of 10,000 of them, what is his expected loss due to deviations of diameter?

Solution:

(a) We set $z = (x - 2)/0.4$, so when $x = 2.8$ inches, $z = (2.8 - 2)/0.4 = 0.8/0.4 = 2$:

$$\int_{-\infty}^{2.8} f(x)\,dx = \dfrac{1}{\sqrt{2\pi}} \int_{-\infty}^{2} e^{-z^2/2}\,dz = \text{Prob}(z \leqslant 2) = 0.9772$$

(b) We can use the results of part (a). Since $\text{Prob}(x \leqslant 2.8 \text{ inches}) = \text{Prob}(z \leqslant 2) = 0.9772$, it follows that $\text{Prob}(x \geqslant 2.8 \text{ inches}) = 1 - 0.9772 = 0.0228$. By the symmetry of the normal density function, we also have that $\text{Prob}(x \leqslant 1.2 \text{ inches}) = 0.0228$. Thus the probability that a ball bearing will deviate more than 0.8 inch from 2 inches is $2(0.0228) = 0.0456$. If 10,000 are manufactured, therefore, we would expect about $0.0456(10,000) = 456$ to deviate more than 0.8 inch from 2 inches. Since he loses 5 cents per bearing, the expected loss is $0.05(456) = \$22.80$.

5.4.6   Example

Suppose that a certain type of television set has a mean life expectancy of 4 years with a standard deviation of $\sigma = 1.5$ years. If a sales outfit guarantees to replace any that break down within the first year, how many can they expect to replace out of the first 1,000 that they sell?

Solution:

Let $z = (x - 4)/1.5$, so when $x = 1$, $z = -3/1.5 = -2$. Thus

$$\int_{-\infty}^{1} f(x) \, dx = \frac{1}{\sqrt{2\pi}} \int_{-\infty}^{-2} e^{-z^2/2} \, dz = \text{Prob}(z \leqslant -2)$$

$$= \text{Prob}(z \geqslant 2) = 1 - \text{Prob}(z \leqslant 2) = 1 - 0.9772$$
$$= 0.0228$$

Thus if we sell 1,000 such sets, we would expect about $0.0228(1,000) \approx 23$ to need to be replaced.

5.4.7   Example in Sociology

The heights of men follow a normal distribution with mean of 5 feet 9 inches and standard deviation of 2.5 inches. Suppose that the policemen of a certain city were required to be over 6 feet tall. What proportion of the men in the population would satisfy this requirement?

Solution:

Since 5 feet 9 inches = 69 inches, we set $z = (x - 69)/2.5$ and have when $x = 6$ feet = 72 inches that $z = (72 - 69)/2.5 = 3/2.5 = 1.2$, so

$$\int_{72}^{\infty} f(x) \, dx = \frac{1}{\sqrt{2\pi}} \int_{1.2}^{\infty} e^{-z^2/2} \, dz = \text{Prob}(z \geqslant 1.2) = 1 - 0.8849$$
$$= 0.1151$$

SUMMARY

In this section we discussed an important set of probability density functions, the *normal density functions*. Although we did not learn an antiderivative for these functions, we did learn that the substitution $z = (x - \mu)/\sigma$ will change any normal density to the *standard normal density function*. The values of the integral of the standard normal density have been tabulated using the techniques of Section 4.7, which involve approximating integrals by sums. We used Table VI, a brief table of these values, to calculate specific probabilities.

EXERCISES 5.4. A

1. Find the mean and variance of the following normal density functions:

   (a) $\dfrac{1}{3\sqrt{2\pi}}\,e^{-x^2/18}$

   (b) $\dfrac{1}{\sqrt{2\pi}}\,e^{-(x-1)^2/2}$

   (c) $\dfrac{1}{4\sqrt{2\pi}}\,e^{-(x-1)^2/32}$

   (d) $\dfrac{2}{\sqrt{2\pi}}\,e^{-x^2/0.5}$

   (e) $\dfrac{1}{\sqrt{2\pi}}\,e^{-(x+3)^2/2}$

2. Find the value of $\displaystyle\int_{-\infty}^{2} f(x)\,dx$ if $f(x)$ is a normal density function having

   (a) $\mu = 0,\ \sigma = 1$
   (b) $\mu = 0,\ \sigma = 2$
   (c) $\mu = 1,\ \sigma = 1$
   (d) $\mu = 2,\ \sigma = 3$

3. If $f(x)$ is a normal density function having $\mu = 1$ and $\sigma = 2$, find the value of

   (a) $\displaystyle\int_{-\infty}^{2} f(x)\,dx$

   (b) $\displaystyle\int_{2}^{5} f(x)\,dx$

   (c) $\displaystyle\int_{-1}^{+2} f(x)\,dx$

4. The lengths of a random selection of tubings were found to have a mean length of 3 feet and a variance of 0.16 foot.
   (a) What proportion of the tubings can we expect to have lengths between 2 and 4 feet?
   (b) If the producer of the tubings loses $0.40 for each tube that deviates from the mean by more than 1 foot and he has contracted to produce 1,000,000, what is his expected loss? (*Hint:* $\sigma = 0.4$.)

5. If the lag in weeks between ordering and receiving a shipment has been found to have a normal density function with $\mu = 3$ and $\sigma = 1$, find the probability that the time lag is over 3.5 weeks for any shipment.

EXERCISES 5.4. B

1. Find the mean and standard deviation of the following density functions.

   (a) $\dfrac{1}{5\sqrt{2\pi}}\,e^{-x^2/50}$

   (b) $\dfrac{1}{\sqrt{2\pi}}\,e^{-(x-2)^2/2}$

   (c) $\dfrac{1}{5\sqrt{2\pi}}\,e^{-(x-2)^2/50}$

   (d) $\dfrac{3}{\sqrt{2\pi}}\,e^{-9x^2/2}$

   (e) $\dfrac{1}{\sqrt{2\pi}}\,e^{-(x+1)^2/2}$

2. Find the value of $\displaystyle\int_{-\infty}^{3} f(x)\,dx$ if $f(x)$ is a normal density function having

   (a) $\mu = 0,\ \sigma = 1$
   (b) $\mu = 0,\ \sigma = 2$
   (c) $\mu = 1,\ \sigma = 1$
   (d) $\mu = 2,\ \sigma = 5$

3. If $f(x)$ is a normal density function having $\mu = 2$ and $\sigma = 1$, find the value of

   (a) $\displaystyle\int_{-\infty}^{2} f(x)\,dx$

   (b) $\displaystyle\int_{2}^{5} f(x)\,dx$

   (c) $\displaystyle\int_{-1}^{2} f(x)\,dx$

4. The gestation period for a random selection of pregnant females of a certain species was found to be 9 months with a standard deviation of $\frac{1}{2}$ month. What proportion of births in this species can be expected to occur after a gestation period of between 8 and 10 months?

5. If the lag in weeks between ordering and receiving a shipment has been found to have a normal density function with $\mu = 2$ and $\sigma = 1$, find the probability that the time lag is over 3.5 weeks for any shipment.

EXERCISES   5.4. C

1. Use Table VI to convince yourself that about 68 percent of a population with a normal density function lies in the interval $\mu - \sigma \leqslant x \leqslant \mu + \sigma$ and about 95 percent lies in the interval $\mu - 2\sigma \leqslant x \leqslant \mu + 2\sigma$.

2. IQ scores have a normal probability distribution with $\mu = 100$ and $\sigma = 16$.
   (a) If a person is said to be mentally retarded when his or her IQ lies below 76, what percentage of the population is mentally retarded?
   (b) If 68 is considered to be the cutoff point for metal retardation, what percentage of the population is mentally retarded?

ANSWERS   5.4. A

1. (a) $\mu = 0$, $\sigma = 3$, var$(x) = 9$ (b) $\mu = 1$, $\sigma = 1$, var$(x) = 1$ (c) $\mu = 1$, $\sigma = 4$, var$(x) = 16$ (d) $\mu = 0$, $\sigma = \frac{1}{2}$, var$(x) = \frac{1}{4}$ (e) $\mu = -3$, $\sigma = 1$, var$(x) = 1$
2. (a) 0.9772 (b) 0.8413 (c) 0.8413 (d) 0.5000
3. (a) 0.6915 (b) 0.2857 (c) 0.5328
4. (a) 0.9876 (b) $4,960
5. 0.3085

ANSWERS   5.4. C

2. (a) 6.7 percent (b) 2.3 percent

SAMPLE QUIZ   ## Sections 5.3 and 5.4

1. (10 pts) $f(x) = \dfrac{1}{4\sqrt{2\pi}} e^{-(x-5)^2/32}$ is a normal density function with mean _____ and standard deviation _____.

2. (20 pts) If the diameter of a certain type of ball bearing has been found to follow a normal density function with $\mu = 1$ inch and $\sigma = 0.2$, find the probability that a ball bearing selected at random will be more than 1.24 inches in diameter.

3. (30 pts) The function $f(x) = \frac{1}{24}x$ if $4 \leqslant x \leqslant 8$ and $f(x) = 0$ otherwise is a probability density function. Find its median and mean.

4. (15 pts) Find the median of the density function $f(x) = 5e^{-5x}$ for $x \geqslant 0$ and $f(x) = 0$ elsewhere.

5. (25 pts) Find the mean and the variance of the uniform density function $f(x) = \frac{1}{4}$ for $4 \leqslant x \leqslant 8$ and $f(x) = 0$ elsewhere.

The answers are at the back of the book.

SAMPLE TEST   # Chapter 5

Do five of the following seven problems. You may want to use Table VI. Each problem counts 20 points.

1. (a) $\displaystyle\int_1^\infty \frac{dx}{x^6} =$ (b) $\displaystyle\int_0^\infty 5e^{-5x}\, dx =$

    (Show your calculations.)

2. Let $f(x)$ be a probability density function. Write the definition of the median, m, and the mean, μ.

3. Suppose that a trolley arrives regularly every 3 minutes.
    (a) Sketch the probability density function describing how long you must wait for it if you arrive at a random time. Label your graph carefully.
    (b) Calculate the mean waiting time, μ, using integrals.

4. Suppose that the electricity, measured in x megawatt-hours, required by a city on a typical day is described by the probability density function:

$$f(x) = \begin{cases} 2e^{-2x} & \text{for } x \geqslant 0 \\ 0 & \text{for } x < 0 \end{cases}$$

    (a) What is the probability that this city will need more than 3 megawatt-hours on a given day?
    (b) Sketch this function, indicating the answer to part (a) on the graph.

5. Suppose that a probability density function is described by

$$f(x) = \begin{cases} kx & \text{for } 3 \leqslant x \leqslant 5 \\ 0 & \text{elsewhere} \end{cases}$$

    (a) Sketch this function.
    (b) Find k.
    (c) Find the median, m.

6. If the lag in weeks between ordering and receiving a shipment has been found to have a normal density function with $\mu = 4$ and $\sigma = 1$, find the probability that the time lag is over 3.5 weeks for any shipment.

7. $\displaystyle\int_4^6 \frac{1}{5\sqrt{2\pi}} e^{-(x-2)^2/50} =$

The answers are at the back of the book.

Chapter 6

FUNCTIONS OF
SEVERAL VARIABLES

6.1 Functions of Several Variables

We have seen that a function is often used as a mathematical model of "real life" in which one variable depends upon another. But usually in real life, and in the business world in particular, the things that matter depend on several variables. For example, the cost of an item will depend on the cost of each material used in its manufacture, labor costs, transportation costs, and so on. If there are several items being produced, each with a marginal cost $m_1, m_2, m_3, \ldots, m_n$, and there is a fixed cost b, the total cost is given by the generalized linear cost-output equation

$$C = m_1x_1 + m_2x_2 + \cdots + m_nx_n + b$$

Figure 6.1–1

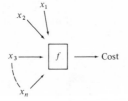

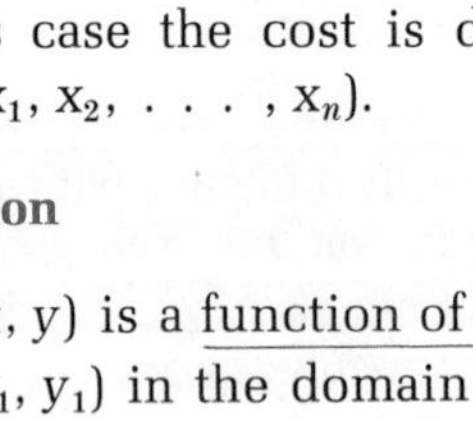

In this case the cost is dependent on n variables, and we can write $C = f(x_1, x_2, \ldots, x_n)$.

6.1.1 Definition

$z = f(x, y)$ is a <u>function of the two variables</u> x and y if for every ordered pair (x_1, y_1) in the domain of f there is a corresponding unique number z_1.

Using the ideas of Section 1.1, we can picture this as shown in Figure 6.1–2. [Note: The order in which we state things counts! It may be true that $f(1, 2) \neq f(2, 1)$.]

Figure 6.1–2

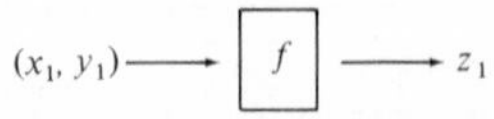

$$(x_1, y_1) \longrightarrow \boxed{f} \longrightarrow z_1$$

More generally, $z = f(x_1, x_2, \ldots, x_n)$ is a function of n variables $x_1, x_2, \ldots, x_n$ if for every ordered n-tuple $(x_1, x_2, \ldots, x_n)$ in the domain of f there is a unique corresponding number z_1. Again, the order counts.

6.1.2   Example

Write the linear cost-output equation of producing two types of shoes if the marginal cost per pair is $3 and $5, respectively, and there is a fixed cost of $1,500 per month.
(a) What is the cost of manufacturing 1,000 pairs of the cheap shoe and 500 of the expensive shoe?
(b) What is the cost of manufacturing 500 pairs of the cheap type and 1,000 of the expensive type?

Solution:

The linear cost-output equation is

$$C = f(x, y) = 3x + 5y + 1,500.$$

where x and y denote the number of pairs of cheap and expensive types, respectively.
(a) $f(1,000, 500) = 3(1,000) + 5(500) + 1,500 = 7,000$
(b) $f(500, 1,000) = 3(500) + 5(1,000) + 1,500 = 8,000$

6.1.3   Example

The demand for a commodity depends on its price per item, x, and the price per item, y, of its competitor, according to the function

$$g(x, y) = 1,250 - 10x + 5y - xy$$

Find
(a) $g(3, 7)$
(b) $g(7, 3)$
(c) $g(x, 3)$
and interpret the meaning of $g(x, 3)$.

Solution:

(a) $g(3, 7) = 1,250 - 10(3) + 5(7) - 3(7) = 1,234$
(b) $g(7, 3) = 1,250 - 10(7) + 5(3) - 7(3) = 1,174$
(c) $g(x, 3) = 1,250 - 10x + 5(3) - x(3) = 1,265 - 13x$

In part (c) we obtain the demand function for x, decreasing as usual, when the price of its competitor is held constant at $y = 3$.

6.1.4   Example

The value V of P dollars after x years invested at $100r$ percent compounded continuously is a function of three variables, P, r, and x.
(a) Write the functional relationship.
Find
(b) $f(100, 2, 0.05)$
(c) $f(100, 0.05, 2)$
(d) $f(100, r, 2)$

Solution:

(a) The functional relationship was found in Chapter 3 to be $V = f(P, r, x) = Pe^{rx}$, where e denotes the base for the natural logarithms, a number near 2.7.
(b) $f(100, 2, 0.05) = 100e^{(2)(0.05)}$, which, by Table III, is approximately $110.52.
(c) $f(100, 0.05, 2) = 100e^{(0.05)(2)}$, which gives the same number as in part (b). Notice, however, that in part (c) we found the value of $100 invested continuously for 2 years at 5 percent, whereas in part (b) we found the value of $100 invested continuously for 0.05 year at 200 percent interest per year.
(d) $V = f(100, r, 2) = 100e^{2r}$. This is a function of a single variable rather than three variables. It gives the value of $100 invested for 2 years at every possible interest rate compounded continuously.

We have shown how graphing can help the understanding of the total behavior of a function of one variable. Making a graph of a function of two variables involves "drawing" surfaces in three-dimensional space, which is complicated at best. Functions of more than two variables are impossible to visualize at all, given the human limitations of living in three-dimensional space.

Under these difficult circumstances we can still get some idea of what a function of several variables looks like by drawing cross sections, the technique used by cartographers when they wish to indicate elevations on a two-dimensional map. We have suggested this possibility in Example 6.1.3(c) and where we set all but two of the variables equal to a constant and considered the relationship described by the remaining variables. By graphing the resulting relationship in two variables for various fixed values of the other variable(s), we get cross sections of the original multivariable function.

6.1.5   Example

Sketch the cross sections obtained by setting $x = 0, 1, 3,$ and 4 in the function

$$C = f(x, y) = 3x + 5y + 1,500$$

Solution:

Figure 6.1–3

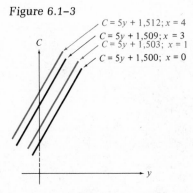

$C = 5y + 1{,}512; x = 4$
$C = 5y + 1{,}509; x = 3$
$C = 5y + 1{,}503; x = 1$
$C = 5y + 1{,}500; x = 0$

6.1.6 Example

Sketch the cross sections obtained by setting $y = 0, 1, 2, 3,$ and 4 in the function

$$d = g(x, y) = 1{,}250 - 10x + 5y - xy$$

Solution:

Figure 6.1–4

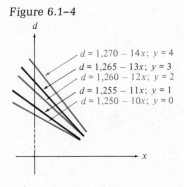

$d = 1{,}270 - 14x; y = 4$
$d = 1{,}265 - 13x; y = 3$
$d = 1{,}260 - 12x; y = 2$
$d = 1{,}255 - 11x; y = 1$
$d = 1{,}250 - 10x; y = 0$

6.1.7 Example

Sketch the cross sections obtained by setting $r = 0.01, 0.02, 0.03, 0.04,$ and 0.05 in the function

$$V = f(P, r, x) = Pe^{rx} \qquad \text{if } P = 100 \text{ is a constant}$$

Solution:

Figure 6.1–5

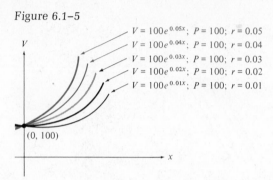

$V = 100e^{0.05x}$; $P = 100$; $r = 0.05$
$V = 100e^{0.04x}$; $P = 100$; $r = 0.04$
$V = 100e^{0.03x}$; $P = 100$; $r = 0.03$
$V = 100e^{0.02x}$; $P = 100$; $r = 0.02$
$V = 100e^{0.01x}$; $P = 100$; $r = 0.01$

Although we shall not put any emphasis on three-dimensional graphing, we shall elaborate here a bit about how functions of two variables, x and y, can be visualized in three-dimensional space. In this case we usually think of the xy-plane as being horizontal (for example, a piece of paper lying on your desk). The z-axis rises up perpendicular to this plane, and for each point on the horizontal xy-plane there is some corresponding distance above or below it [the value of z for the given pair (x, y)] at which there is a point on the surface which describes the function. Each point in space is identified with an ordered triple of numbers (x, y, z), only some of which appear on the graph (a surface) of a function of two variables.

Figure 6.1–6

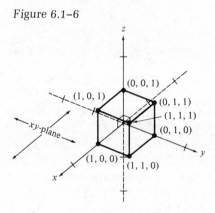

Graphing such figures is not easy, but the technique of cross sections may help. Setting the z-variable equal to a constant is equivalent, geometrically, to intersecting the surface with a given plane parallel to the xy-plane. Think of cutting it with a sharp knife at precisely the level of the given z-value above or below the xy-plane. The following example shows how this might be done.

6.1.8   Example

A production function $z = f(x, y) = x^2 + y^2$ depends upon the input of
x units of one factor and y units of another, according to the given formula. Give the cross sections describing the possible inputs of x and y
corresponding to the production levels of $z = 1, 4, 9,$ and 16 units, then
use this to sketch the production surface.

Solution:

Figure 6.1–7

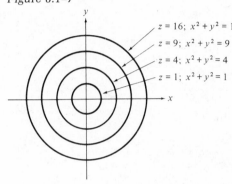

Figure 6.1–8

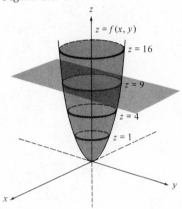

SUMMARY

In this section the definition and some uses of multivariable functions
(functions of more than one variable) were discussed. Often these functions can be helpfully pictured by use of cross sections.

EXERCISES   **6.1. A**

1. Evaluate the given functions at the given points.
    (a) $f(x, y) = x^2 + y^3 + 2x + y + 1$ at $(3, 2)$
    (b) $f(x, y, z) = 2x + 3y + 4z + xyz$ at $(2, -1, 1)$
    (c) $g(x, y, z) = (x + y + z)/x$ at $(5, -2, 1)$
    (d) $g(x, y) = (x + y)^2 - x^3 + y^3$ at $(3, 2)$
2. (a) Write the linear cost-output equation of producing two types of hats if
         the marginal cost is $2.55 for one type and $4.00 for the other and there
         is a fixed cost of $244 per month.
    (b) What is the cost of producing 100 of the less expensive hat and 50 of
         the more expensive hat?
    (c) What is the cost of manufacturing 50 of the cheap hat and 100 of the
         other?

3. (a) Write the linear cost-output equation of producing machine parts at three locations where, because of varying local conditions, the marginal cost per part is $0.25, $0.30, and $0.40, respectively, and there is an overall fixed operating cost of $2,000 for all three plants.
    (b) What is the cost of producing 2,000, 2,500, and 350 parts at the three plants in the order given?

4. The demand for a commodity depends on its price, x, and that of its competitor, y, according to the function $f(x, y) = 1,000 - x^2 + 2y - xy$. Find the quantity demanded when the prices are given by
    (a) $x = 5, y = 2$ (b) $x = 2, y = 5$ (c) $x = x_1, y = 2$

5. Sketch cross sections for each of the following functions by setting, in turn, $x = 1$, $x = 2$, and $x = 2.5$.
    (a) $f(x, y) = z = x + y$ (c) $f(x, y) = z = x^2 - x - y^2$
    (b) $f(x, y) = z = xy$ (d) $f(x, y) = z = ye^x$

EXERCISES   6.1. B

1. Evaluate the given functions at the given points.
    (a) $f(x, y) = x^3 + y^2 + 2y + x + 1$ at $(2, 3)$
    (b) $f(x, y, z) = 2y + 3x + 4z + xyz$ at $(-1, 2, 1)$
    (c) $g(x, y, z) = (x + y + z)/y$ at $(-2, 5, 1)$
    (d) $g(x, y) = (x + y)^2 - y^3 + x^3$ at $(2, 3)$

2. (a) Write the linear cost-output equation of producing two kinds of vests if the marginal cost is $3 for one type and $4.50 for the other and there is a fixed cost of $350 per month.
    (b) What is the cost of producing 100 of the less expensive and 200 of the more expensive vest?
    (c) What is the cost of manufacturing 200 of the less expensive and 100 of the more expensive vest?

3. (a) Write the linear cost-output equation of producing machine parts at three locations where the marginal cost is $0.25, $0.50, and $0.60 at each location, respectively, and there is a fixed cost of $1,000 for all three plants.
    (b) What is the cost of producing 2,000, 2,500, and 1,000 parts at the three plants in the order given?

4. The demand for a commodity depends on its price, x, and that of its competitor, y, according to the function $f(x, y) = 2,000 - 2x + y^2 - xy$. Find that quantity demanded when the prices are given by
    (a) $x = 2, y = 3$ (b) $x = 3, y = 2$ (c) $x = 2, y = y$

5. Sketch cross sections of the following functions by setting $y = 1$, $y = 2$, and $y = 3$.
    (a) $f(x, y) = z = 2x + y$ (c) $f(x, y) = z = x^2 - y^2$
    (b) $f(x, y) = z = x^2y$ (d) $f(x, y) = z = ye^x$

ANSWERS   6.1. A

1. (a) 26 (b) 3 (c) 0.8 (d) 6
2. (a) $C(x, y) = 2.55x + 4y + 244$ (b) $699 (c) $771.50

3. (a) $C(x, y, z) = 0.25x + 0.3y + 0.4z + 2,000$ (b) $3,390
4. (a) $969 (b) $996 (c) $q(x_1) = 1,004 - x_1^2 - 2x_1$
5. (a)   (b)

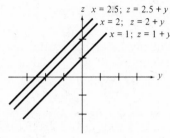

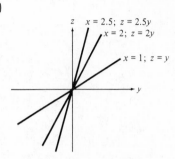

(c)   (d)

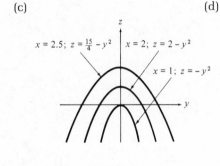

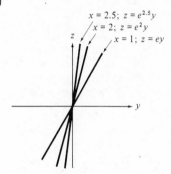

6.2   Partial Derivatives

To consider the rate of change of a function of several variables, we use the approach suggested in Section 6.1; we keep all but one of the independent variables constant. Then we can differentiate as usual with respect to the one variable that is considered to be changing.

For example, let us reconsider the demand function $f(x, y) = 1,250 - 10x + 5y - xy$ of Example 6.1.3. If the competitor's price y is kept constant at y_1, then

$$f(x, y_1) = (1,250 + 5y_1) - (10 + y_1)x$$

is a demand function of the single variable x and can be differentiated as usual, obtaining $-(10 + y_1)$. This expression gives the rate of change of $f(x, y)$ with respect to x at a constant value y_1 of y.

6.2.1   Definition

If $z = f(x, y)$ is a function of two variables,

$$\lim_{\Delta x \to 0} \frac{f(x + \Delta x, y) - f(x, y)}{\Delta x}$$

if it exists, is called the <u>partial derivative</u>, or simply <u>partial</u>, of the function f with respect to x at (x, y). It is denoted by

$$f_1(x, y) \quad \text{or} \quad f_x(x, y) \quad \text{or} \quad \frac{\partial f}{\partial x} \quad \text{or} \quad \frac{\partial z}{\partial x}$$

A similar definition holds for the partial derivative of f with respect to y: If $z = f(x, y)$ is a function of two variables, then

$$\lim_{\Delta y \to 0} \frac{f(x, y + \Delta y) - f(x, y)}{\Delta y}$$

if this limit exists, is called the partial derivative of f with respect to y at (x, y) and is denoted by

$$f_2(x, y) \quad \text{or} \quad f_y(x, y) \quad \text{or} \quad \frac{\partial f}{\partial y} \quad \text{or} \quad \frac{\partial z}{\partial y}$$

Clearly this definition can be generalized to functions of more than two variables; we just consider all but one of the variables to be constant. Then we apply the rules we learned for differentiating functions of a single variable.

6.2.2   Example

Find the first partial derivatives, $\dfrac{\partial z}{\partial x}$ and $\dfrac{\partial z}{\partial y}$, for each of the following functions.

(a) $z = 3x + 5y + 2$
(b) $z = 3x^2 + 2y^2 + 5xy$

Solution:

To find $\dfrac{\partial z}{\partial x}$, we pretend that y is a constant (for example, 2, π, or e). Then we differentiate with respect to x as we did in Chapters 2 and 3.

(a) $\dfrac{\partial z}{\partial x} = 3$ (since if y is a constant, so is 5y)

(b) $\dfrac{\partial z}{\partial x} = 6x + 5y$ (since the constant 5y times the variable x has the derivative of 5y)

To find $\dfrac{\partial z}{\partial y}$, we pretend that x is a constant and differentiate with respect to y:

(a) $\dfrac{\partial z}{\partial y} = 5$

(b) $\dfrac{\partial z}{\partial y} = 4y + 5x$

If we want to find the partial derivatives for specific values of the independent variables, we first find the partial derivatives in the general case and then plug in the given specific values.

6.2.3 Example

Evaluate $f_x(1, 2)$ in parts (a) and (b) of Example 6.2.2. In other words, find $\frac{\partial z}{\partial x}$ when $x = 1$ and $y = 2$. Use the results of Example 6.2.2.

Solution:

(a) Since $\frac{\partial z}{\partial x} = 3$, this partial derivative is 3 for every value of x and y. So, in particular, $f_x(1, 2) = 3$.

(b) $f_x(x, y) = \frac{\partial z}{\partial x} = 6x + 5y$, so $f_x(1, 2) = 6(1) + 5(2) = 16$.

6.2.4 Example

Find all first partial derivatives of $z = 3xe^y + 7ye^x + xw$.

Solution:

To find $\frac{\partial z}{\partial x}$, we hold both y and w constant and differentiate with respect to x, getting $\frac{\partial z}{\partial x} = 3e^y + 7ye^x + w$.

To find $\frac{\partial z}{\partial y}$, we hold x and w constant and differentiate with respect to y, getting $\frac{\partial z}{\partial y} = 3xe^y + 7e^x$.

To find $\frac{\partial z}{\partial w}$, we hold x and y constant and differentiate with respect to w, getting merely $\frac{\partial z}{\partial w} = 0 + 0 + x = x$.

6.2.5 Example

The production function for a certain operation has been found to be $z = 2x + 3y^2 + xy + w$, where z denotes the number of units output for an input of x units of labor, y units of materials, and w units of power. Find the marginal product of increasing labor by one unit if the current input is $x = 3$, $y = 1$, and $w = 2$.

Solution:

The marginal product with respect to labor, x, is

$$\frac{\partial z}{\partial x} = 2 + y$$

Hence at the given point, $\frac{\partial z}{\partial x} = 2 + 1 = 3$.

Thus the increase of 1 unit of labor will produce an approximate increase of 3 units in production at the production level $x = 3$, $y = 1$, and $w = 2$.

Higher-order partial derivatives may be defined by letting all but one of the variables be constant for each differentiation. When there is more than one variable, however, we need not differentiate with respect to the same variable each time. So there are an increasing number of partial derivatives as the number of differentiations increases. For example, a function of two variables has four second partial derivatives and eight third partial derivatives, since we can do each differentiation with respect to either variable. A function of three variables will have 3 first partial derivatives, 9 second partial derivatives, and 27 third partial derivatives.

The 4 second partial derivatives of a function of two variables, $z = f(x, y)$, are (in several possible notations)

$$\frac{\partial}{\partial x}\left(\frac{\partial z}{\partial x}\right) = \frac{\partial^2 z}{\partial x^2} = f_{xx} = f_{11}$$

$$\frac{\partial}{\partial x}\left(\frac{\partial z}{\partial y}\right) = \frac{\partial^2 z}{\partial x\, \partial y} = f_{yx} = f_{21}$$

$$\frac{\partial}{\partial y}\left(\frac{\partial z}{\partial x}\right) = \frac{\partial^2 z}{\partial y\, \partial x} = f_{xy} = f_{12}$$

$$\frac{\partial}{\partial y}\left(\frac{\partial z}{\partial y}\right) = \frac{\partial^2 z}{\partial y^2} = f_{yy} = f_{22}$$

Notice that when using the f_{xy} notation, the variables with respect to which we are taking the derivatives are listed from left to right, but that in the $\frac{\partial^2 z}{\partial y\, \partial x}$ notation, they are listed from right to left. Both f_{xy} and $\frac{\partial^2 z}{\partial y\, \partial x}$ indicate that we first take the derivative with respect to x and then take the derivative with respect to y.

6.2.6   Example

Find all the second-order partial derivatives of $z = f(x, y) = x^2 y^3 + x^3 + y^2$ at $x = 1$, $y = 3$.

Solution:

We first find the two first partials and then we take both partial derivatives of each of these to get the four second partials. Finally, we shall evaluate these at the specified point.

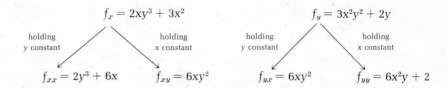

$$f_x = 2xy^3 + 3x^2 \qquad\qquad f_y = 3x^2y^2 + 2y$$

holding y constant   holding x constant   holding y constant   holding x constant

$$f_{xx} = 2y^3 + 6x \qquad f_{xy} = 6xy^2 \qquad f_{yx} = 6xy^2 \qquad f_{yy} = 6x^2y + 2$$

Therefore,

$$f_{xx}(1,\ 3) = 2(3)^3 + 6(1) = 54 + 6 = 60$$
$$f_{xy}(1,\ 3) = f_{yx}(1,\ 3) = 6(1)(3)^2 = 54$$
$$f_{yy}(1,\ 3) = 6(1)^2(3) + 2 = 20$$

6.2.7   Example

Find all the second-order partial derivatives of the function $f(x,\ y) = z = xy^2 + 3x^2 + 5y^3 + x(\ln y)$ at the point $x = 1$, $y = 4$.

Solution:

We first note that
(a) $f_x(x,\ y) = y^2 + 6x + \ln y$
(b) $f_y(x,\ y) = 2xy + 15y^2 + x/y$
    Then the second partial derivatives are

$$\frac{\partial^2 z}{\partial x^2} = f_{xx} = 6 \qquad \text{[y is held constant while (a) is differentiated]}$$

$$\frac{\partial^2 z}{\partial x\,\partial y} = f_{yx} = 2y + \frac{1}{y} \qquad \text{[y is held constant while (b) is differentiated]}$$

$$\frac{\partial^2 z}{\partial y\,\partial x} = f_{xy} = 2y + \frac{1}{y} \qquad \text{[x is held constant while (a) is differentiated]}$$

$$\frac{\partial^2 z}{\partial y^2} = f_{yy} = 2x + 30y - \frac{x}{y^2} \qquad \text{[x is held constant while (b) is differentiated]}$$

Therefore, at the given point, we have

$$f_{xx}(1,\ 4) = 6 \qquad\qquad f_{yx}(1,\ 4) = 2(4) + \tfrac{1}{4} = 8\tfrac{1}{4}$$
$$f_{xy}(1,\ 4) = 2(4) + \tfrac{1}{4} = 8\tfrac{1}{4} \qquad f_{yy}(1,\ 4) = 2 + 120 - \tfrac{1}{16} = 121\tfrac{15}{16}$$

You may have noticed that in both the examples above, $f_{xy} = f_{yx}$. Under the conditions that you will encounter, this is generally true. We say: "The cross-partials are equal."

Geometric Interpretation

The geometric interpretation of the partial derivative is what you might expect. The partial derivatives are the slope of the tangent line to a cross section of the function.

6.2.8   Example

Interpret the partial derivative $\dfrac{\partial w}{\partial x}$ of the function $w = x^2 + 2xy + zy$ at the point $(1, -2, 0, -3)$.

Solution:

We have $\dfrac{\partial w}{\partial x} = 2x + 2y$. Hence, at $(1, -2, 0, -3)$, $\dfrac{\partial w}{\partial x} = 2(1) + 2(-2) = -2$. An interpretation is given in an xw-plane, where, if we let $y = -2$ and $z = 0$, $w = x^2 + 2x(-2) + (0)(-2) = x^2 - 4x$. The line tangent to this parabola, $w = x^2 - 4x$, at the point $(x, w) = (1, -3)$, has slope $\dfrac{\partial w}{\partial x} = -2$.

Figure 6.2–1

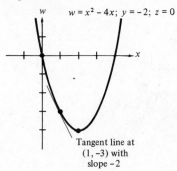

For functions of only two variables we can visualize the geometry of partial derivatives in space. This is because a function $z = f(x, y)$ will have as its graph a *surface*. At the point (x, y, z) on this surface there may be one or more tangent lines in space. Although lines in space do not have a "slope" as we have defined it, for certain lines the idea of slope makes sense. In a plane $y = c$ parallel to the xz-plane or in a plane $x = c$ parallel to the yz-plane, a line can be said to have slope $\dfrac{\Delta z}{\Delta x}$ or $\dfrac{\Delta z}{\Delta y}$, respectively, since the "missing" variable is always constant. Hence we have the following interpretation.

6.2.9   Rule

If $z = f(x, y)$ has partial derivatives f_x and f_y at the point (x_1, y_1):

1. $f_x(x_1, y_1)$ is the slope of the line in the plane $y = y_1$ tangent to the surface $z = f(x, y)$ at the point (x_1, y_1, z_1).
2. $f_y(x_1, y_1)$ is the slope of the line in the plane $x = x_1$ tangent to the surface $z = f(x, y)$ at the point (x_1, y_1, z_1).

The simultaneous equations of these lines are

$$y = y_1 \quad \text{and} \quad z - z_1 = f_x(x_1, y_1)(x - x_1)$$

and

$$x = x_1 \quad \text{and} \quad z - z_1 = f_y(x_1, y_1)(y - y_1)$$

respectively.

Notice that in three-dimensional space two equations are required to describe a line. This is because one equation describes a plane, and a line is the intersection of two planes.

6.2.10   Example

Find the slope of the lines tangent to the surface $z = 3x^2 + y^2$ at the point $(1, 2, 7)$ in the planes
(a) $x = 1$
(b) $y = 2$
Interpret your results using cross sections.

Solution:

(a) We consider first the intersection with the plane $x = 1$, which is parallel to the yz-plane. $f_y(x, y) = 2y$, so $f_y(1, 2) = 4$. If $x = 1$, the cross section is $z = 3 + y^2$ and the tangent line in that plane is given by $(z - 7) = 4(y - 2)$; so the line tangent to that cross section can be described by the simultaneous equations $x = 1$ and $z = 4y - 1$.

Figure 6.2–2

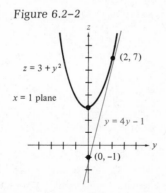

(b) Now we turn to the other case. In the plane $y = 2$ we are interested in the line tangent to the cross section $z = 3x^2 + 4$ at the point $(1, 7)$. Computing $f_x(x, y) = 6x$, we get that $f_x(1, 2) = 6$, so the equation of the tangent line in the $y = 2$ plane is given by $(z - 7) = 6(x - 1)$ or $z = 6x + 1$. Thus the tangent line can be described by the simultaneous equations $y = 2$ and $z = 6x + 1$.

Figure 6.2–3

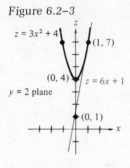

Figure 6.2–4

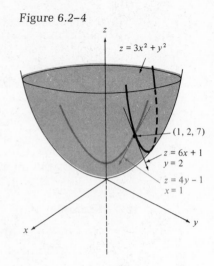

SUMMARY

The partial derivatives of a function of more than one variable are the rate of change of the function with respect to one variable while the other variables are held constant. Geometrically, these amount to the slope of the tangent line to one of the cross sections of the function. In the case of a function of two variables they can also be thought of as the slope of a line tangent, in a certain plane, to the given surface. In this section we practiced finding partial derivatives of functions of several variables, as well as interpreting our results geometrically. We also defined higher-order (second, third, etc.) partial derivatives in a manner analogous to that for functions of a single variable.

EXERCISES 6.2. A

1. Find the first partial derivatives of the following functions and evaluate them at the given values of the independent variables.
   (a) $f(x, y) = 3x - 2y + 7$ at $x = 2, y = -4$
   (b) $f(x, y) = 3x^2 - 2y^3 + 4xy^2$ at $x = 1, y = -2$
   (c) $f(x, y) = x^3 e^y$ at $x = 2, y = 0$
   (d) $f(x, y) = x^5 - y^{-2} + x^2 y - \ln y$ at $x = 0, y = 1$
   (e) $f(x, y, u, v) = xyuv$ at $x = 1, y = 2, u = 3, v = 4$
2. Find all the second-order partial derivatives of the functions in parts (a)–(d) of problem 1. How many second-order partials will there be in part (e)? How many third-order partials in part (e)?
3. The cost function for the two products manufactured by a certain company is $C(x, y) = x - 0.02x^2 + 10y + 0.03y^3 + 1,000$. Find the marginal cost of increasing y one unit if $x = 15$ and $y = 10$.
4. The revenue obtained by a firm is given by $R(x, y) = 5x - 100x^{-1} + 4y - 0.07y^2 + 2xy$, where x and y are the amounts in thousands of dollars spent

on newspaper and radio advertising, respectively. If $x = 100$ and $y = 10$, what is the effect of increasing newspaper advertising by one unit?

5. The demand for a commodity depends upon its unit price x and that of its two closest competitors, y and z, according to the formula $f(x, y, z) = 1,000 - 3x + y + z - xyz$. Find the marginal demand for this commodity when $x = 15$, $y = 10$, and $z = 20$.

EXERCISES 6.2. B

1. Find the first partial derivatives of the following functions and evaluate them at the given values of the independent variables.
    (a) $f(x, y) = -2x + 3y + 7$ at $x = -4, y = 2$
    (b) $f(x, y) = -2x^3 + 3y^2 + 4x^2y$ at $x = -2, y = 1$
    (c) $f(x, y) = y^3 e^x$ at $x = 0, y = 2$
    (d) $f(x, y) = -x^{-2} + y^5 + xy^2 - \ln(x)$ at $x = 1, y = 0$
    (e) $f(x, y, z) = xy^2z^3$ at $x = 2, y = 3, z = 1$
2. Find all the second-order partial derivatives of the functions in parts (a)–(d) of problem 1. How many second-order partials will there be in part (e)? How many third-order partials in part (e)?
3. The cost function for the two products manufactured by a certain company is $C(x, y) = 10x + 0.03x^3 + y + 0.02y^2 + 1,000$. Find the marginal cost of increasing y one unit if $x = 10, y = 15$.
4. The revenue obtained by a firm is given by $R(x, y) = 4x - 0.07x^2 + 5y - (100/y) + 2xy$, where x and y are the amounts in thousands of dollars spent on newspaper and radio advertising, respectively. If $x = 100$ and $y = 10$, what is the effect of increasing newspaper advertising by one unit?
5. The demand for a commodity depends upon its unit price x and that of its two closest competitors, y and z, according to the formula $f(x, y, z) = 1,000 - x + 3y + z - xyz$. Find the marginal demand for this commodity when $x = 10$, $y = 15$, and $z = 15$.

EXERCISES 6.2. C

1. If a substance is injected into an artery or vein, the concentration of the substance will vary with time t and with the position x, where we think of the x-axis as parallel to the axis of an ideally cylindrical vein. Under certain conditions this concentration can be described by a function of the form

    $$C(x, t) = e^{ax+bt}$$

    where a and b are constants. This function will satisfy the diffusion equation:

    $$\frac{\partial C}{\partial t} = \delta \frac{\partial^2 C}{\partial x^2}$$

    where δ is called a diffusion constant. Show by differentiation that in this situation, $b = \delta a^2$.
2. Suppose that the two demand functions for widgets and gizmos are given by

    $$d_1 = F(p_1, p_2) \quad \text{and} \quad d_2 = G(p_1, p_2)$$

where p_1 is the price of widgets and p_2 is the price of gizmos. If the partial derivatives F_2 and G_1 are both positive, then the demand for widgets increases as the price for gizmos rises (and vice versa), and we say that the goods are competitive. If both these partial derivatives are negative, then the demand for each good changes in the opposite direction as the price of the other, and we say that the goods are complementary. Which of the following demand functions are for complementary and which are for competitive goods?

(a) $F(x, y) = -x - y + 100$; $G(x, y) = -y - x + 100$
(b) $F(x, y) = -3x + y + 100$; $G(x, y) = 2x - 2y + 100$
(c) $F(x, y) = -x + y + 100$; $G(x, y) = -x - y + 100$

ANSWERS   6.2. A

1.   (a) $\dfrac{\partial f}{\partial x} = 3$, $\dfrac{\partial f}{\partial y} = -2$; $f_1(2, -4) = 3$, $f_2(2, -4) = -2$ (b) $\dfrac{\partial f}{\partial x} = 6x + 4y^2$, $\dfrac{\partial f}{\partial y} =$

$-6y^2 + 8xy$; $f_1(1, -2) = 22$, $f_2(1, -2) = -40$ (c) $\dfrac{\partial f}{\partial x} = 3x^2 e^y$, $\dfrac{\partial f}{\partial y} = x^3 e^y$,

$f_1(2, 0) = 12$, $f_2(2, 0) = 8$ (d) $\dfrac{\partial f}{\partial x} = 5x^4 + 2xy$, $\dfrac{\partial f}{\partial y} = 2y^{-3} + x^2 - \dfrac{1}{y}$,

$f_x(0, 1) = 0$, $f_y(0, 1) = 2 - 1 = 1$ (e) $\dfrac{\partial f}{\partial x} = yuv$, $\dfrac{\partial f}{\partial y} = xuv$, $\dfrac{\partial f}{\partial u} = xyv$,

$\dfrac{\partial f}{\partial v} = xyu$, $f_1(1, 2, 3, 4) = 24$, $f_2(1, 2, 3, 4) = 12$, $f_3(1, 2, 3, 4) = 8$,

$f_4(1, 2, 3, 4) = 6$
2.   (a) $f_{xx} = f_{yy} = f_{xy} = f_{yx} = 0$ (b) $f_{xx} = 6$, $f_{xy} = 8y = f_{yx}$, $f_{yy} = -12y + 8x$
(c) $f_{xx} = 6xe^y$, $f_{yy} = x^3 e^y$, $f_{xy} = 3x^2 e^y = f_{yx}$ (d) $f_{xx} = 20x^3 + 2y$, $f_{yy} =$
$-6y^{-4} + y^{-2}$, $f_{xy} = 2x = f_{yx}$
There will be 16 second-order partials and 64 third-order partials in 1(e).
3. $C_y(x, y) = 10 + 0.09y^2$, so $C_y(15, 10) = 19$
4. $R_x(x, y) = 5 + 100x^{-2} + 2y$, so $R_x(100, 10) = 5 + 0.01 + 20 = 25.01$
5. $f_x(x, y, z) = -3 - yz$, so $f_x(15, 10, 20) = -203$

ANSWER   6.2. C

2.   (a) complementary (b) competitive (c) neither

6.3   Extreme Values of Functions of Two Variables

In this section we shall limit our discussion to functions of two variables because the analogous discussion for functions of more than two variables is extremely involved. Even the two-variable case is considerably more complicated than the single-variable situation, as we see in Rule 6.3.1 (although the first idea [part (a) of Rule 6.3.1] is the same as for the single-variable case). In order for a function of two variables to have a local maximum at a point (that is, a peak of the surface there), its cross sections must have a local maximum when cut by a plane parallel to the xz-

plane and when cut by a plane parallel to the yz-plane. This suggests, correctly, that Rule 6.3.1(a) is a necessary condition for a local extremum.
Part (b) is a new feature of the two-variable case and is essential for
this test. Conditions 1 and 2 correspond to the second derivative test
for single-variable functions. (*Note:* In what follows, "relative" = "local.")

6.3.1   Rule

If (a) $f_x(x_1, y_1) = 0$ and $f_y(x_1, y_1) = 0$, and
    (b) $[f_{xy}(x_1, y_1)]^2 - [f_{xx}(x_1, y_1)][f_{yy}(x_1, y_1)] < 0$, then

1. $f(x, y)$ has a relative maximum at (x_1, y_1) if $f_{xx}(x_1, y_1) < 0$.
2. $f(x, y)$ has a relative minimum at (x_1, y_1) if $f_{xx}(x_1, y_1) > 0$.

   Notice that if condition (b) fails, the test fails. But if (b) holds,
$f_{xx}(x_1, y_1)$ and $f_{yy}(x_1, y_1)$ must have a positive product since f_{xy}^2 is always
nonnegative; thus f_{xx} and f_{yy} must have the same sign. It follows that f_{yy}
may be used instead of f_{xx} in conditions 1 and 2 to obtain the same
result.
   The complexity of this test gives some indication of the problems involved in treating more than one variable.

6.3.2   Example

Find the extreme values of the function

$$f(x, y) = x^2 + 2x + y^2 + 4y + 10$$

Solution:
We first find the possible relative extrema of the function by taking its
first partial derivatives and setting them equal to zero.

$$f_x = 2x + 2 = 0 \text{ implies that } x = -1$$

and

$$f_y = 2y + 4 = 0 \text{ implies that } y = -2$$

Next we evaluate the second partial derivatives and consider condition (b):

$$f_{xx} = 2 \qquad f_{yy} = 2 \qquad f_{xy} = 0$$

so $0 - 2 \cdot 2 < 0$ and condition (b) is satisfied.
   Since condition 2 is also satisfied, f must have a relative minimum
at $(-1, -2)$.

6.3.3   Example

Find the extreme values of the function

$$f(x, y) = 2x - x^2 + 4y - y^2 + 10$$

Solution:

We first locate the possible relative extrema by taking first partial derivatives and setting them equal to zero.

$$f_x = 2 - 2x = 0 \text{ implies that } x = 1$$
$$f_y = 4 - 2y = 0 \text{ implies that } y = 2$$

Evaluating the second partial derivatives, we get

$$f_{xx} = -2 \qquad f_{yy} = -2 \qquad f_{xy} = 0$$

$0 - (-2)(-2) < 0$ and condition (b) again is satisfied.

Since condition 1 is satisfied this time, f must have a relative maximum at $(1, 2)$.

A nonhorizontal plane in three dimensions is given by an equation of the form $z = ax + by + c$, where a and b are not both zero. It is intuitively clear that a nonhorizontal plane should have no minima or maxima.

6.3.4   Example

Show that a plane has no extreme values.

Solution:

We use the equation given above for an arbitrary plane:

$$z = ax + by + c$$

Taking first and second partial derivatives, we get $\dfrac{\partial z}{\partial x} = a$ and $\dfrac{\partial z}{\partial y} = b$.

These clearly do not become zero at any value for x and y (except the trivial case when $a = b = 0$), so there cannot be any critical point; hence there are no extrema.

There are other surfaces in three dimensions that do not have extreme values. Sometimes condition (a) of Rule 6.3.1 is satisfied, but the critical point thus determined gives neither a local minimum nor a local maximum. If f_{xx} and f_{yy} have opposite signs, the point on the surface will be a <u>saddle point</u>, as shown in Figure 6.3–1.

Figure 6.3–1

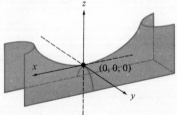

The following rule extends the ideas of Rule 6.3.1.

6.3.5   Rule

If $f_x(x_1, y_1) = 0$ and $f_y(x_1, y_1) = 0$ and $f^2_{xy}(x_1, y_1) - f_{xx}(x_1, y_1)f_{yy}(x_1, y_1) > 0$, there is a saddle point at (x_1, y_1).
If $f_x(x_1, y_1) = 0$ and $f_y(x_1, y_1) = 0$ and $f^2_{xy}(x_1, y_1) - f_{xx}(x_1, y_1) f_{yy}(x_1, y_1) = 0$, the test fails, and we have no information about the point.

6.3.6   Example

Show that the surface $f(x, y) = x^2 - y^2$ has no relative extrema and one saddle point.

Solution:

We begin by finding the first partial derivatives and setting them equal to zero:

$$f_x = 2x = 0 \text{ implies that } x = 0$$
$$f_y = -2y = 0 \text{ implies that } y = 0$$

Taking the second partial derivatives, we get $f_{xx} = 2, f_{yy} = -2$, and $f_{xy} = 0$, so $(f_{xy})^2 - f_{xx}f_{yy} = 0^2 - (2)(-2) > 0$ everywhere and condition (b) is not satisfied. Notice that $f_{xx} = 2 > 0$ and $f_{yy} = -2 < 0$ indicates that the critical point $(0, 0, 0)$ must be a relative maximum with respect to y and a relative minimum with respect to x, so we have a saddle point, as shown in Figure 6.3–1.

6.3.7   Example

A refinery produces two grades of steel in quantities of x and y tons, respectively. The total cost is $C(x, y) = x^2 + 100x + y^2 + 200y - xy$ and the revenue is $R(x, y) = 1{,}000x - x^2 + 2{,}000y - 2y^2 + xy$. Find the production level (x, y) that maximizes profit.

Solution:

Since profit $=$ revenue $-$ cost, we have

$$P(x, y) = -2x^2 - 3y^2 + 900x + 1{,}800y + 2xy$$
$$P_x = -4x + 900 + 2y \qquad P_y = 2x + 1{,}800 - 6y$$
$$P_{xx} = -4 \qquad P_{xy} = 2 \qquad P_{yx} = 2 \qquad P_{yy} = -6$$

Setting the first partials equal to zero and solving the resulting equations simultaneously for x and y, we find that $x = 450$ and $y = 450$ gives the only possible extremum. Since

$$(P_{xy})^2 - (P_{xx})(P_{yy}) = 2^2 - (-4)(-6) = 4 - 24 = -20 < 0$$

condition (b) of Rule 6.3.1 is satisfied. Since $P_{xx} = -4 < 0$ we have a local maximum at the critical point (450, 450). Thus $P(450, 450) = 607,500$ is the maximum possible profit.

In such problems we often have the added restriction that x and y must both be positive—since it is impossible to sell fewer than zero tons of any grade of steel! In Example 6.3.7 such a restriction would have made no difference in the answer, since the maximum occurred where both x and y are positive. But if the solution had included a negative x or y (or both), we would have been forced to examine the boundary to find the absolute maximum in the region of interest.

The situation in more than one variable is analogous to that of one variable; the absolute maximum of a continuous function will occur either at a critical point or at a boundary point of the region of interest. If the critical point does not occur in the permissible region, we must test the boundary of that region. If the region is specified by $x \geq 0$, $y \geq 0$, the boundary will consist of the two half-lines

$$x = 0 \quad \text{and} \quad y \geq 0 \qquad \text{and} \qquad y = 0 \quad \text{and} \quad x \geq 0$$

Thus we must test for absolute maxima of the functions $f(0, y)$ and $f(x, 0)$ for $y \geq 0$ and $x \geq 0$, respectively. To see an example of how this is done, look at the solution to problem 2(b) in Exercises 6.3.C.

SUMMARY

A function of two variables has a local minimum or maximum only when both first partial derivatives are zero. But to show we have a local minimum or maximum we must use other tests, too, as given in Rule 6.3.1. This section included examples on how to use this rule and a brief discussion of what may happen at critical points where the rule's assumptions are not valid. To find absolute minima and maxima we must test boundary points as well as critical points, as in the single-variable case.

EXERCISES 6.3. A

1. Find the relative maximum and minimum values of the following functions.
   (a) $f(x, y) = x^2 + y^2$
   (b) $f(x, y) = x^2 + 4x + y^2 - 8y + 3$
   (c) $f(x, y) = 6x - x^2 - 10y - y^2 + 7$
   (d) $f(x, y) = 2y - 6x - x^2 - y^2$
   (e) $f(x, y) = x^3 - 12xy + 8y^3$
2. The total profit obtained by selling a product on two different markets is given by $P(x, y) = 1,000 - x^2 - y^2 + 100x + 120y$, $x \geq 0$, $y \geq 0$. Find the maximum profit.
3. The cost of repairs at two inspection points on an assembly line is given by

$C(x, y) = 4x^2 + 2y^2 - 24x - 12y + 90$. What is the number of inspections that should be made to minimize the cost of repairs?

4. Suppose that the cost of producing x gizmos and y widgets is $C(x, y) = x^2 - 6x + 2y^2 - 5y + 500$. If each gizmo sells for \$4 and each widget sells for \$7, find the number of each that gives maximum profit.

EXERCISES 6.3. B

1. Find the relative minimum and maximum values of the following functions.
   (a) $f(x, y) = 2x^2 + y^2$
   (b) $f(x, y) = x^2 + 6x + y^2 - 4y + 5$
   (c) $f(x, y) = 8x - x^2 + 2y - y^2 + 3$
   (d) $f(x, y) = 4x - x^2 - 10y - y^2 + 7$
   (e) $f(x, y) = x^3 + y^3 - 6xy$
2. The total profit obtained by selling a product on two different markets is given by $P(x, y) = 2,000 - 2x^2 - 2y^2 + 240x + 200y$, $x \geq 0$, $y \geq 0$. Find the maximum profit.
3. The cost of repairs at two inspection points on an assembly line is given by $C(x, y) = x^2 + 2y^2 - 6x - 12y + 40$. What is the number of inspections that should be made to minimize the cost of repairs?
4. Suppose that if we sell x gizmos and y widgets, the total cost is $C(x, y) = -8x + 2x^2 - 7y + y^2 + 500$. If each gizmo is sold for \$8 and each widget is sold for \$7, find the number of gizmos and widgets that produce maximum profit.

EXERCISES 6.3. C

1. Find the relative minimum and maximum points of
   (a) $f(x, y) = xy - 6x - x^2 - y^2$ (b) $f(x, y) = x^3 - 12xy + 24y^3$
2. Find the absolute extreme values, if any, for $x \geq 0$, $y \geq 0$ of
   (a) $f(x, y) = 8x - x^2 + 2y - y^2 + 3$ (b) $f(x, y) = x^2 + 6x + y^2 - 4y + 5$
   (Hint: Use the results of problem 1 in Exercises 6.3.B.)
3. Show that the function $f(x, y) = xy$ has a saddle point at $(0, 0)$.

ANSWERS 6.3. A

1. (a) relative minimum at $(0, 0)$ (b) relative minimum at $(-2, 4)$
   (c) relative maximum at $(3, -5)$ (d) relative maximum at $(-3, 1)$
   (e) relative minimum at $(2, 1)$
2. maximum profit of \$7,100 at $(50, 60)$
3. three inspections at each station
4. 5 gizmos and 3 widgets

ANSWERS 6.3. C

1. (a) relative maximum at $(-4, -2)$ (b) relative minimum at $(6 \cdot 9^{-2/3}, 9^{-1/3})$
2. (a) Since the only relative maximum is at $(4, 1)$, which is in the region of interest, the absolute maximum will occur there. It will be $f(4, 1) =$

$32 - 16 + 2 - 1 + 3 = 20$. As x and/or y becomes very large, $f(x, y) \rightarrow -\infty$, so there is no absolute minimum.

(b) As x and/or y becomes very large, $f(x, y) \rightarrow \infty$, so there is no absolute maximum. Since the only local minimum occurs at the point $(-3, 2)$, which is not in the region, we must look at the boundary to find the absolute minimum. The boundary consists of points where $x = 0$ or $y = 0$. If $x = 0$, $f(0, y) = y^2 - 4y + 5$. Taking the derivative of this, we get $f' = 2y - 4$, which equals zero when $y = 2$, which is a local minimum by the second derivative test. Thus $f(0, 2) = 2^2 - 4(2) + 5 = 1$ is one possible test point.

If $y = 0$, then $f(x, 0) = x^2 + 6x + 5$, which has a derivative $f' = 2x + 6$, which equals zero when $x = -3$. But the point $(-3, 0)$ is not within the permitted region ($x \geqslant 0, y \geqslant 0$). Thus we must test the endpoint, $(0, 0)$. Since $f(0, 0) = 5$, and $1 < 5$, the absolute minimum in the specified region is 1; it is taken at the point $(0, 2)$.

3. $f_x = y, f_y = x, f_{xx} = 0, f_{yy} = 0, f_{xy} = 1$, so the only critical point is $(0, 0)$. Since $(f_{xy})^2 - f_{xx}f_{yy} = 1 - 0 > 0$, the critical point is a saddle point.

SAMPLE QUIZ   ## Sections 6.1, 6.2, and 6.3

1. (20 pts) The cost function for two products manufactured by a certain company is $C(x, y) = 10x + 0.01x^2 + 100 + 0.5xy + 5y - 0.02y^2$. Find the marginal cost of the second product (the one corresponding to y) at the production level $x = 10, y = 5$.

2. (20 pts) Find any local minima or maxima of

$$z = \tfrac{1}{2}x^2 - 6x + y^2 + 8y - 3$$

3. (40 pts) Find all first and second partial derivatives of

$$f(x, y) = z = x^2 e^y + \frac{\ln (x)}{y} + e^x$$

4. (20 pts) Graph the cross sections of $z = xe^y$ for $x = 1, 2$.

The answers are at the back of the book.

6.4   Applications to Least-Squares Models

Except for determining the linear and quadratic models for two and three points, respectively, in Section 1.4, this book has not attempted to motivate or derive the cost, profit, or demand functions that we have used so frequently. This section will introduce an important method for deriving such functions.

6.4.1   Illustrations

If we seek a function $y = f(x)$ that approximately describes past events and, hopefully, predicts the future, the first step is a careful tabulation of observed data. Then we can plot the data on a graph, getting a scatter diagram. For example, in Figure 6.4–1 we have plotted cumulative productivity versus time over a period of 5 months for some manufacturing process. This yields five points, not all on one straight line. Several lines may be drawn, as in Figure 6.4–2, each of which approximates the data in some fashion. The method of least squares asserts that the "best" such function is the one that minimizes the sum, S, of the squares of the vertical distances between the data values of y_i and those values of y corresponding to x_i on the straight-line graph:

$$S = \sum_{i=1}^{5} [y_i - (mx_i + b)]^2$$

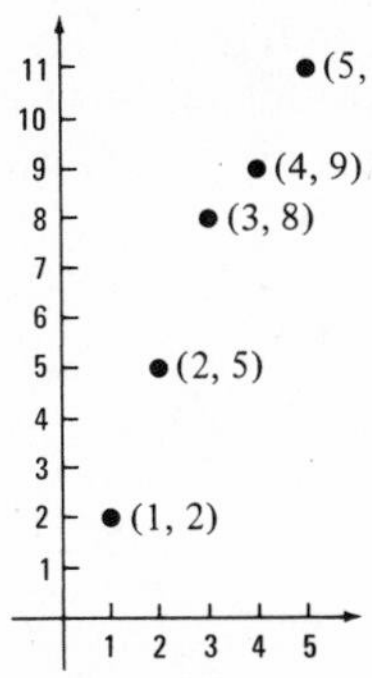

Figure 6.4–1 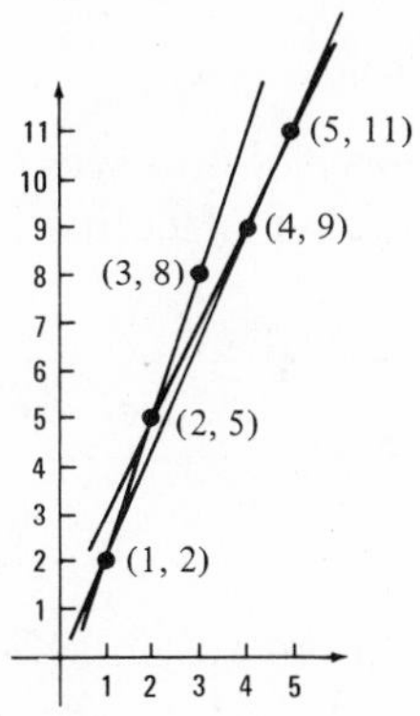   Figure 6.4–2 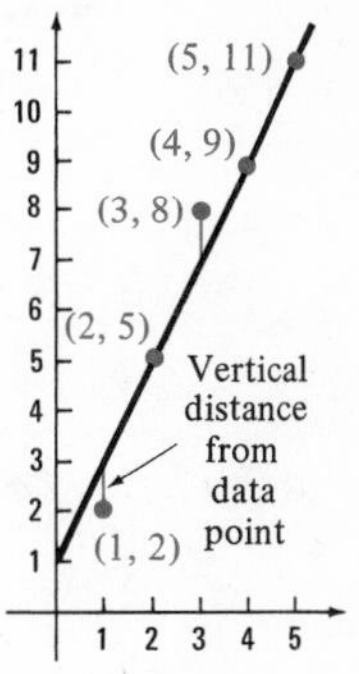   Figure 6.4–3

This line, as in Figure 6.4–3, is called the regression line of the data.

6.4.2   Example

Find the regression line for the data given in the scatter diagram of Figure 6.4–1. Use it to predict the cumulative productivity for the sixth month.

Solution:

The sum

$$S(m, b) = \sum_{i=1}^{5} [y_i - (mx_i + b)]^2 = [2 - (m + b)]^2 + [5 - (2m + b)]^2$$
$$+ [8 - (3m + b)]^2 + [9 - (4m + b)]^2 + [11 - (5m + b)]^2$$

is a function of the two independent variables, m and b. To minimize S, we set its first two partial derivatives equal to zero. We get

$$\frac{\partial S}{\partial m} = 2(-1)[2 - (m + b)] + 2(-2)[5 - (2m + b)]$$
$$+ 2(-3)[8 - (3m + b)] + 2(-4)[9 - (4m + b)]$$
$$+ 2(-5)[11 - (5m + b)] = 0$$

$$\frac{\partial S}{\partial b} = 2(-1)(2 - (m + b)) + 2(-1)[5 - (2m + b)]$$
$$+ 2(-1)[8 - (3m + b)] + 2(-1)[9 - (4m + b)]$$
$$+ 2(-1)[11 - (5m + b)] = 0$$

Collecting terms and dividing by 2, these become

$$55m + 15b = 127$$
$$15m + \ 5b = \ 35$$

which have the simultaneous solution $m = 2.2$ and $b = 0.4$. Thus the regression line is $y = 2.2x + 0.4$, and for $x = 6$ we predict $y = 2.2(6) + 0.4 = 13.2 + 0.4 = 13.6$.

6.4.3   Example

A supervisor observes that the productivity of his subordinates rises with the number of coffee breaks they have during the day according to the table

x (number of coffee breaks)	0	1	2	3
y (number of jobs completed)	2	2.5	3.5	4

Write the regression line for these data.

Solution:

Let

$$S(m, b) = \sum_{i=1}^{4} [y_i - (mx_i + b)]^2 = [2 - (0m + b)]^2$$
$$+ [2.5 - (m + b)]^2 + [3.5 - (2m + b)]^2 + [4 - (3m + b)]^2$$

Then

$$\frac{\partial S}{\partial m} = 0 - 2[2.5 - (m + b)] + 2(-2)[3.5 - (2m + b)]$$
$$+ 2(-3)[4 - (3m + b)] = 0$$
$$\frac{\partial S}{\partial b} = 2(-1)(2 - b) - 2[2.5 - (m + b)] - 2[3.5 - (2m + b)]$$
$$-2[4 - (3m + b)] = 0$$

Again dividing by 2 and collecting terms, this time we get

$$14m + 6b = 21.5$$
$$6m + 4b = 12$$

The simultaneous solution to these equations is $m = \frac{7}{10}$, $b = \frac{39}{20}$, so the equation of the regression line is $y = \frac{7}{10}x + \frac{39}{20}$.

The process we have used in the preceding two examples is straight-forward enough, but it would be boring if we had many points, and many points are generally required for an accurate prediction. It would be help-ful to have a formula that we can program on a computer. In the next example we shall derive such a formula.

But first let us emphasize that regression lines are not fortune-tellers; they must be used with common sense. The previous problem, for ex-ample, would predict that if the supervisor gives his employees 100 coffee breaks a day, they will complete 72 jobs! If we believe this conclusion, the fault is not with the regression line, of course, but with ourselves for using it improperly.

6.4.4   Example

Find the regression line fitting the points (x_1, y_1), (x_2, y_2), . . . , (x_n, y_n) of a scatter diagram.

Solution:

We seek a function $y = mx + b$ such that $\sum_{i=1}^{n} [y_i - (mx_i + b)]^2 = S(m, b)$

is a minimum. Setting $\dfrac{\partial S}{\partial m} = 0 = \dfrac{\partial S}{\partial b}$, we have

$$\frac{\partial S}{\partial m} = 2 \sum_{i=1}^{n} (-x_i)[y_i - (mx_i + b)] = 0$$

so

$$\sum_{i=1}^{n} y_i x_i = b \cdot \sum_{i=1}^{n} x_i + m \cdot \sum_{i=1}^{n} x_i^2$$

and

$$\frac{\partial S}{\partial b} = 2 \sum_{i=1}^{n} (-1)[y_i - (mx_i + b)] = 0$$

so

$$\sum_{i=1}^{n} y_i = nb + m \sum_{i=1}^{n} x_i$$

A computer can be programmed to solve the equations above. If we de-fine $\bar{x} = (1/n) \sum_{i=1}^{n} x_i$ and $\bar{y} = (1/n) \sum_{i=1}^{n} y_i$, the solution will be

$$m = \left(\sum_{i=1}^{n} x_i y_i - n\bar{x}\bar{y} \right) \Big/ \left(\sum_{i=1}^{n} x_i^2 - n\bar{x}^2 \right) \qquad \text{and} \qquad b = \bar{y} - m\bar{x}$$

Some regression-line problems are simplified by assuming that the line must go through the point $(0, 0)$. In Section 1.3 we saw that this was equivalent to having an equation of the form $y = mx$. In this case we say that y is proportional to x.

6.4.5   Example

In Example 6.4.2 suppose that the regression line must be of the form $y = mx$. Find the regression line of this form that minimizes the sum of the squares as before. Compare the new line with the old in terms of the data.

Solution:

Since we assume that the new regression line has the form $y = mx$, we minimize

$$S = \sum_{i=1}^{5} (y_i - mx_i)^2 = (2 - m)^2 + (5 - 2m)^2 + (8 - 3m)^2$$

$$+ (9 - 4m)^2 + (11 - 5m)^2$$

This is a function of one variable, m, so we merely take the first derivative and set it equal to zero.

$$\frac{dS}{dm} = 2(-1)(2 - m) + (-4)(5 - 2m) + (-6)(8 - 3m)$$

$$+ (-8)(9 - 4m) - 10(11 - 5m)$$

$$= 110m - 254 = 0$$

Hence $m = \frac{254}{110}$ is a solution. The second derivative is $110 > 0$, and so this gives a minimum. The required equation is $y = \frac{254}{110}x \approx 2.3x$.

To determine which line gives the best fit, we can calculate the sum of the squares of the differences for each function. The results are shown in Table 6.4–1.

Table 6.4–1 Sum of the Squares

x	y	$y_1 = 2.2x + 0.4$	$y_2 = 2.3x$	$(y - y_1)^2$	$(y - y_2)^2$
1	2	2.6	2.3	0.36	0.09
2	5	4.8	4.6	0.04	0.16
3	8	7.0	6.9	1.00	1.21
4	9	9.2	9.2	0.04	0.04
5	11	11.4	11.5	0.16	0.25
				1.60	1.75

Thus y_1 indeed gives a better "fit" in terms of least squares.

SUMMARY

In this section we considered an extremely useful optimization problem involving functions of several variables: find the best line or curve "fitting" a set of empirical data (scatter points). These lines (or curves) are called "regression lines." They can be found (1) by solving a minimization problem (least squares) or (2) by using a formula derived by the first method. The first technique is useful to know, even though we have a formula, since it can be applied to find other types of regression curves. (See problem 1 of Exercises 6.4.C.)

EXERCISES   6.4. A

1. Find the regression lines $y = mx + b$ for the following sets of data, using either the formula developed in Example 6.4.4 or using partial derivatives as in Examples 6.4.2 and 6.4.3. Plot both a scatter diagram and the regression line in each case.

(a)

x	1	2	3	4
y	3	5	7	9

(b)

x	1	2	3	4	5
y	1	4	5	4	6

2. Find the best least-squares line of the form $y = mx$ which fits each set of data in problem 1.
3. Compare the goodness of fit in terms of the sums of the squares of the deviations of the predictions from the data in problems 1 and 2.
4. The average demand q for a commodity has been observed to depend on the commodity's unit price, p, according to the accompanying table. Find the demand law given by a regression line $q = mp + b$ fitting the data.

p	25	50	75	100
q	1,005	760	825	610

5. The effect of feeding a certain number of units of a special feed to hogs at an experimental farm is expected to give a proportional increase in weight from the average of 100 pounds. Find the best line $y = mx$ fitting the observed data.

x (quantity of feed)	15	30	45
y (weight increase)	2	5	6

6. The average life expectancy of a female living in the United States is given in the following table for three different years. Find the least-squares line of the form $y = mx + b$ that fits these data, and use it to predict the life expectancy of the American female in the year 2000. (You may think of 1950 as year 1, 1960 as year 2, etc.)

Year, x	(1) 1950	(2) 1960	(3) 1970	(6) 2000
life expectancy of U.S. female, y	72	73	75	

EXERCISES   6.4. B

1. Find the regression lines $y = mx + b$ for the following sets of data, using either the formula developed in Example 6.4.4 or using partial derivatives as

in Examples 6.4.2 and 6.4.3. Plot both a scatter diagram and the regression line in each case.

(a)

x	1	2	3	4
y	1	2	4	6

(b)

x	2	4	6	8	10
y	2	8	10	8	12

2. Find the best least-squares line of the form $y = mx$ which fits each of the above sets of data.

3. Compare the goodness of fit in terms of the sums of the squares of the deviations of the predictions from the data in problems 1 and 2.

4. The average demand q for a commodity has been observed to depend on the commodity's unit price, p, according to the accompanying table. Find the demand law given by a regression line $q = mp + b$ fitting the data.

p	50	100	150	200
q	2,010	1,520	1,650	1,220

5. The following scores of a training group were observed by their personnel manager after 6 months of training. Find the least-squares line of the form $y = mx$ that best fits these data.

x (aptitude)	6	7	8	9
y (achievement)	8	8	9	10

6. It has been observed that crickets chirp more rapidly on warm nights than on cool nights. Find a least-squares line that fits the following data relating the number of cricket chirps per minute to temperature in degrees Fahrenheit. Use your least-squares model to predict the temperature at which the crickets will chirp at a rate of 80 chirps per minute.

chirps per minute	60	76	84	96
degrees Fahrenheit	55	59	61	64

EXERCISES 6.4. C

1. Find the best curve of the form $y = ax^2 + bx + c$ fitting the data in problem 1(a) of Exercises 6.4. A in terms of least squares.

2. Show that the average data points $(\bar{x}, \bar{y})$ always lie on the regression line.

3. Show that the sum of the signed deviations of the data from the regression line $\left(\text{that is, } \sum_{i=1}^{n} [y_i - (mx_i + b)]\right)$ is always zero.

4. Show that the best least-squares line of the form $y = mx$ fitting points $((x_1, y_1), (x_2, y_2), \ldots, (x_n, y_n))$ is such that
$$m = \frac{x_1y_1 + x_2y_2 + \cdots + x_ny_n}{x_1^2 + x_2^2 + \cdots + x_n^2}$$

HC 5.   The 1940, 1950, 1960, and 1970 death rates from lung cancer for American males (per hundred thousand) are given below. Use the formula derived in Example 6.4.4 to find the best least-squares line of the form $y = mx + b$ to fit the data given. We let 1940 be time $x = 0$: 1940, 8; 1950, 18; 1960, 33; and 1970, 45.

	x_i	y_i	x_i^2	$x_i y_i$
	0	8		
	10	18		
	20	33		
	30	45	900	1,350
sums	60			

ANSWERS 6.4. A

1.   (a) $y = 2x + 1$ (b) $y = x + 1$
2.   (a) $y = \frac{7}{3}x$ (b) $y = \frac{14}{11}x$
3.   (a) The sum of squares for number 1 is 0; for number 2 the sum is $\frac{2}{3}$. Therefore, the first is the better fit. (b) The sum of squares for number 1 is 4; for number 2 the sum is $4\frac{10}{11}$. Therefore, the first is the better fit.
4. $q = -\frac{112}{25}p + 1,080$
5. $y = \frac{450}{3,150}x$
6. $y = \frac{3}{2}x + 70\frac{1}{3}$; $79\frac{1}{3}$ in 2000 A.D.

ANSWERS 6.4. C

1.   Solving three equations of the form $\frac{\partial S}{\partial a} = \frac{\partial S}{\partial b} = \frac{\partial S}{\partial c} = 0$, obtain $a = 0$, $b = 2$, and $c = 1$. Thus $y = 0 \cdot x^2 + 2x + 1$.
2. $y = \bar{x}\dfrac{\sum x_i y_i - n\bar{x}\bar{y}}{\sum x_i^2 - n\bar{x}^2} + \left(\bar{y} - \dfrac{\sum x_i y_i - n\bar{x}\bar{y}}{\sum x^2 - n\bar{x}^2}\bar{x}\right) = \bar{y}$
3. $\sum [y_i - (mx_i + b)] = \sum y_i - \sum (mx_i + b) = n\bar{y} - nm\bar{x} - nb = 0$ because $\bar{y} = m\bar{x} + b$ by problem 2 of Exercises 6.4. C.
4.   Differentiate $S = \sum_{i=1}^{n} (mx_i - y_i)^2$ with respect to m; set $\dfrac{dS}{dm} = 0$ and solve for m.
5. $y = 1.26x + 7.1$

6.5   Constrained Extrema Using Lagrange Multipliers

Often we want to find the extreme values of a function $f(x, y)$ (for example, the maximum of a profit function) where there are special conditions, called *constraints*, on the variables x and y. (For example, we might have to limit our total production of the two products while maximizing the profit.) The constraint can usually be expressed in the form $g(x, y) = 0$. For example, if the constraint is $x + y = 10$, we can write $g(x, y) = x + y - 10 = 0$.

One method that can sometimes be used for solving such problems is algebraic substitution. We solve for one variable in terms of the other in the constraining equation, obtaining $y = h(x)$. Then we find the extrema of $f(x, h(x))$, a function of one variable, using the techniques of Chapter 2.

6.5.1   Example

The cost of paying for the inspection of an assembly-line operation depends upon the number of inspections, x and y, at each of two sites according to the formula $C(x, y) = 2x^2 + xy + y^2 + 100$. How many inspections should be made at each site to minimize the cost if the total number of inspections must be 16?

Solution:

Using the constraint $x + y = 16$, we can write $y = 16 - x$. Then we can write the cost function

$$C(x, 16 - x) = 2x^2 + x(16 - x) + (16 - x)^2 + 100$$
$$= 2x^2 + 16x - x^2 + 256 - 32x + x^2 + 100$$
$$= 2x^2 - 16x + 356 = f(x)$$

Thus $f'(x) = 4x - 16$, which equals zero when $x = 4$. Since $f''(x) = 4 > 0$, the function takes a minimum (of 324) at $x = 4$. At this point $y = 16 - 4 = 12$. Hence we make four inspections at the first station and 12 at the second.

6.5.2   Example

Find all the extreme values of the function $f(x, y) = x^2 + xy + y^2 + 100$ subject to the constraint $x^2 + y^2 = 18$.

Solution:

In this case we have $x = \pm\sqrt{18 - y^2}$, so we must substitute both of these values into the given equation.

$$f(\pm\sqrt{18 - y^2}, y) = 18 - y^2 \pm (\sqrt{18 - y^2})y + y^2 + 100$$

so

$$f' = -2y \pm \tfrac{1}{2}(18 - y^2)^{-1/2}(-2y)y \pm \sqrt{18 - y^2} + 2y$$
$$= \mp y^2(18 - y^2)^{-1/2} \pm \sqrt{18 - y^2}$$

Setting this equal to zero, we get $y^2 = 18 - y^2$, or $2y^2 = 18$. Thus $y^2 = 9$, giving $y = \pm 3$. Substituting back into $x = \pm\sqrt{18 - y^2}$, we get that $x = \pm 3$, so there are four critical points of $f(x, y)$: $(3, 3)$, $(-3, -3)$, $(-3, 3)$, and $(3, -3)$. The first two are local maxima and the second two are local minima. We conclude this, for example, by noting that $f(3, 3) = 127$, while for the nearby constrained point $(4, \sqrt{2})$, we have that $f(4, \sqrt{2}) = 16 + 4\sqrt{2} + 2 + 100 \approx 123.6 < 127$.

To use the substitution method shown in the previous two examples, we must be able to solve for one variable in terms of the other in the constraining equation. Sometimes this is difficult or impossible. Moreover, there may be more than one constraint, complicating matters further. In these situations one can use *Lagrange multipliers*, named after Joseph Louis Lagrange (1736–1813). We state here the method of Lagrange multipliers for the case of two variables and one constraining equation; later we shall indicate how it can be generalized.

6.5.3   Rule

The relative extrema of the function $f(x, y)$ subject to the condition $g(x, y) = 0$ are among those values of x and y simultaneously satisfying the equations

$$f_x(x, y) + L \cdot g_x(x, y) = 0$$
$$f_y(x, y) + L \cdot g_y(x, y) = 0$$
$$g(x, y) = 0$$

provided that all the indicated partial derivatives exist.

The variable L is called the Lagrange multiplier.

6.5.4   Example

Solve Example 6.5.1 using the method of Lagrange multipliers.

Solution:

We wish to minimize $f(x, y) = 2x^2 + xy + y^2 + 100$ subject to the constraint $x + y = 16$ [that is, $g(x, y) = x + y - 16 = 0$]. Following Rule 6.5.3, we compute $f_x = 4x + y$, $f_y = x + 2y$, $g_x = 1$, and $g_y = 1$, and then write the three equations:

$$4x + y + L \cdot 1 = 0$$
$$x + 2y + L \cdot 1 = 0$$
$$x + y - 16 = 0$$

Subtracting the second equation from the first, we get $3x - y = 0$, which gives $y = 3x$. Substituting this into the third equation, we get $x + 3x = 16$, which yields $x = 4$. Since $y = 3x$, $y = 12$. In this case it happens that $L = -28$.

6.5.5   Example

Solve Example 6.5.2 using Lagrange multipliers.

Solution:

Since $f(x, y) = x^2 + xy + y^2 + 100$ and $g(x, y) = x^2 + y^2 - 18$, we have $f_x = 2x + y$, $f_y = x + 2y$, $g_x = 2x$, and $g_y = 2y$. Thus the three equations are

(1) $2x + y + L \cdot 2x = 0$
(2) $x + 2y + L \cdot 2y = 0$
(3) $x^2 + y^2 = 18$

Multiply equations (1) and (2) by y and $-x$, respectively, and add, eliminating the L terms; we get

(4) $2xy + y^2 - x^2 - 2xy = 0$ or $y^2 - x^2 = 0$

Adding equations (3) and (4), we get $2y^2 = 18$, so $y = \pm 3$. Equation (4) then tells us that $x = \pm 3$.

It is not hard to extend the method of Lagrange multipliers to more than two independent variables. For each additional variable, we simply take another partial derivative and add another equation. If we wish to optimize $f(x, y, z, w)$, for example, subject to the constraint $g(x, y, z, w) = 0$, we add the two equations $f_z(x, y, z, w) + L \cdot g_z(x, y, z, w) = 0$ and $f_w(x, y, z, w) + L \cdot g_w(x, y, z, w) = 0$ to the set since there are two additional variables.

6.5.6   Example

Maximize the function $f(x, y, z) = xyz$ subject to the constraint $x + y + z = 60$.

Solution:

We have $f_x = yz$, $f_y = xz$, $f_z = xy$, and $g_x = g_y = g_z = 1$. Thus

(1) $yz + L = 0$
(2) $xz + L = 0$
(3) $xy + L = 0$
(4) $x + y + z = 60$

From equations (1) and (2) we get $yz = xz$, so $y = x$. [If $z = 0$, $f(x, y, 0) = 0$, which is clearly not a maximum.] From equations (2) and (3) we get $xz = xy$, so $z = y$. Substituting these two results into equation (4), we get $x + x + x = 60$, so $x = 20$. It follows that $y = z = 20$.

To see if $x = 20$, $y = 20$, $z = 20$ really gives a maximum, not a minimum, we might compare $f(20, 20, 20) = 8,000$ with the value of f at a nearby point that also satisfies the constraint. Since $f(20, 19, 21) = 7,980 < 8,000$, we conclude that f does indeed take a maximum at $(20, 20, 20)$.

We close this section with the remark that it is possible to extend the method of Lagrange multipliers to cases where there is more than one constraining equation. If we wish to optimize $f(x, y)$, for example, subject to the constraints $g(x, y) = 0$ and $h(x, y) = 0$, we use two Lagrange multipliers, solving the equations $f_x + L \cdot g_x + K \cdot h_x = 0$, $f_y + L \cdot g_y + K \cdot h_y = 0$, $g(x, y) = 0$, and $h(x, y) = 0$ simultaneously for the variables x, y, L, and K.

6.5.7   Example

Solve Example 6.5.6 with the added constraint that $y = x + z$.

Solution:

We include an extra Lagrange multiplier K, and an extra equation in our calculations. Since $h(x, y, z) = -x + y - z$, we have $h_x = -1$, $h_y = 1$, and $h_z = -1$. Thus our equations are

(1) $yz + L - K = 0$
(2) $xz + L + K = 0$
(3) $xy + L - K = 0$
(4) $x + y + z = 60$
(5) $-x + y - z = 0$

Adding equations (4) and (5), we get $2y = 60$, so $y = 30$. Substituting this

into equations (1) and (3) and then subtracting equation (3) from equation (1), we get

$$30z - 30x = 0 \qquad \text{so } z = x$$

Since $y = x + x = 2x$, we have $x = 15$ and thus $z = 15$.

SUMMARY

The problem considered in the present section is that of optimizing a function of several variables where the variables are subject to some additional conditions called constraints. Two methods were discussed: substitution and the powerful method of Lagrange multipliers. This method involves introducing new variables, one for each constraint, and solving a system of simultaneous equations, one for each variable and one for each constraint. Lagrange multipliers are widely used since in real life constraints (or limiting conditions) are the rule rather than the exception.

EXERCISES   **6.5. A**

1. Maximize $f(x, y) = 6xy - x^3 - y^3$ subject to $x + y = 4$.
2. Find all extrema of $f(x, y) = x + y$ subject to $x^2 + y^2 = 50$.
3. Maximize $f(x, y, z) = xyz$ subject to $x + y + z = 12$.
4. Minimize $f(x, y, z) = x^2 + y^2 + 2z^2$ subject to $x + y = 20$ and $y + z = 20$.
5. The Multiwealth Corporation has three sales offices, X, Y, and Z, which

sell an average of x, y, and z units per year, respectively. If the goal for the year is to sell a total of 200 units, and the profit from selling such units is measured by the function $P(x, y, z) = 2xy + yz + xz$, find the maximum profit.

6. A certain profligate decides to spend his million dollars on wine, people, and song in amounts x, y, and z. He measures his pleasure according to the function $P(x, y, z) = x + yz - xz$, but he is constrained by the fact that he must have precisely as much wine as people and song combined. Maximize his pleasure!

EXERCISES 6.5. B

1. Maximize $f(x, y) = xy$ subject to $x + y = 4$.
2. Find all extrema of $f(x, y) = x + y$ subject to $3x^2 + y^2 = 48$.
3. Minimize $f(x, y, z) = x^2 + y^2 + 2z^2$ subject to $x + y + z = 30$.
4. Find all extrema of the function $f(x, y, z) = x^2 + y^2 + z^2$ subject to the constraints $x + y = 1$, $y = 2z$.
5. Suppose that the revenue received from selling a certain item is $R = x^2 + 3xy - 6y + 100$, where x is the amount sold of the given item and y is the amount sold of its competitor. Find the maximum revenue if the total amount sold of the item and its competitor will be 50.
6. A medical researcher is planning a diet for a group of experimental animals that are to be fed at a fixed level of 196 kilocalories. He can choose from two kinds of food, the first of which costs $2 per unit and the second of which costs $4 per unit. If the number of kilocalories provided by x units of the first and y units of the second is 8xy, how many units of each type of food will give a minimal cost?

EXERCISES 6.5. C

1. If the market price of widgets is given in terms of its quantity, x, and that of its competitor, y, according to the formula $D = y^2 + 3xy - 6x + 100$, and the sum of x and y is always 50, find the quantity of both items that makes the price for widgets a maximum.
2. An interpretation of the Lagrange multiplier L is $L = -df/dg$, where f is the function to be optimized and g is the constraining function. Sometimes L has the special interpretation of "marginal change in utility with respect to income" or "marginal utility of income"; a unit change in income produces $-L$ units change in utility.
    (a) What is the marginal utility of an income of $100 which is to be completely used up on food, shelter, clothing, and whiskey, measured in dollars by x, y, z, and w, respectively (x and w are independent, of course), if utility is measured by $f(x, y, z, w) = xyz + w$?
    (b) What is the marginal utility of an income of $100, as in part (a), if utility is measured instead by the function $f(x, y, z, w) = xyzw^2$?

ANSWERS 6.5. A

1. $f(2, 2) = 8$
2. $f(5, 5) = 10$ is the maximum. $f(-5, -5) = -10$ is the minimum

3. $f(4, 4, 4) = 64$
4. $f(5, 15, 5) = 300$
5. $P(100, 100, 0) = 20,000$
6. $x = 500,000, y = 500,000, z = 0, P_{max} = 500,000$

ANSWERS   6.5. C

1. $x = 11$ and $y = 39$
2. (a) $L = -1$ (b) $L = -640,000$

6.6   Multiple Iterated Integrals

Integration, like differentiation, can be generalized from one to several variables. In a manner similar to that of taking partial derivatives, we consider only one variable at a time while pretending the others are constants. In this section, for simplicity, we shall restrict ourselves to continuous functions of two variables and shall assume that all integrals in sight exist. (See Exercises 6.6.C for the three-variable case.)

6.6.1   Definition

Let $f(x, y)$ be a continuous function over the rectangle $a \leqslant x \leqslant b, c \leqslant y \leqslant d$. Then the symbol

$$\int_c^d \int_a^b f(x, y) \, dx \, dy$$

is called the <u>double iterated integral of $f(x, y)$</u> over the described region.

The double integral is evaluated by writing

$$\int_c^d \int_a^b f(x, y) \, dx \, dy = \int_c^d \left\{ \int_a^b f(x, y) \, dx \right\} dy$$

where the integral in braces is evaluated first while treating y as a constant and the resulting expression is then integrated with respect to y.

6.6.2   Example

Evaluate the following double iterated integrals.

(a) $\displaystyle\int_3^4 \int_1^2 6xy^2 \, dx \, dy$

(b) $\displaystyle\int_1^2 \int_3^4 6\,xy^2 \, dy \, dx$

Solution:

(a) $\int_3^4 \int_1^2 6xy^2 \, dx \, dy = \int_3^4 \left\{ \int_1^2 6xy^2 \, dx \right\} dy = \int_3^4 \left\{ 3x^2y^2 \Big|_{x=1}^{x=2} \right\} dy$

$\qquad = \int_3^4 (12y^2 - 3y^2) \, dy = \int_3^4 9y^2 \, dy = 3y^3 \Big|_3^4 = 111$

(b) $\int_1^2 \int_3^4 6xy^2 \, dy \, dx = \int_1^2 \left\{ \int_3^4 6xy^2 \, dy \right\} dx = \int_1^2 \left\{ 2xy^3 \Big|_{y=3}^{y=4} \right\} dx$

$\qquad = \int_1^2 (128x - 54x) \, dx = \int_1^2 74x \, dx = 37x^2 \Big|_1^2$

$\qquad = 37(4 - 1) = 111$

Example 6.6.2 illustrates the fact that whenever $f(x, y)$ is a continuous function, the order of integration may be reversed. That is,

$$\int_c^d \int_a^b f(x, y) \, dx \, dy = \int_a^b \int_c^d f(x, y) \, dy \, dx$$

6.6.3   Example

Evaluate the following double iterated integrals.

(a) $\int_1^2 \int_0^1 e^x y \, dx \, dy$

(b) $\int_0^1 \int_1^2 e^x y \, dy \, dx$

Solution:

(a) $\int_1^2 \left\{ \int_0^1 e^x y \, dx \right\} dy = \int_1^2 \left\{ e^x y \Big|_{x=0}^{x=1} \right\} dy = \int_1^2 (ey - (1)y) \, dy$

$\qquad = \int_1^2 (e - 1)y \, dy = \frac{e-1}{2} y^2 \Big|_1^2 = \frac{e-1}{2} (4 - 1)$

$\qquad = \frac{3}{2}(e - 1)$

(b) $\int_0^1 \int_1^2 e^x y \, dy \, dx = \int_0^1 \left\{ \int_1^2 e^x y \, dy \right\} dx = \int_0^1 \left\{ e^x \frac{y^2}{2} \Big|_{y=1}^{y=2} \right\} dx$

$\qquad = \int_0^1 e^x \left(\frac{4-1}{2} \right) dx = \frac{3}{2} \int_0^1 e^x \, dx = \frac{3}{2} e^x \Big|_0^1 = \frac{3}{2}(e - 1)$

Double iterated integrals may be used to evaluate joint density functions in statistics, that is, population density functions that depend upon two variables. As in Section 5.2, $f(x, y)$ is a density function if

1. $f(x, y) \geqslant 0$ for all x and y.

2. $\int_c^d \int_a^b f(x, y) \, dx \, dy = 1$, where all nonzero values of f are such that

   $a \leqslant x \leqslant b$ and $c \leqslant y \leqslant d$.

Then the probability that $\alpha \leqslant x \leqslant \beta$ and $\gamma \leqslant y \leqslant \delta$ is given by

$$\int_\gamma^\delta \int_\alpha^\beta f(x, y) \, dx \, dy$$

In other words, this is the probability that x and y will both satisfy the given conditions simultaneously. The following example shows how this might be used.

6.6.4   Example

The joint probability density of x, the length of a telephone call, and y, the number of calls made on a random day, by executives of a certain business firm is given by

$$f(x, y) = \begin{cases} ke^{-x}y^{-2} & \text{for } x \geqslant 0,\, y \geqslant 1 \\ 0 & \text{elsewhere} \end{cases}$$

(a) Determine the value of k.
(b) What is the probability that a randomly selected executive will make between 5 and 10 phone calls on a certain day that last between 1 and 2 minutes?

Solution:

(a) $1 = \displaystyle\int_1^\infty \int_0^\infty ke^{-x}y^{-2} \, dx \, dy = \int_1^\infty \left\{ -ke^{-x}y^{-2} \Big|_{x=0}^{x=\infty} \right\} dy$

$= \displaystyle\int_1^\infty \left\{ 0 - [-k(1)y^{-2}] \right\} dy = k \int_1^\infty y^{-2} \, dy = k(-1)y^{-1} \Big|_1^\infty$

$= k(-1)(0 - 1) = k$

Therefore, $k = 1$.

(b) The probability is given by

$$\int_5^{10} \int_1^2 e^{-x}y^{-2} \, dx \, dy = \int_5^{10} (e^{-1} - e^{-2})y^{-2} \, dy = (e^{-1} - e^{-2})(-y^{-1}) \Big|_5^{10}$$

$$= (e^{-1} - e^{-2})(\tfrac{1}{5} - \tfrac{1}{10}) = \tfrac{1}{10}(e^{-1} - e^{-2})$$

6.6.5   Example

A joint probability function is given by

$$f(x, y) = \begin{cases} ke^{-x-2y} & \text{for } x \geqslant 0,\, y \geqslant 0 \\ 0 & \text{otherwise} \end{cases}$$

(a) Determine k.

(b) Find the probability that x ≤ 1 and y ≤ 2.

Solution:

(a) $1 = \int_0^\infty \int_0^\infty ke^{-x-2y} \, dx \, dy = \int_0^\infty \left\{ -ke^{-x-2y} \Big|_{x=0}^{x=\infty} \right\} dy = \int_0^\infty ke^{-2y} \, dy =$

$(-\frac{1}{2})ke^{-2y} \Big|_0^\infty = \frac{1}{2}k$, so we conclude that $k = 2$.

(b) The given probability is $\int_0^2 \int_0^1 2e^{-x-2y} \, dx \, dy = \int_0^2 (-2e^{-x-2y}) \Big|_0^1 \, dy =$

$\int_0^2 (2e^{-2y} - 2e^{-1-2y}) \, dy = (-e^{-2y} + e^{-1-2y}) \Big|_0^2 = -e^{-4} + e^{-5} + e^0 - e^{-1} =$

$e^{-5} + 1 - e^{-4} - e^{-1}$.

Just as a definite integral of a function of one variable can be pictured as an *area* under the curve $y = f(x)$, the double iterated integral of $f(x, y) \geq 0$ can be thought of as the volume under the surface $z = f(x, y)$.

6.6.6   Example

Find the volume under the surface $z = -x + y$ over the region where $0 \leq x \leq 1$ and $2 \leq y \leq 4$.

Solution:

$V = \int_2^4 \int_0^1 (-x + y) \, dx \, dy = \int_2^4 \left(\frac{-x^2}{2} + xy \right) \Big|_{x=0}^{x=1} \, dy = \int_2^4 (-\frac{1}{2} + y) \, dy$

$= \left(\frac{-y}{2} + \frac{y^2}{2} \right) \Big|_2^4 = (-2 + 8) - (-1 + 2) = 5$

The volume thus calculated resembles a box; it has its bottom on the xy-plane and a sloping top, as shown in Figure 6.6–1.

Figure 6.6–1

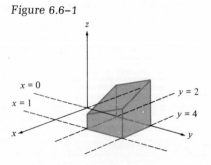

SUMMARY

A continuous function of more than one variable can be integrated by integrating over each variable in turn. This is called "iterated integration." It is useful for evaluating joint density functions and finding the volume under curved surfaces.

EXERCISES 6.6. A

1. Evaluate the following iterated integrals.

(a) $\int_1^0 \int_0^1 (x + y^2) \, dx \, dy$

(e) $\int_0^1 \int_2^3 (3 + y) \, dx \, dy$

(b) $\int_0^1 \int_1^0 (x + y^2) \, dy \, dx$

(f) $\int_2^3 \int_0^1 (3 + y) \, dy \, dx$

(c) $\int_0^1 \int_2^3 x^2 y \, dx \, dy$

(g) $\int_1^0 \int_0^1 e^{x+2y} \, dx \, dy$

(d) $\int_2^3 \int_0^1 x^2 y \, dx \, dy$

(h) $\int_0^1 \int_1^0 e^{x+2y} \, dy \, dx$

2. If $f(x, y) = ke^{-ax-by}$ for $x \geq 0$, $y \geq 0$, $f(x, y) = 0$ elsewhere, is a joint density function of x and y, evaluate k. (See Examples 6.6.4 and 6.6.5.)
3. If $f(x, y) = kx^2 y^3$, $-1 \leq x \leq 1$, $0 \leq y \leq 2$ is a joint density function, evaluate k.
4. For each of the functions in problems 2 and 3, find the probability that $0 \leq x \leq 1$ and $1 \leq y \leq 2$.
5. Find the volume of the region under the surface $f(x, y) = x^2 + y^2$, where $-1 \leq x \leq 1$ and $-1 \leq y \leq 1$.
6. Find the volume of the region under the surface $f(x, y) = xy$, where $0 \leq x \leq 1$, $1 \leq y \leq 2$.

EXERCISES 6.6. B

1. Evaluate the following iterated integrals.

(a) $\int_0^2 \int_2^0 (x^2 + y) \, dx \, dy =$

(e) $\int_2^3 \int_0^1 (3 + x) \, dy \, dx =$

(b) $\int_2^0 \int_0^2 (x^2 + y) \, dy \, dx =$

(f) $\int_0^1 \int_2^3 (3 + x) \, dx \, dy =$

(c) $\int_1^3 \int_0^1 xy^2 \, dx \, dy =$

(g) $\int_0^1 \int_1^0 e^{3x+y} \, dx \, dy =$

(d) $\int_0^1 \int_1^3 xy^2 \, dy \, dx =$

(h) $\int_1^0 \int_0^1 e^{3x+y} \, dy \, dx =$

2. If $f(x, y) = kye^{-x}$ for $x \geq 0$, $0 \leq y \leq 1$, $f(x, y) = 0$ elsewhere, and $f(x, y)$ is a joint density function, find k.
3. If $f(x, y) = kx^3 y^2$, $0 \leq x \leq 2$, $-1 \leq y \leq 3$, $f(x, y) = 0$ elsewhere, and $f(x, y)$ is a joint density function, find k.
4. For each of the functions in problems 2 and 3, find the probability that $1 \leq x \leq 2$ and $0 \leq y \leq 1$.

5. Find the volume of the region under the surface $f(x, y) = x + y$, where $0 \leqslant x \leqslant 1$ and $0 \leqslant y \leqslant 2$.

6. Find the volume of the region under the surface $f(x, y) = x^2y^3$, where $0 \leqslant x \leqslant 1$ and $0 \leqslant y \leqslant 2$.

EXERCISES   6.6. C

Extending Definition 6.6.1 in the natural way for a function of three variables, evaluate

1. $\int_0^1 \int_3^4 \int_1^2 (x + 2y + z) \, dx \, dy \, dz$ 3. $\int_1^2 \int_0^1 \int_3^4 (x + 2y + z) \, dy \, dz \, dx$

2. $\int_3^4 \int_1^2 \int_0^1 (x + 2y + z) \, dz \, dx \, dy$

ANSWERS   6.6. A

1. (a) $-\frac{5}{6}$ (b) $-\frac{5}{6}$ (c) $\frac{19}{6}$ (d) $\frac{5}{6}$ (e) $\frac{7}{2}$ (f) $\frac{7}{2}$ (g) $\frac{1}{2}(e + e^2 - e^3 - 1)$
   (h) $\frac{1}{2}(e + e^2 - e^3 - 1)$
2. $k = ab$
3. $k = \frac{3}{8}$
4. (a) $e^{-b} - e^{-a-b} - e^{-2b} + e^{-a-2b}$ (b) $\frac{15}{32}$
5. $\frac{8}{3}$
6. $\frac{3}{4}$

ANSWERS   6.6. C

1. 9   3. 9
2. 9

SAMPLE QUIZ   **Sections 6.4, 6.5, and 6.6**

1. (30 pts)

   (a) $\int_1^2 \int_0^1 (xy + e^x) \, dx \, dy =$

   (b) $\int_0^1 \int_1^2 (xy + e^x) \, dy \, dx =$

2. (30 pts) Maximize the function $f(x, y) = x^2 + 2xy + 2y$ subject to the constraint $x + y = 10$.

3. (30 pts) Using the least-squares method, find the regression line for the data points $(1, 0)$, $(2, 2)$, and $(3, 5)$.

4. (10 pts) Find the line going through the origin that gives the least-squares difference for the above points.

The answers are at the back of the book.

SAMPLE TEST   Chapter 6

Do 100 of the 110 available points.

1. (a) (5 pts) Evaluate the function $f(x, y, z) = x^2 + xy + yz^3 + 10$ at $x = 1$, $y = 2$, $z = -1$.

    (b) (5 pts) Write the linear cost-output equation of producing x widgets and y gizmos if each widget costs \$1.29, each gizmo costs \$0.98, and the fixed overhead is \$100.

    (c) (5 pts) Sketch cross sections of the function $z = 2x^2 + 3xy + 5$ in the yz-plane, letting $x = 0$ and 1.

2. Let $f(x, y) = x^2 + y^3 + xy$.

    (a) (8 pts) Find $f_x(1, 2)$ and $f_y(1, 2)$.

    (b) (8 pts) Find f_{xx} and $\dfrac{\partial^2 f}{\partial y\, \partial x}$.

    (c) (4 pts) Write the equations of the line tangent to the surface above at the point $(1, 2, 10)$ in the plane $x = 1$.

3. Let $C(x, y) = x^2 + 6x + y^2 + 2y + 40$.

    (a) (10 pts) What is the minimum value of $C(x, y)$? Where does it occur?

    (b) (5 pts) Show that your answer is indeed a relative minimum using the second derivative test.

    (c) (5 pts) If $C(x, y)$ is a production cost function and we include the conditions $x \geqslant 0$, $y \geqslant 0$, what is now the answer to part (a)?

4. (a) (10 pts) Find the regression (least-squares) line for the points $(1, 2)$, $(3, 3)$, and $(5, 8)$ using partial derivatives.

    (b) (10 pts) Find the best line of the form $y = mx$ for the above data.

    (c) (5 pts) Which line gives a better fit?

5. (15 pts) Find all extreme values of the function $C(x, y) = x^2 + y^2$ subject to the condition $x + y = 4$.

6. (10 pts) Evaluate $\displaystyle\int_0^1 \int_1^2 (2x + 3y^2)\, dx\, dy$.

7. (5 pts) Evaluate $\displaystyle\int_1^2 \int_0^1 (2x + 3y^2)\, dy\, dx$.

The answers are at the back of the book.

Chapter 7

TRIGONOMETRY

7.1 Periodic Functions: Sine and Cosine

Many of the activities of man and nature repeat themselves in a more or less cyclical way. If a phenomenon as described by a function $f(t)$ repeats itself every time t changes by p units so that $f(t + p) = f(t)$ for all t, then mathematicians say that the phenomenon or function is <u>periodic with period p</u>.

Many phenomena are truely periodic in this strict mathematical way; others are only approximately periodic. For example, the earth makes exactly one revolution on its axis each day: a function that describes this motion must, therefore, have a period of one day. Consequently, many social and biological patterns, such as waking and sleeping, conform to a 24-hour cycle of activity. But these activities are only approximately periodic, as are business cycles, seasonal variations in precipitation, heartbeats, breathing patterns, population fluctuations, and the like. Nevertheless, any adequate mathematical model of these phenomena must take their periodic nature into account.

Figure 7.1-1, for example, shows graphs of the yearly numbers of lynx and snowshoe hare pelts collected by the Hudson's Bay Company in

Figure 7.1–1

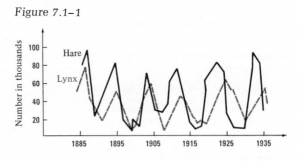

Canada over several years. Both graphs are roughly periodic, with a period of about 10 years. A strongly cyclical interrelationship is indicated between the predator and prey populations, in addition to the periodic rising and falling supply of pelts available for sale by the company.

A somewhat more regular periodic pattern is sometimes obtained from an electrocardiogram, as it records heartbeats translated into elec-

Figure 7.1–2

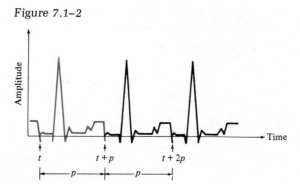

trical impulses. An idealized version of this pattern is given in Figure
7.1–2.

Notice that the graph in Figure 7.1–2 has one *complete* cycle every *p*
units. In other words, the fundamental shape of the curve is repeated in
this interval but in no smaller subinterval. For this reason, we say that
the period of *p* units is a <u>fundamental period</u>. One cannot obtain the
characteristic shape (from *t* to *t* + *p*) in any shorter interval.

Figure 7.1–3

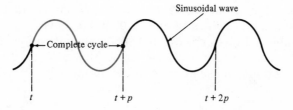

For a strikingly regular example of a periodic function, there is the
familiar undulating pattern of the sine (or sinusoidal) wave induced by
an AC electric current, as it is often seen on an oscilloscope. Indeed, the
sine curve, or function, and its complement, the cosine, are the most important tools mathematicians have for modeling periodic phenomena.
We now turn our attention to these functions, sine and cosine.

Sine and Cosine

The definitions of the sine and the cosine of an angle which we often
learn in elementary algebra have ostensibly little to do with periodic
functions. Instead, they are given in terms of ratios between lengths of
the sides of a right triangle. If an angle *A* of a right triangle has a measure
of *t* degrees, the trigonometric definitions of the sine and cosine of *t*°
(abbreviated as sin *t*° and cos *t*°, respectively) are

$$\sin t° = \frac{\text{length of side opposite } A}{\text{length of hypotenuse}}$$

$$\cos t° = \frac{\text{length of side adjacent to } A}{\text{length of hypotenuse}}$$

Figure 7.1–4

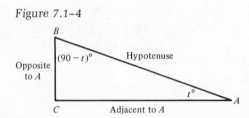

Two angles of a triangle are <u>complementary</u> if the sum of their angles is 90°. It is clear from Figure 7.1–4 that $\sin t° = \cos (90 - t)°$ and $\cos t° = \sin (90 - t)°$. That is, the sine of an angle equals the cosine of the complement of that angle.

7.1.1   Example

The value of $\sin 30° = \frac{1}{2}$. The value of $\cos 30° = \sqrt{3}/2$. These facts can be obtained, for instance, from the right triangle ABC with hypotenuse of length s, as in Figure 7.1–5. Or we may resort to Table VII, where the sines and cosines of angles 0° to 90° have been compiled. (Notice that Table VII gives $\cos 30° = 0.8660$, which is a four-place approximation to the exact value $\sqrt{3}/2$.) Even in this form, $\sin t°$ and $\cos t°$ are useful functions whose domain of definition is the interval $0 \leq t \leq 90$.

Figure 7.1–5

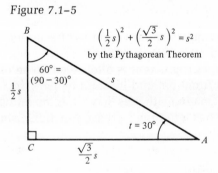

7.1.2   Example in Biology

Experimental data show that the guppy (*Lebistes reticulatis*) uses forces L and G induced by light and gravitation, respectively, to adjust its upright position in the water. (See Figure 7.1–6.) We may think of these directed quantities (or vectors) as the arrows in Figure 7.1–6. They are

Figure 7.1-6

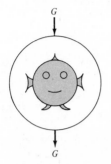 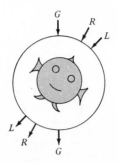 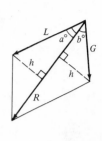

parallel to the rays of light entering water in the case of L, or to the vertical force of gravity in the case of G. Moreover, they generate a resultant force, R, the diagonal of the parallelogram whose sides are L and G, which determines the position of the guppy. Using our definition of sine and Figure 7.1-6, we find that $\sin a° = h/L$ and $\sin b° = h/G$. Dividing these, we get

$$\frac{\sin a°}{\sin b°} = \frac{h/L}{h/G} = \frac{G}{L} \qquad \text{or simply} \qquad \frac{\sin a°}{\sin b°} = \frac{G}{L}$$

where $a°$ and $b°$ are the angles between L and R and between G and R, respectively.

For example, if we know that G is always 32 units, and we observe that all the guppies in our pond are tilted at such an angle that $a° = 45°$ and $b° = 30°$, it is possible to compute the intensity, L, of the force induced by the light rays on each guppy that day. Using Table VII, a hand calculator, or diagrams such as Figure 7.1-5, we can find that $\sin 30° = 0.5$ and $\sin 45° = 0.7071$. Thus since

$$\frac{\sin a°}{\sin b°} = \frac{G}{L} \qquad \text{we have} \qquad \frac{0.7071}{0.5} = \frac{32}{L}$$

which implies that $L \approx 22.6276$.

It is desirable to extend the domains of $\sin t°$ and $\cos t°$ so that t may be any real number. The following definition does this and more. It makes the sine and cosine into periodic functions, each having a fundamental period of 360°.

7.1.3   Definition

Given a point $P:(x, y)$ on the unit circle $x^2 + y^2 = 1$, we define

$$\sin t° = y \qquad \text{and} \qquad \cos t° = x$$

where t is the number of degrees in the angle measured counterclockwise from the x-axis to the unit radius OP. (See Figure 7.1-7.)

Figure 7.1-7

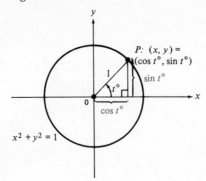

If t is negative, the angle is measured clockwise.

Notice that sin $t° = y/1 = y$ and cos $t° = x/1 = x$, so that our new definition is consistent with the old version. Further, $\cos^2 t° + \sin^2 t° = 1$, since $x^2 + y^2 = 1$. [Note that we write $\sin^2 t°$ when we mean $(\sin t°)^2$.]

We see that the values of the functions cos $t°$ and sin $t°$ vary continuously and periodically between 1 and -1 as the angle t sweeps continuously around the circle. Indeed,

$$\cos (t + 360)° = \cos t° \qquad \text{and} \qquad \sin (t + 360)° = \sin t° \quad \text{for all } t$$

so the sine and cosine both have a period of 360°. It is easy to see why the sine and cosine are called "circular" functions: to evaluate them, we locate points on the unit circle.

7.1.4   Example

Find the sine and cosine of t, where (a) $t = 0°$, 90°, 180°, 270°; (b) $t = -90°$, 450°. (c) Find $\cos^2 150° + \sin^2 150°$.

Solution:

(a) The given angles are obtained when P is at the points $(1, 0)$, $(0, 1)$, $(-1, 0)$, and $(0, -1)$, respectively. (See Figure 7.1-8.) It follows, for example, that cos 270° $= 0$ and sin 270° $= -1$.

(b) To find cos $(-90°)$, we use the periodicity of the cosine function. Thus,

$$\cos (-90°) = \cos (-90° + 360°) = \cos 270° = 0$$

Similarly, sin $(-90°) = \sin 270° = -1$, cos 450° $= \cos 90° = 0$, and sin 450° $= \sin 90° = 1$.

(c) This example does not require any observation other than $\cos^2 t° + \sin^2 t° = 1$. Hence, in particular, $\cos^2 150° + \sin^2 150° = 1$.

Of course, if you happen to know (see below) that $\cos 150° = -\cos 30° = -\sqrt{3}/2$ and $\sin 150° = \sin 30° = \frac{1}{2}$, you can verify directly that $\cos^2 150° + \sin^2 150° = (-\sqrt{3}/2)^2 + (\frac{1}{2})^2 = \frac{3}{4} + \frac{1}{4} = 1$.

Figure 7.1–8

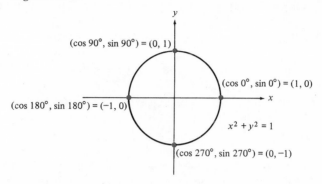

It is possible, as suggested above, to always find the values of $\sin t°$ and $\cos t°$ using Table VII even if t is outside the interval $0 \leq t \leq 90$. We simply use certain relationships obtained from the symmetry of the unit circle, as in Figure 7.1–9(a), (b), and (c).

The following rule, therefore, should be committed to memory. (Equivalently, Figure 7.1–9 below should be remembered.)

7.1.5   Rule

(a) $\cos (180 - t)° = -\cos t°$, $\sin (180 - t)° = \sin t°$.
(b) $\cos (180 + t)° = -\cos t°$, $\sin (180 + t)° = -\sin t°$.
(c) $\cos (360 - t)° = \cos t°$, $\sin (360 - t)° = -\sin t°$.

(See also problem 1 in Exercises 7.1.C.)

Part (a) of Rule 7.1.5 is used for angles between 90° and 180°; part (b) is used for angles between 180° and 270°; and part (c) is used for angles between 270° and 360°.

Figure 7.1–9

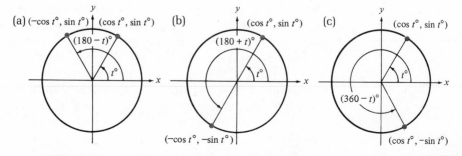

7.1.6   Example

Find (a) sin 150°, (b) cos 240°, (c) sin 315°, (d) cos (−45)°.

Solution:

(a) sin 150° = sin (180 − 30)° = sin 30° = 0.5000, using Table VII and Rule 7.1.5(a).
(b) cos 240° = cos (180 + 60)° = −cos 60° = −0.5000, using Table VII and Rule 7.1.5(b).
(c) sin 315° = sin (360 − 45)° = −sin 45° = −0.7071, using Table VII and Rule 7.1.5(c).
(d) cos (−45)° = cos (−45 + 360)° = cos 315° = cos 45° = 0.7071, using Table VII, Rule 7.1.5(c), and the periodicity of the cosine function.

We can now sketch the graphs of $y = \sin t°$ and $y = \cos t°$ quite easily. If we do so on the same coordinate axes, we notice that the graphs follow the same pattern of rising and falling, each having a fundamental period of 360°, but they are "out of phase" by 90°. That is, the graph of $y = \sin t°$ lags behind that of $y = \cos t°$ by 90°. (See Figure 7.1–10.) Hence, we may write sin $(t + 90)° = \cos t°$ for all t. We say that the graph of $y = \sin (t + 90)°$ is a <u>horizontal translation</u> (or "shift") of the graph of $y = \sin t°$ by 90°.

Figure 7.1–10

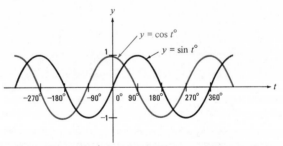

There are other important ways to modify the graphs of $y = \sin t°$ and $y = \cos t°$ without changing their characteristic "shapes." (See Figure 7.1–11.)

1. The graph of $y = A \sin t°$ (A > 0) has the same period as $y = \sin t°$ but an <u>amplitude</u> of A units. That is, for each t the corresponding y value is multiplied by the number A, so $−A \le A \sin t° \le A$. [See Figure 7.1–11(b).] We might say that A "stretches" the graph vertically.
2. The graph of $y = \sin (Rt)°$ has the same amplitude (1) as that of $y = \sin t°$, but a <u>fundamental period</u> of 360°/R (R > 0). We might say that the constant R "stretches" the graph horizontally [Figure 7.1–11(c)].
3. The graph of $y = V + \sin t°$ has the same period as $y = \sin t°$, but

Figure 7.1–11

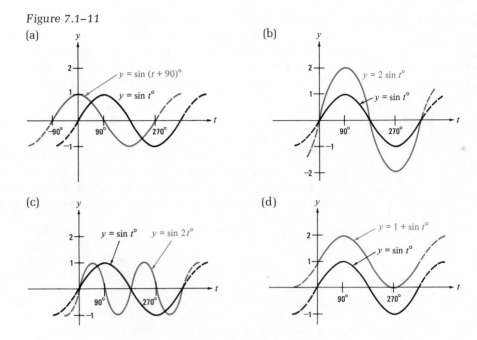

(a)
$y = \sin(t + 90)°$
$y = \sin t°$

(b)
$y = 2 \sin t°$
$y = \sin t°$

(c)
$y = \sin t°$ $y = \sin 2t°$

(d)
$y = 1 + \sin t°$
$y = \sin t°$

each point is <u>translated vertically</u> by V units. [See Figure 7.1–11(d).] We might say that the constant V "lifts" the graph.

One can also combine these modifications, as in

$$y = V + A \sin(Rt + H)°$$

The cosine function can be modified in a similar way.

7.1.7   Example in Biology

In Figure 7.1–12 a modified sine curve is fitted to telemetric measurements of intraperitoneal temperatures of a female rat. The function is of

Figure 7.1–12

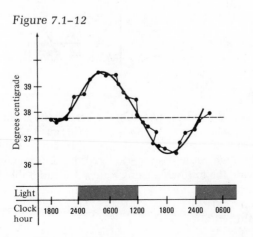

the form $F(t) = A + B \sin(Ct + D)$, where the constants A, B, C, and D are used to "lift," "stretch," and "shift" the basic sine function into the appropriate size and position. Notice that the horizontal axis refers to units of time, not degrees. Nevertheless, time is cyclical as well as linear, and we can let 24 hours correspond to 360° in a natural way.

SUMMARY

In our brief discussion of the functions sine and cosine, we have sketched out the basic properties of these periodic circular functions and suggested some applications. Further properties can be found in the exercises, as well as in the later sections, where we shall investigate the calculus of the sine, cosine, and the other circular functions.

EXERCISES   7.1. A

1. If a right triangle ABC has a right angle at C, while sides AB, BC, and AC are 5, 4, and 3 units, respectively, and the measures of angles A and B are $a°$ and $b°$, respectively, find (a) $\sin a°$, (b) $\cos b°$. (c) If $x > 0$, $y > 0$, and $x + y = 90$, is it always true that $\sin x° = \cos y°$? (d) That $\cos x° = \sin y°$?
2. If a right triangle ABC has a right angle at C, the angle at A measures 29°, and side AB measures 10 meters, find the length of BC and AC. (Use Table VII at the back of the book.)
3. Estimate the fundamental period of each of the following graphs:

(a)

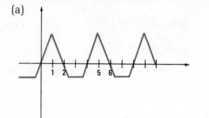

(b)

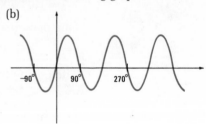

(c)

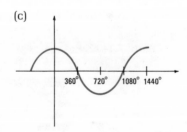

4. Evaluate the following, using the periodicity of sine and cosine:
   (a) $\sin 390°$ (b) $\cos(-270)°$
   (c) $\sin 720°$ (d) $\cos(-3{,}690)°$
5. Explain why sine has a period 720° but a fundamental period of only 360°.
6. Evaluate the following using Rule 7.1.5 and Table VII:
   (a) $\sin 120°$ (b) $\cos 240°$
   (c) $\cos 300°$ (d) $\sin 480°$
   (e) $\sin(-135)°$

7. A ladder 10 meters long leans against a house so that it just reaches the ledge of a window. If the angle that the ladder makes with the ground is 65°, find the distance from the ground to the window's ledge. (*Hint*: Make a diagram with a right triangle.)

8. Referring to Example 7.1.2, suppose that the force G induced by gravity is 8 units, the force L induced by light is 10 units, and the angle measure a° between the light rays and the guppy is 45°. Find the angle between the guppy and the vertical force G.

9. What is the fundamental period and amplitude of the function $y = \cos(t + 90)°$. Sketch the graph.

10. What is the fundamental period and amplitude of the function $y = 3 \sin 4t°$. Sketch the graph.

11. Without Table VII find the value of $2 \sin^2 49° + 2 \cos^2 49°$.

12. Sketch the following sequence of graphs; use each graph to obtain the subsequent graph:
    (a) $y = \sin t°$
    (b) $y = \sin(t + 180)°$
    (c) $y = \sin 3(t + 180)°$
    (d) $y = 2 \sin 3(t + 180)°$
    (e) $y = 4 + 2 \sin 3(t + 180)°$

EXERCISES   7.1. B

1. If a right triangle RST has a right angle at T while sides RS, ST, and RT are 13, 12, and 5 units, respectively, and the measures of angles R and S are r° and s°, respectively, find (a) sin r°, (b) cos s°. (c) If $a + b = 90$, is it always true that $\sin a° = \cos b°$? That $\cos a° = \sin b°$?

2. If the right triangle RST has a right angle at T, and R measures 64°, and side RS measures 100 units, find the lengths of sides ST and RT.

3. Estimate the fundamental period of each of the following graphs:

(a)

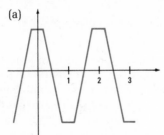

(b)

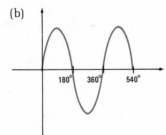

(c)

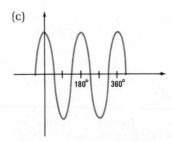

4. Evaluate the following, using the periodicity of sine and cosine:
   (a) $\cos 405°$
   (b) $\sin(-180)°$
   (c) $\sin 300°$
   (d) $\cos 420°$

5. Explain why cosine has a period of 1,080°, but a fundamental period of only 360°.
6. Evaluate the following using Rule 7.1.5:
   (a) cos 120° (b) sin 240°
   (c) sin 300° (d) cos 480°
   (e) cos (−135)°
7. A straight tree is toppled against a tall building in a storm, making an angle of 52° with the ground. If the base of the tree is 5 meters from the bottom of the building, how tall was the tree? (*Hint:* Draw a right triangle.)
8. Referring to Example 7.1.2, suppose that the force G induced by gravity is 5 units, the force L induced by light is 4 units, and the angle measure $a°$ between the guppy and the light ray is 60°. Find the angle between the guppy and the vertical force G.
9. What is the fundamental period and amplitude of the function $y = \cos (t - 90)°$? Sketch the graph.
10. What is the fundamental period and amplitude of the function $y = 4 \cos 3t°$? Sketch the graph.
11. Without using Table VII, find the value of $-10 \sin^2 38° - 10 \cos^2 38°$.
12. Sketch the following sequence of graphs; use each graph to obtain the subsequent graph:
    (a) $y = \cos t°$ (b) $y = \cos (t + 180)°$
    (c) $y = \cos 3(t + 180)°$ (d) $y = 2 \cos 3(t + 180)°$
    (e) $y = 4 + 2 \cos 3(t + 180)°$

EXERCISES 7.1. C

1. Four useful formulas involving sums and differences are:

$$\sin (a \pm b)° = \sin a° \cos b° \pm \cos a° \sin b°$$
$$\cos (a \pm b)° = \cos a° \cos b° \mp \sin a° \sin b°$$

(See Appendix VI.)
(a) Use these formulas to prove the validity of Rule 7.1.5.
(b) Use these formulas to prove that $\cos (90 - t)° = \sin t°$, that is, that the sine of an angle equals the cosine of its complement.
(c) Prove that $\sin (-t)° = -\sin t°$ and $\cos (-t)° = \cos t°$, that is, that the sine is an odd function and the cosine is an even function.

HC 2. Use your hand calculator to fill in the following table. That is, find the value of the functions listed on the left for the values of t listed at the top.

	Degrees				
t	1	0.5	0.2	0.1	0.05
$\sin t°$					
$\cos t°$					

3. The other circular functions are defined by the quotients $\tan t° = \sin t°/\cos t°$, $\cot t° = \cos t°/\sin t°$, $\sec t° = 1/\cos t°$, and $\csc t° = 1/\sin t°$. (a) Where are these functions not defined? (b) What are their periods?
4. What are the periods of (a) $\sin t° + \cos 2t°$ (b) $\sin^2 t°$?

ANSWERS 7.1. A

1. (a) $\frac{4}{5}$ (b) $\frac{4}{5}$ (c) yes (d) yes
2. $BC \approx 4.848$ meters; $AC \approx 8.746$ meters
3. (a) 4 (b) 180° (c) 1,440°
4. (a) $\frac{1}{2}$ (b) 0 (c) 0 (d) 0
5. The sine function repeats every 720°, so 720° is a period; there is no smaller period than 360°, so 360° is the fundamental period.
6. (a) 0.866 (b) −0.5 (c) 0.5 (d) 0.866 (e) −$\sqrt{2}$/2 ≈ −0.7071
7. 9.063 meters
8. 62°
9. fundamental period: 360°   amplitude: 1

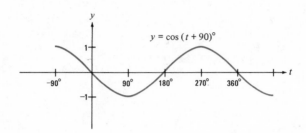

10. fundamental period: 90°   amplitude: 3

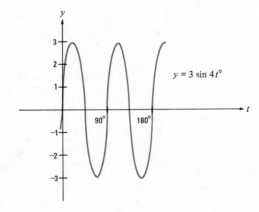

11. 2
12.

(a)

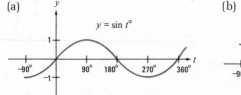

(b)

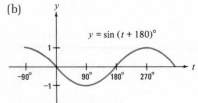

(c)

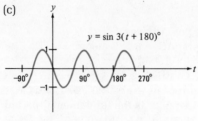

$y = \sin 3(t + 180)°$

(d)

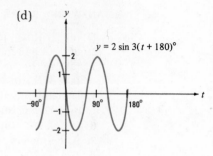

$y = 2 \sin 3(t + 180)°$

(e)

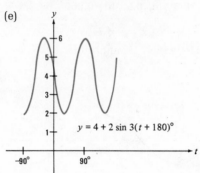

$y = 4 + 2 \sin 3(t + 180)°$

ANSWERS   7.1. C

1. (a) $\cos (180 - t)° = \cos 180° \cos t° + \sin 180° \sin t° = -\cos t°$
         $\sin (180 - t)° = \sin 180° \cos t° - \cos 180° \sin t° = -(-1) \sin t°$
                          $= \sin t°$
         $\cos (180 + t)° = \cos 180° \cos t° - \sin 180° \sin t° = -\cos t°$
         $\sin (180 + t)° = \sin 180° \cos t° + \cos 180° \sin t° = -\sin t°$
         $\cos (360 - t)° = \cos 360° \cos t° + \sin 360° \sin t° = \cos t°$
         $\sin (360 - t)° = \sin 360° \cos t° - \cos 360° \sin t° = -\sin t°$
    (b) $\cos (90 - t)° = \cos 90° \cos t° + \sin 90° \sin t° = \sin t°$
    (c) $\sin (-t)° = \sin (0 - t)° = \sin 0° \cos t° - \cos 0° \sin t° = -\sin t°$
         $\cos (-t)° = \cos (0 - t)° = \cos 0° \cos t° + \sin 0° \sin t° = \cos t°$

HC 2.

	1	0.5	0.2	0.1	0.05
$\sin t°$	0.017452	0.008727	0.003491	0.001745	0.000873
$\cos t°$	0.999848	0.999962	0.999994	0.999998	1.000000

3. (a) $\tan t°$ and $\sec t°$ are not defined whenever $\cos t°$ is 0—that is, for
         $90° + 180n°$ for any integer n. Similarly, $\cot t°$ and $\csc t°$ are not defined
         whenever $\sin t°$ is 0—that is, at $180n°$ for every integer n.
    (b) The periods of $\tan t°$ and $\cot t°$ are 180°; the periods of $\sec t°$ and $\csc t°$
         are 360°.
4. (a) 360°   (b) 180°

7.2   Derivatives of Sine and Cosine

Radian Measure

The degree, an invention that is usually attributed to the ancient Baby-
lonians, is a useful and venerable unit of angle measurement. And, yet,

for the purposes of calculus, the degree has been found wanting. The reason is simple. One degree does not correspond to one standard unit of length. Consequently, the use of degrees often leads to excessively complicated formulas when finding derivatives.

It is possible, however, to measure angles with units called *radians*. Since the *unit* circle has a circumference of 2π units in length, corresponding to a full rotation of 360°, the following definition is useful.

7.2.1   Definition

A <u>radian</u> is the length of an arc on the unit circle, as well as the measure of the corresponding central angle, <u>where 2π radians is equivalent to 360°.</u>

Figure 7.2–1

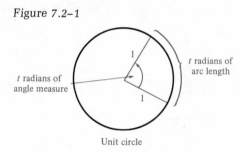

Unit circle

It follows that 180° is π radians and 30° is $\pi/6$ radians.

7.2.2   Example

Convert (a) $\pi/2$ radians of angle measure to degrees, and (b) 450° to radians. (c) Find the value of $\dfrac{\sin 5\pi/2}{5\pi/2}$ where $5\pi/2$ is given in radians.

Solution:

(a) Since 2π radians $= 360°$, it follows that 1 radian $= (360/2\pi)°$. Hence, on the unit circle,

$$\frac{\pi}{2} \text{ radians} = \left(\frac{\pi}{2}\right)\left(\frac{360}{2\pi}\right) \text{ degrees} = 90°$$

(b) Since $360° = 2\pi$ radians, $1° = 2\pi/360$ radians. Hence,

$$450° = (450)\left(\frac{2\pi}{360}\right) \text{ radians} = \frac{5\pi}{2} \text{ radians}$$

(c) First, observe that $\dfrac{\sin 5\pi/2}{5\pi/2}$ makes sense as a quotient since $\sin 5\pi/2$ and $5\pi/2$ are both units of length. Moreover, $\sin 5\pi/2 = \sin 450° = \sin (360° + 90°) = \sin 90° = 1$. Hence,

$$\frac{\sin 5\pi/2}{5\pi/2} = \frac{1}{5\pi/2} = \frac{2}{5\pi}$$

From now on, when we write an expression such as sin t, it will be understood that t is measured in radians. All the rules given thus far can be restated in terms of radians. For example, "sin $(t + 2\pi) = $ sin t" means that the period of the function $f(t) = $ sin t is 2π, where t is a certain number of units of length.

Derivatives

Let us begin by finding the derivatives of sin t and cos t where $t = 0$. This is easy (for the cosine at any rate) if you remember only that the derivative is the slope of the tangent to the graph. In the case of the cosine function the tangent line is horizontal to the graph at the point where $t = 0$. (See Figure 7.2–2.) Thus, the derivative of cosine at $t = 0$ is — you guessed it! — zero.

Figure 7.2–2

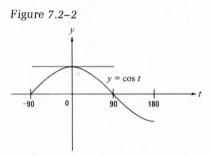

We proceed the same way for the sine function. But we congratulate you if you can see that the slope of the tangent line to $y = $ sin t at $t = 0$ is exactly 1. Not that this is implausible! (See Figure 7.2–3.)

Table 7.2–1 gives further evidence that at $t = 0$ the definition of the derivative yields

$$\frac{d}{dt}\,(\sin t) = \lim_{\Delta t \to 0} \frac{\sin \Delta t - \sin 0}{\Delta t} = \lim_{\Delta t \to 0} \frac{\sin \Delta t}{\Delta t} = 1$$

Figure 7.2–3

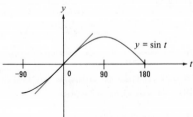

A similar calculation (with a hand calculator) should convince you that at $t = 0$,

$$\frac{d}{dt}(\cos t) = \lim_{\Delta t \to 0} \frac{\cos \Delta t - \cos 0}{\Delta t} = \lim_{\Delta t \to 0} \frac{\cos \Delta t - 1}{\Delta t} = 0$$

Table 7.2–1

	t (radians)				
	0.1	0.03	0.007	0.0004	0.00005
sin t	0.099833	0.029996	0.007000	0.000400	0.000050
cos t	0.995004	0.999550	0.999976	1.000000	1.000000
cos t − 1	−0.004996	−0.000450	−0.000024	0.000000	0.000000
(sin t)/t	0.998330	0.999850	1.000000	1.000000	1.000000
(cos t − 1)/t	−0.049960	−0.015000	−0.003429	0.000000	0.000000

The numbers in this table are correct to six places; if you have a hand calculator, check this.

7.2.3   Rule

(a) $\lim_{t \to 0} \dfrac{\sin t}{t} = 1$ (b) $\lim_{t \to 0} \dfrac{\cos t - 1}{t} = 0$

See Table 7.2–1 for justification of this rule. For the next rule we shall need the following trigonometric formula. (See Appendix VI for a proof.)

$$* \quad \sin (a + b) = \sin a \cos b + \cos a \sin b$$

7.2.4   Rule

For all real numbers t:

(a) $\dfrac{d}{dt}(\sin t) = \cos t$ (b) $\dfrac{d}{dt}(\cos t) = -\sin t$

Proof of part (a):

$$\frac{d}{dt}\sin t = \lim_{\Delta t \to 0} \frac{\sin (t + \Delta t) - \sin t}{\Delta t} \qquad \text{(by definition)}$$

$$= \lim_{\Delta t \to 0} \frac{(\sin t \cos \Delta t + \cos t \sin \Delta t) - \sin t}{\Delta t} \qquad \text{(by starred trigonometric formula above)}$$

$$= \lim_{\Delta t \to 0} \cos t \, \frac{\sin \Delta t}{\Delta t} + \lim_{\Delta t \to 0} \sin t \, \frac{\cos \Delta t - 1}{\Delta t} \qquad \text{(by algebra)}$$

$$= \cos t \lim_{\Delta t \to 0} \frac{\sin \Delta t}{\Delta t} + \sin t \lim_{\Delta t \to 0} \frac{\cos \Delta t - 1}{\Delta t} \qquad \text{(since } \cos t \text{ and } \sin t \text{ are constant as } \Delta t \to 0\text{)}$$

$$= (\cos t)(1) + (\sin t)(0) = \cos t \qquad \text{(by Rule 7.2.3)}$$

We postpone the proof of part (b) until Example 7.2.9(a), where it is proved using the chain rule.

7.2.5   Example

Find the second derivatives of sin t and cos t.

Solution:

Since $\dfrac{d}{dt}$ (sin t) = cos t, we have $\dfrac{d^2}{dt^2}$ (sin t) = $\dfrac{d}{dt}$ (cos t) = −sin t.

Similarly, $\dfrac{d}{dt}$ (cos t) = −sin t implies that $\dfrac{d^2}{dt^2}$ (cos t) = −cos t.

7.2.6   Example

Find the derivatives of (a) 2 sin t − 3 cos t, (b) (sin t)(cos t), and (c) sin t/cos t.

Solution:

(a) $\dfrac{d}{dt}$ (2 sin t − 3 cos t) = 2 $\dfrac{d}{dt}$ (sin t) − 3 $\dfrac{d}{dt}$ (cos t)

$$= 2 \cos t + 3 \sin t$$

(b) $\dfrac{d}{dt}$ (sin t)(cos t) = (sin t) $\dfrac{d}{dt}$ (cos t) + (cos t) $\dfrac{d}{dt}$ (sin t)

$$= -\sin^2 t + \cos^2 t$$

(c) $\dfrac{d}{dt} \left(\dfrac{\sin t}{\cos t} \right) = \dfrac{(\cos t)(\cos t) - (\sin t)(-\sin t)}{\cos^2 t}$

$$= \dfrac{\cos^2 t + \sin^2 t}{\cos^2 t} = \dfrac{1}{\cos^2 t}$$

[*Note:* The function $f(t) = \sin t / \cos t$ is called "tan t." Hence, $\dfrac{d}{dt}$ (tan t) = $1/\cos^2 t$. This function will be discussed further in Section 7.4.]

7.2.7   Example in Biology

A bee's cell is a regular hexagonal prism with one open end and one trihedral apex, the latter consisting of three quadrilaterals all of whose sides are equal. (See Figure 7.2–4.) It can be shown that the surface area of this cell, which is made of bee's wax, is given by

$$A(t) = 6hs + \tfrac{3}{2}s^2 \left(-\dfrac{\cos t}{\sin t} + \dfrac{\sqrt{3}}{\sin t} \right)$$

where t is the measure of the variable angle BAC, and h and s are constants. Find the angle t at the apex that *minimizes* the amount of wax used, that is, the surface area.

Figure 7.2–4

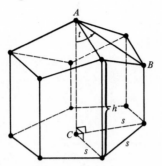

Solution:

To minimize the surface area $A(t)$, we set the derivative

$$\frac{dA}{dt} = \tfrac{3}{2}s^2 \left(\frac{1}{\sin^2 t} - \frac{\sqrt{3} \cos t}{\sin^2 t} \right) = 0$$

Hence, $1 - \sqrt{3} \cos t = 0$, and $\cos t = 1/\sqrt{3} \approx 0.5774$. Using Table VII, we find that $t \approx 55°$. The first derivative test (Rule 2.5.1) easily shows that this angle gives a minimum. (It is a fact that bees tend to use this angle, on the average! How this came to pass is an interesting question . . . for biologists.)

We may apply our curve-sketching techniques of Section 2.7 to functions that are combinations of sine and cosine. Such functions, particularly sums of powers of the sine and cosine functions called "trigonometric polynomials" or "Fourier series," are of great importance in approximating periodic data.

7.2.8   **Example**

(a) Show that the function $f(x) = 1 + \cos x + \sin x$ has a period of 2π.
(b) Using its derivative, sketch the graph of $y = f(x)$ in the interval $0 \leq x \leq 2\pi$.

Solution:

(a) Using the fact that sine and cosine have a period of 2π, we have $f(x + 2\pi) = 1 + \cos (x + 2\pi) + \sin (x + 2\pi) = 1 + \cos x + \sin x = f(x)$. Hence, $f(x)$ also has a period of 2π.
(b) Setting $f'(x) = -\sin x + \cos x = 0$, we discover that $\sin x = \cos x$. Hence, $x = \pi/4$ and $\tfrac{5}{4}\pi$ are the critical points, since this is where the graphs of $y = \sin x$ and $y = \cos x$ intersect. Since $f''(x) = -\cos x - \sin x$, Table VII gives us $f''(\pi/4) < 0$ and $f''(\tfrac{5}{4}\pi) > 0$, indicating, by Rule 2.6.7, a relative maximum at $x = \pi/4$ and a relative minimum at $x = \tfrac{5}{4}\pi$. As for inflection points, $f''(x) = -\sin x - \cos x = 0$, if and only if $\sin x = -\cos x$. You can verify that this equation is

true when $x = \frac{3}{4}\pi$ and $\frac{7}{4}\pi$. Plotting a few more points yields the graph in Figure 7.2–5.

Figure 7.2–5

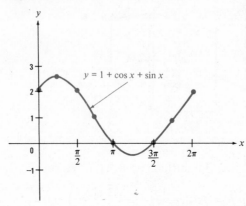

The chain rule, too, can be applied to composite functions of the sine and cosine. Indeed, we may use this rule (3.1.4), together with the fact that sine and cosine are "cofunctions," to derive the formula for the derivative of $\cos t$, as in part (a) of the following.

7.2.9   Example

Find the derivatives of the following composite functions:
(a) $\sin (\pi/2 - t)$, (b) $\cos^2 t$, (c) $\sin 2t$, (d) $\cos (t^2)$.

Solution:

(a) $\dfrac{d}{dt} \sin (\pi/2 - t) = (-1) \cos (\pi/2 - t) = -\cos (\pi/2 - t)$

$\left[\text{Since } \cos t = \sin (\pi/2 - t) \text{ and } \sin t = \cos (\pi/2 - t), \text{ all this amounts to is } \dfrac{d}{dt} \cos t = -\sin t. \right]$

(b) $\dfrac{d}{dt} \cos^2 t = 2(\cos t)(-\sin t) = -2(\cos t)(\sin t)$

(c) $\dfrac{d}{dt} \sin 2t = 2 \cos 2t$

(d) $\dfrac{d}{dt} \cos (t^2) = (2t)[-\sin (t^2)] = -2t \sin (t^2)$

7.2.10   Example

A certain profit cycle is approximated by the function $P(t) = 10 + \sin 2\pi t - \cos 2\pi t$. Calculate (a) the period of this cycle, (b) the maximum value of $P(t)$, and (c) sketch the graph of $y = P(t)$ from $t = 0$ to $t = 2$.

Solution:

(a) To find the period, notice that

$$P(t + 1) = 10 + \sin 2\pi(t + 1) - \cos 2\pi(t + 1)$$
$$= 10 + \sin (2\pi t + 2\pi) - \cos (2\pi t + 2\pi)$$
$$= 10 + \sin 2\pi t - \cos 2\pi t = P(t)$$

The period of $P(t)$ is 1.

(b) We find the maximum of $P(t)$ by setting $P'(t) = 2\pi \cos 2\pi t + 2\pi \sin 2\pi t = 0$. Hence, $\sin 2\pi t = -\cos 2\pi t$. And the last equation can only be true if $2\pi t = \frac{3}{4}\pi$ or $\frac{7}{4}\pi$ (see Figure 7.1–10 on page 334); that is, if $t = \frac{3}{8}$ or $t = \frac{7}{8}$, for $0 \leq t \leq 1$. Since $P''(t) = -4\pi^2 \sin 2\pi t + 4\pi^2 \cos 2\pi t$, $P''(\frac{3}{8}) < 0$ and $P''(\frac{7}{8}) > 0$. Hence, $P(\frac{3}{8}) \approx 11.4142$ is a relative maximum and $P(\frac{7}{8}) \approx 8.5858$ is a relative minimum. By Rule 2.4.10, they are also absolute extrema on the closed interval $0 \leq t \leq 1$, since $P(0) = P(1) = 9$. Hence, the maximum profit is about 11.4142.

(c) The only additional information we need to sketch the graph is the location of inflection points. Since $P''(t) = 0$ only if $\sin 2\pi t = \cos 2\pi t$, we must have $2\pi t = \pi/4$ or $5\pi/4$, so the inflection points are $(\frac{1}{8}, 10)$ and $(\frac{5}{8}, 10)$. Clearly, $y = P(t)$ is concave downward if $\frac{1}{8} < t < \frac{5}{8}$ and concave upward if $0 \leq t < \frac{1}{8}$ or $\frac{5}{8} < t < 1$ on the interval $0 \leq t \leq 1$. For $1 < t \leq 2$, we just repeat the first cycle of the graph. (See Figure 7.2–6.)

Figure 7.2–6

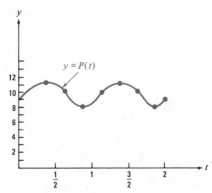

SUMMARY

The need for radian measure becomes evident when one calculates the derivatives of sine and cosine. We can use these derivatives in the usual ways, such as finding extreme values of functions and sketching graphs. In fact, it is now an easy matter to find the antiderivatives of the sine and cosine functions. We shall do this in the next section.

EXERCISES   7.2. A

1. Convert the following angle measures from degrees to radians:
   (a) 30°   (b) 60°   (c) 180°
   (d) 360°   (e) 135°   (f) −240°

2. Convert the following radian measures of central angles of the unit circle to degrees:
   (a) $\pi/4$ (b) $3\pi/4$ (c) 1
   (d) $-\pi$ (e) $5\pi/2$

3. Evaluate
   (a) $\sin \dfrac{\pi}{3}$ (b) $\cos(-\pi)$ (c) $\cos \frac{5}{2}\pi$

   (d) $\cos 1$ (e) $\frac{1}{2}\cos \frac{5}{2}\pi$

4. Find the first derivatives of the following:
   (a) $2 \sin t + \cos t$ (b) $2 \cos t$
   (c) $\sin t - t^2$ (d) $t \sin t$

5. Find the second derivatives of the functions in problem 4.

6. If the cost of producing x items is given in dollars by $C(x) = 100 - 10(\sin x + \cos x)$, find the minimum cost of producing between 0 and 10 items.

7. Use the chain rule to differentiate the following:
   (a) $\cos 3t$ (b) $\sin(2t + \pi)$ (c) $\sin 4t$
   (d) $\cos(3t + \pi/2)$ (e) $\sin(t^3)$

8. Find an equation of the tangent line to the graph of $y = \sin t$ at $t = \pi/6$.

9. Sketch the graph of $y = \sin^2 t$ for $-2\pi \leq t \leq 2\pi$.

HC 10. Use your hand calculator to complete the following table:

	t (radians)				
	0.2	0.04	0.008	0.0001	0.00003
sin t					
cos t					
cos t − 1					
(sin t)/t					
(cos t − 1)/t					

11. Find the critical points of the following and determine which give relative maxima and which give relative minima:
    (a) $y = \sin 2t$ (b) $y = \sin^2 t$

12. Find all the critical points of $y = 3 \sin(2t + \pi)$ and determine which give relative maxima and which give relative minima.

EXERCISES   7.2. B

1. Convert the following angle measures from degrees to radians:
   (a) 45°   (b) 90°   (c) 270°
   (d) 720°   (e) 300°   (f) −135°

2. Convert the following radian measures of central angles of the unit circle to degrees:

(a) $\pi/6$ (b) $4\pi/3$ (c) -3π
(d) $\frac{1}{2}$ (e) $\pi/2$

3. Evaluate
   (a) $\cos \pi/3$ (b) $\sin 3\pi/2$
   (c) $\sin (-\pi/2)$ (d) $\sin 1$

4. Find the first derivatives of the following:
   (a) $\sin t + 3 \cos t$ (b) $3 \sin t$
   (c) $t^2 - \cos t$ (d) $t \cos t$

5. Find the second derivatives of the functions in problem 4.

6. If a profit cycle varies according to the formula $P(t) = 1{,}500 + 100(\sin 3\pi t - \cos 3\pi t)$, where t is measured in months and $P(t)$ in dollars:
   (a) Find the fundamental period of this cycle.
   (b) Find the maximum profit.

7. Use the chain rule to differentiate the following:
   (a) $\sin 3t$ (b) $\cos (\pi - 2t)$ (c) $\cos 5t$
   (d) $\sin (2t - \frac{3}{2}\pi)$ (e) $\cos (t^3)$

8. Find an equation of the tangent line to the graph of $y = \cos t$ at $t = \pi/6$.

9. Sketch the graph of the function $y = \cos^2 t$ for $-2\pi \leqslant t \leqslant 2\pi$.

HC 10. Use your hand calculator to complete the following table:

		t (radians)				
		0.3	0.02	0.005	0.0006	0.00001
$\sin t$						
$\cos t$						
$\cos t - 1$						
$(\sin t)/t$						
$(\cos t - 1)/t$						

11. Find all the critical points of the following and determine which give maxima and which give minima:
    (a) $y = \cos 2t$ (b) $y = \cos^2 t$

12. Find all the critical points of $2 \cos (3t - \pi)$ and determine which give relative maxima and which give relative minima.

EXERCISES   7.2. C

1. Recall that in Example 7.2.7 $A(t) = 6hs + \frac{3}{2}s^2 \left(-\dfrac{\cos t}{\sin t} + \dfrac{\sqrt{3}}{\sin t}\right)$. Verify that the formula for $A'(t)$ was given correctly and that $t \approx 55°$ gives a minimum.

2. If we define $\cot x = \cos x/\sin x$, $\sec x = 1/\cos x$, $\csc x = 1/\sin x$, verify that
$$\frac{d}{dx}(\cot x) = -\csc^2 x; \quad \frac{d}{dx}(\sec x) = (\tan x)(\sec x); \quad \frac{d}{dx}(\csc x) = -(\cot x)(\csc x).$$

3. Evaluate the following limits using Rule 7.2.3:
   (a) $\displaystyle\lim_{t \to 0} \frac{\sin t}{2t}$ (b) $\displaystyle\lim_{t \to 0} \frac{\sin 2t}{2t}$

   (c) $\displaystyle\lim_{t \to 0} \frac{\sin 2t}{t}$ (d) $\displaystyle\lim_{t \to 0} \frac{1 - \cos t}{2t}$

ANSWERS 7.2. A

1. (a) $\pi/6$ (b) $\pi/3$ (c) π (d) 2π (e) $3\pi/4$ (f) $-4\pi/3$

2. (a) $45°$ (b) $135°$ (c) $\dfrac{180°}{\pi} \approx 57°$ (d) $-180°$ (e) $450°$

3. (a) $\sqrt{3}/2 \approx 0.866$ (b) -1 (c) 0 (d) 0.54 (e) 0

4. (a) $2\cos t - \sin t$ (b) $-2\sin t$ (c) $\cos t - 2t$ (d) $\sin t + t\cos t$

5. (a) $-2\sin t - \cos t$ (b) $-2\cos t$ (c) $-\sin t - 2$ (d) $2\cos t - t\sin t$

6. The minimum is $100 - 10\sqrt{2}$. It is attained at $\pi/4$ and $5\pi/4$, which clearly cannot be the number of items produced. If integer quantities are needed, try both adjacent integers, but if a continuous quantity is to be produced, the numbers as given may be useful.

7. (a) $-3\sin 3t$ (b) $2\cos(2t + \pi)$ (c) $4\cos 4t$

   (d) $-3\sin\left(3t + \dfrac{\pi}{2}\right)$ (e) $3t^2\cos(t^3)$

8. $y = \dfrac{\sqrt{3}}{2}t + \dfrac{1}{2} - \dfrac{\pi\sqrt{3}}{12}$

9.

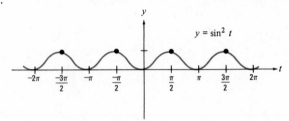

10.

	0.2	0.04	0.008	0.0001	0.00003
$\sin t$	0.198669	0.039989	0.008000	0.000100	0.000030
$\cos t$	0.980067	0.999200	0.999968	1.000000	1.000000
$\cos t - 1$	-0.019933	-0.000800	-0.000032	0.000000	0.000000
$(\sin t)/t$	0.993345	0.999725	1.000000	1.000000	1.000000
$(\cos t - 1)/t$	-0.099665	-0.020000	-0.004000	0.000000	0.000000

t (radians)

11. (a) Critical points are $t = \pi/4 + (\pi/2)n$ for all integers n; when n is even, there is a local maximum, and when n is odd, there is a local minimum.

    (b) Critical points are $(\pi/2)n$ for all integers n; when n is even, there is a local minimum and when n is odd, there is a local maximum.

12. The critical points are $t = (\pi/2)n + \pi/4$ for all integers n. There is a local maximum if n is odd. There is a local minimum if n is even.

ANSWERS 7.2. C

3. (a) $\frac{1}{2}$ (b) 1 (c) 2 (d) 0

Sections 7.1 and 7.2

1. (10 pts) Convert to radians:
   (a) 360°   (b) 45°
2. (10 pts) Convert to degrees:
   (a) π (b) $-\pi/6$
3. (5 pts) If the function is given in degrees, what is the period of $y = \sin 3x$?
4. (30 pts) Given that $\sin 30° = 0.5$, find:
   (a) $\cos 60°$ (b) $\sin(-30°)$ (c) $\cos(-60°)$
   (d) $\sin 390°$ (e) $\sin 7\pi/6$ [part (e) in radians]
   (f) If a ladder leaning against a house makes a 30° angle with the ground and the distance from the bottom of the ladder to the house is 2 meters, how long is the ladder?
5. (10 pts) Write the equation of the tangent line to the curve $y = \sin x$ at the point where $x = \pi$.
6. (15 pts) Graph
   (a) $y = \cos x$ (b) $y = 2 \cos x$ (c) $y = \cos^2 x$
7. (20 pts) Find the derivatives:
   (a) $y = 3 \cos x + 2 \sin x$ (b) $y = \cos 3x$
   (c) $y = \sin(x^2)$ (d) $y = \cos^4 x$

The answers are at the back of the book.

7.3   Integration of Sine and Cosine

Having learned the derivatives of the functions $\sin x$ and $\cos x$ in Section 7.2, we can easily find their antiderivatives. In fact, we state the following rule, which you should commit to memory.

7.3.1   Rule

(a) $\displaystyle\int \sin x \, dx = -\cos x + k.$

(b) $\displaystyle\int \cos x \, dx = \sin x + k$, where k is an arbitrary constant.

Part (b) follows immediately from the equation $\dfrac{d}{dx}(\sin x) = \cos x$.

For part (a), we can write $\dfrac{d}{dx}(-\cos x) = -\dfrac{d}{dx}(\cos x) = -(-\sin x) = \sin x$, which is the integrand of $\displaystyle\int \sin x \, dx$.

Warning! It is a common error among students to write "$\displaystyle\int \sin x \, dx = +\cos x + k$." Don't forget the minus sign.

Effective use of Rule 7.3.1 often depends upon the technique of substitution, which we studied in Section 4.2. This technique is illustrated in the next few examples.

7.3.2   Example

Evaluate the following integrals:

(a) $\displaystyle\int \cos (x + 1)\, dx$ (b) $\displaystyle\int \sin 2x\, dx$ (c) $\displaystyle\int 3x \cos (x^2)\, dx$

Solution:

(a) To find the antiderivatives of $\cos (x + 1)$, just let $u = x + 1$. Then, $du = dx$, and we may write

$$\int \cos (x + 1)\, dx = \int \cos u\, du = \sin u + k = \sin (x + 1) + k$$

(b) To evaluate $\displaystyle\int \sin 2x\, dx$, let $u = 2x$. Then $du = 2\, dx$, and we can write

$$\begin{aligned}
\int \sin 2x\, dx &= \tfrac{1}{2} \int 2 \sin 2x\, dx && \text{[since } \tfrac{1}{2}(2) = 1] \\
&= \tfrac{1}{2} \int \sin 2x\, (2\, dx) && \text{(getting the integrand into the right form)} \\
&= \tfrac{1}{2} \int \sin u\, du && \text{(making the substitutions)} \\
&= \tfrac{1}{2} (-\cos u) + k && \text{[using Rule 7.3.1(a)]} \\
&= -\tfrac{1}{2} \cos 2x + k && \text{(substituting } 2x = u)
\end{aligned}$$

[*Careful*: $\tfrac{1}{2} \cos 2x \neq \cos \tfrac{1}{2}(2x)$.]

(c) To evaluate $\displaystyle\int 3x \cos (x^2)\, dx$, let $u = x^2$. Then $du = 2x\, dx$, and we have

$$\begin{aligned}
\int 3x \cos (x^2)\, dx &= \tfrac{3}{2} \int 2x \cos (x^2)\, dx && \text{[since } \tfrac{3}{2}(2) = 3] \\
&= \tfrac{3}{2} \int \cos (x^2)(2x\, dx) && \text{(getting the integrand into the right form)} \\
&= \tfrac{3}{2} \int \cos u\, du && \text{(making the substitutions)} \\
&= \tfrac{3}{2} \sin u + k && \text{[using Rule 7.3.1(b)]} \\
&= \tfrac{3}{2} \sin (x^2) + k && \text{(substituting } x^2 = u)
\end{aligned}$$

Sometimes you will find it useful to treat an entire circular function as variable u, for example, letting $u = \sin x$. This technique is illustrated below.

7.3.3   Example

Evaluate the following integrals: (a) $\displaystyle\int \sin^2 x \cos x\, dx$, (b) $\displaystyle\int \frac{\sin x}{\cos x}\, dx$.

Solution:

(a) To evaluate $\int \sin^2 x \cos x\, dx$, we will let $u = \sin x$. This is a good choice, since it gives us $du = \cos x\, dx$, which appears in the integrand. Thus,

$$\int \sin^2 x \cos x\, dx = \int u^2\, du = \frac{u^3}{3} + k = \frac{\sin^3 x}{3} + k$$

Remember: $\sin^2 x = (\sin x)^2$; that is, the value of the function is squared at each x.

(b) To evaluate $\int \dfrac{\sin x}{\cos x}\, dx$, let $u = \cos x$. This is the correct choice, since $du = -\sin x\, dx$ is just the negative of the numerator in the integral. Hence,

$$\int \frac{\sin x}{\cos x}\, dx = -\int \frac{-\sin x\, dx}{\cos x} = -\int \frac{du}{u}$$
$$= -\ln |u| + k = -\ln |\cos x| + k$$

Let us observe once again (as in Example 7.2.6) that $\sin x/\cos x$ is an important circular function in its own right, which you may recognize as the tangent of x, written "$\tan x = \sin x/\cos x$." We shall consider this function in more detail in Section 7.4, but we have already proved the next rule.

7.3.4   Rule

$$\int \tan x\, dx = -\ln |\cos x| + k$$

The technique of integration by parts (Section 4.3) is often effective when the integrand is a product of a circular function and another type of function.

7.3.5   Example

Evaluate the integral $\int x \cos x\, dx$.

Solution:

Let $f(x) = x$ and $g'(x) = \cos x$. Then $f'(x) = 1$ and $g(x) = \sin x$, so

$$\int x \cos x\, dx = x \sin x - \int 1 \sin x\, dx$$
$$= x \sin x + \cos x + k$$

by Rule 4.3.5.

As we pointed out in the beginning of this chapter, much that occurs in nature takes place seasonally or periodically. This is often also true about the man-made world. Prices in the business world, for example, tend to follow certain cycles determined by the weather, the seasons, and other factors. And when they do, these cycles can often be modeled using the circular functions, as the following example suggests.

7.3.6   Example

Suppose that the rate of change of the price of a commodity is proportional to the difference between demand and supply. Thus, $\frac{dp}{dt} = k(D - S)$. To be specific, we will let $k = 3\pi$. Further, let us assume that both the supply and demand functions are periodic, each having a 1-year cycle. In fact, let them be given by $S(t) = 100(1 + \sin 2\pi t)$ and $D(t) = 100(1 + \cos 2\pi t)$, respectively. Find the price function if $p(0) = 1{,}000$.

Solution:

We merely have to solve the separable differential equation

$$\frac{dp}{dt} = 3\pi[100(1 + \cos 2\pi t) - 100(1 + \sin 2\pi t)]$$

$$dp = 300\pi(\cos 2\pi t - \sin 2\pi t)\, dt$$

Integrating both sides, we obtain

$$p(t) = 150(\sin 2\pi t + \cos 2\pi t) + k$$

To evaluate k, we write $p(0) = 150(\sin 0 + \cos 0) + k = 1{,}000$. Hence, $k = 1{,}000 - 150 = 850$, and $p(t) = 150(\sin 2\pi t + \cos 2\pi t) + 850$. Notice that this is a function whose period is also 1 year in length, since, as t varies from 0 to 1, $2\pi t$ varies from 0 to 2π. Sketches of the graphs of these functions are given in Figures 7.3–1 and 7.3–2.

Figure 7.3–1

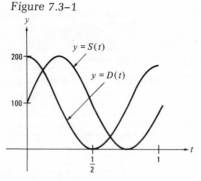

Figure 7.3–2

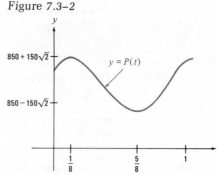

7.3.7   Example

Find the <u>total change</u> in the price function $p(t)$ from time $t = 5$ to time $t = 10$ if the marginal price $p'(t) = 300\pi(\cos 2\pi t - \sin 2\pi t)$.

Solution:

Since we have already found one function $p(t)$ that solves the given differential equation (see Example 7.3.6), we could, of course, just write $p(10) - p(5)$. Instead, let us recall that $p(10) - p(5) =$

$$\int_5^{10} p'(t)\,dt = 150(\sin 2\pi t + \cos 2\pi t)\Big|_5^{10}$$

$$= 150[(0-1)-(0-1)] = 0$$

Hence, the total change in *any* such $p(t)$ is zero, regardless of the value of $p(0)$.

When finding the area "under" the graph of a circular function, it is always wise to keep in mind a clear picture of the graph and its periodic nature. This is illustrated by the next example.

7.3.8   **Example**

Find the area between the x-axis and the graph of $y = \sin x$ between $x = 0$ and $x = 2\pi$.

Solution:

From a sketch of the graph of $y = \sin x$ (see Figure 7.3–3) we recall that the graph is above the x-axis from $x = 0$ to $x = \pi$; but from $x = \pi$ to

Figure 7.3–3

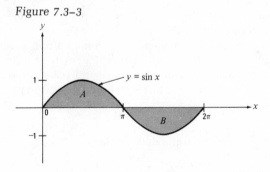

$x = 2\pi$, it is below the x-axis. We must, therefore, calculate the two areas, A and B, and add them. Hence,

$$A = \int_0^\pi \sin x\,dx = -\cos x\Big|_0^\pi = -(\cos \pi - \cos 0)$$

$$= -(-1-1) = 2$$

$$B = -\int_\pi^{2\pi} \sin x\,dx = -(-\cos x)\Big|_\pi^{2\pi}$$

$$= +(\cos 2\pi - \cos \pi) = (1+1) = 2$$

And the total area $= 2 + 2 = 4$.

(We also could have noticed that the graph from π to 2π is merely the "mirror image" of the graph from 0 to π; hence, $A = B$ without further calculation.)

SUMMARY

In this section we discussed integrals involving the sine and cosine functions. We then used these integrals to calculate areas and to model the business cycle of a price function. In the following section we shall study the integrals (and derivatives) of other circular functions.

EXERCISES   7.3. A

1. Find all the antiderivatives of the functions
   (a) $y = \cos 2x$ (b) $\sin 3x$ (c) $\sin (x + 4)$
   (d) $\cos \frac{1}{2}x$ (e) $3 \cos x$ (f) $\sin (-4x)$
2. Find all the antiderivatives of the functions
   (a) $y = 3 \sin 4x$ (b) $4 \cos (-2x)$ (c) $5 \cos 4x$
   (d) $3 \cos \frac{1}{2}x$ (e) $8 \sin 3x$
3. A certain population has a rate of change given by $\dfrac{dF}{dt} = 50\pi \sin[(\pi/2)(t + 1)]$,

   where $F(t)$ is the size of the population at time t. Find the total change in population between times $t = 1$ and $t = 10$.
4. If $F(1) = 250$, find a particular solution to the differential equation given in problem 3.
5. Find the value of the definite intergral $\displaystyle\int_0^{\pi/2} \sin x \, dx$.
6. Find the area of the region between the graph of $y = \sin 2x$ and the x-axis from $x = 0$ to $x = 2\pi$.
7. Suppose that the demand for a commodity is given by $D(t) = 50\pi \sin 10\pi t$ and its supply is given by $S(t) = 50\pi \cos 10\pi t$ at time t. If the rate of change of the price of the commodity is given by the difference between demand and supply and $p(0) = 200$, find the price function $p(t)$.
8. Evaluate $\displaystyle\int x \cos x^2 \, dx$.
9. Find the value of the definite integral $\displaystyle\int_{\pi/2}^{\pi} \cos x \sin x \, dx$.
10. Evaluate $\displaystyle\int \cos^3 x \sin x \, dx$.
11. Evaluate $\displaystyle\int \dfrac{\sin x}{\cos^2 x} \, dx$.
12. Integrate $y = \tan 2x$. (See Rule 7.3.4.)

EXERCISES   7.3. B

1. Find all antiderivatives of the functions
   (a) $\sin (x + 1)$ (b) $\sin 2x$ (c) $\cos (2x + 1)$
   (d) $\cos 5x$ (e) $4 \sin x$ (f) $6 \cos \frac{1}{3}x$
2. Find all antiderivatives of the functions
   (a) $4 \sin 3x$ (b) $6 \cos (-2x)$ (c) $4 \cos 5x$
   (d) $3 \sin (-2x)$ (e) $7 \sin 2x$

3. Find the value of the definite integral $\int_{\pi}^{\pi/2} \cos x \, dx$.

4. If the rate of change of a population is given by $\dfrac{dF}{dt} = 110\pi \cos \pi(t + 1)$, where $F(t)$ is the size of the population at time t, find the total change in population from times $t = 2$ to $t = 4$.

5. If $F(0) = 100$, solve the differential equation in problem 4.

6. Find the area between the graphs of $y = \sin x$ and $y = \cos x$ from $x = \pi/2$ to $x = \pi$.

7. The rate of change in population $\dfrac{dP}{dt}$ of a certain community is equal to the difference of the birth rate $\dfrac{dB}{dt} = 10\pi \sin \pi t$ and the mortality rate $\dfrac{dM}{dt} = -10\pi \cos \pi t$. If $P(0) = 100$, find the population function $P(t)$.

8. Evaluate $\int 2x^3 \sin x^4 \, dx$.

9. Find the value of the definite integral $\int_{0}^{\pi/2} \sin 2x \, dx$.

10. Evaluate $\int \sin^4 x \cos x \, dx$.

11. Evaluate $\int \dfrac{\cos x}{\sqrt{\sin x}} \, dx$.

12. Integrate $y = \tan 3x$. (See Rule 7.3.4.)

EXERCISES   7.3. C

1. Evaluate the following integrals:

   (a) $\int (\sin^2 x - 1) \sin x \, dx$ (b) $\int \dfrac{\cos x \, dx}{(1 - \cos^2 x)^2}$

   (c) $\int [\sin(\cos x)] \sin x \, dx$ (d) $\int \dfrac{[\ln (\cos x)] \sin x \, dx}{\cos x}$

   (e) $\int e^{\sin x} \cos x \, dx$ (f) $\int x \sin x \, dx$

2. If $\cot x = \dfrac{\cos x}{\sin x}$, find $\int \cot x \, dx$.

ANSWERS   7.3. A

1. (a) $\frac{1}{2} \sin 2x + k$ (b) $-\frac{1}{3} \cos 3x + k$ (c) $-\cos (x + 4) + k$ (d) $2 \sin \frac{1}{2}x + k$
   (e) $3 \sin x + k$ (f) $\frac{1}{4} \cos(-4x) + k$
2. (a) $-\frac{3}{4} \cos 4x + k$ (b) $-2 \sin (-2x) + k$ (c) $\frac{5}{4} \sin 4x + k$
   (d) $6 \sin (\frac{1}{2}x) + k$ (e) $-\frac{8}{3} \cos 3x + k$
3. -100
4. $F(t) = -100 \cos (\pi/2)(t + 1) + 150$
5. 1
6. 4
7. $p(t) = -5 \cos 10\pi t - 5 \sin 10\pi t + 205$
8. $\frac{1}{2} \sin x^2 + k$

9. $-\frac{1}{2}$ 10. $-\frac{1}{4}\cos^4 x + k$

11. $\dfrac{1}{\cos x} + k$

12. $-\frac{1}{2}\ln|\cos x| + k$

ANSWERS 7.3. C

1.   (a) $\frac{1}{3}\cos^3 x + k$ (b) $\dfrac{-1}{3\sin^3 x} + k$ (c) $\cos(\cos x) + k$

(d) $-\frac{1}{2}[\ln(\cos x)]^2 + k$ (e) $e^{\sin x} + k$ (f) $-x\cos x + \sin x + k$

2. $\ln|\sin x| + k$

7.4   The Other Circular Functions

The four other circular functions are the tangent, cotangent, secant, and cosecant. As with the sine and cosine functions, the tangent, cotangent, secant, and cosecant (abbreviated tan, cot, sec, and csc, respectively) may be defined for positive angles less than $\pi/2$ radians using the sides of a right triangle. Indeed, with respect to right triangle ABC in Figure 7.4.1:

$$\tan t = \frac{\text{length side opposite } A}{\text{length side adjacent } A} \qquad \cot t = \frac{\text{length side adjacent } A}{\text{length side opposite } A}$$

$$\sec t = \frac{\text{length hypotenuse}}{\text{length side adjacent } A} \qquad \csc t = \frac{\text{length hypotenuse}}{\text{length side opposite } A}$$

We immediately notice that the slope of the line AB is $\sin t/\cos t = \tan t$.

Figure 7.4–1

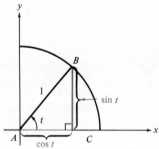

7.4.1   Example in Biology

A spermatozoon consists of a head and cylindrical tail. If we assume that its motion is planar, then at any time t the tail forms a sinusoidal wave, called a lateral displacement wave. At a given time t the points (x, y) on the tail are described by the equation $y = a\sin k(x + ct)$, a function of horizontal position x and time t. At any fixed time t, the tangent line at

Figure 7.4–2

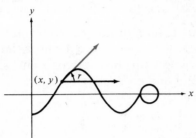

point (x, y) has a slope $\dfrac{dy}{dx} = ka \cos k(x + ct)$. As we saw in Figure 7.4–1, the slope of the line is the tangent of the angle of inclination. Thus, if the angle of inclination to the x-axis is r radians,

$$\tan r = \frac{dy}{dx} = ka \cos k(x + ct)$$

It is easy to see that the traditional trigonometric formulas, given above for $0 \leqslant t \leqslant \pi/2$, are consistent with the following more general definitions.

7.4.2   Definition

For all t such that the denominators are not zero,

(a) $\tan t = \dfrac{\sin t}{\cos t}$ (c) $\sec t = \dfrac{1}{\cos t}$

(b) $\cot t = \dfrac{\cos t}{\sin t}$ (d) $\csc t = \dfrac{1}{\sin t}$

Since $\sin t = 0$ when $t = k\pi$, for all integers k, we see that the functions $\cot t$ and $\csc t$ are not defined where $t = k\pi$. Similarly, the functions $\tan t$ and $\sec t$ are not defined where $t = (k + \frac{1}{2})\pi$, for all whole numbers k, since the denominators are zero there.

Moreover, the functions defined in Definition 7.4.2 are periodic. Both the secant and cosecant obviously have the same period as cosine and sine, respectively, namely 2π. And, while it is true that $\tan (t + 2\pi) = \tan t$, the period of $\tan t$ and $\cot t$ is actually π. (This is the fundamental period, since it gives precisely one cycle of the function.)

7.4.3   Example

Show that $\tan (t + \pi) = \tan t$ for all real numbers t.

Solution:

We can apply Rule 7.1.5(b) in radian form to obtain

$$\tan{(t + \pi)} = \frac{\sin{(t + \pi)}}{\cos{(t + \pi)}} = \frac{-\sin{t}}{-\cos{t}} = \frac{\sin{t}}{\cos{t}} = \tan{t}$$

We can evaluate our four new functions, if we wish, by using Definition 7.4.2 directly. It is useful to remember, however, that the signs of $\sec{t}$ and $\csc{t}$ agree with those of $\cos{t}$ and $\sin{t}$, respectively, while $\tan{(\pi - t)} = -\tan{t}$ and $\cot{(\pi - t)} = -\cot{t}$.

7.4.4   Example

Evaluate the following: (a) $\sec{\frac{2}{3}\pi}$, (b) $\csc{\frac{2}{3}\pi}$, (c) $\tan{\frac{2}{3}\pi}$, (d) $\cot{\frac{2}{3}\pi}$, (e) $\cot{\frac{3}{2}\pi}$.

Solution:

(a) $\sec{\frac{2}{3}\pi} = \dfrac{1}{\cos{\frac{2}{3}\pi}} = \dfrac{1}{-\frac{1}{2}} = -2$

(b) $\csc{\frac{2}{3}\pi} = \dfrac{1}{\sin{\frac{2}{3}\pi}} \approx \dfrac{1}{0.8660}$, by Table VII

(c) $\tan{\frac{2}{3}\pi} = -\tan{(\pi - \frac{2}{3}\pi)} = -\tan{\frac{1}{3}\pi} = -\tan{60°} \approx -1.7321$, by Table VII

(d) $\cot{\frac{2}{3}\pi} = \dfrac{1}{\tan{\frac{2}{3}\pi}} \approx \dfrac{1}{-1.7321}$

(e) $\cot{\frac{3}{2}\pi}$ is not defined, since $\cos{\frac{3}{2}\pi} = 0$.

It is now possible to plot the graphs of our "new" functions. Observe the points where the functions are not defined.

Figure 7.4–3

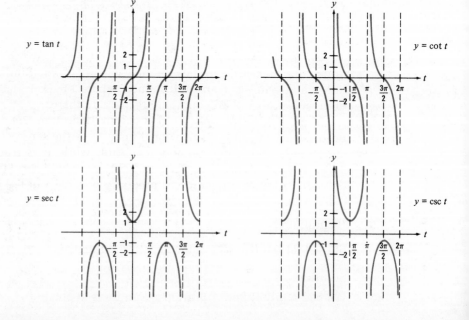

The derivatives of the tangent, cotangent, secant, and cosecant functions can be found quite easily by applying our basic rules of differentiation (2.3.3, especially) to the quotients of Definition 7.4.2. In fact, we already found that of tan t in Example 7.2.6(c). But we will state them here for easy reference.

7.4.5   Rule

(a) $\dfrac{d}{dt}$ (tan t) $=$ sec^2 t (c) $\dfrac{d}{dt}$ (sec t) $=$ (sec t)(tan t)

(b) $\dfrac{d}{dt}$ (cot t) $= -$csc^2 t (d) $\dfrac{d}{dt}$ (csc t) $= -$(csc t)(cot t)

7.4.6   Example

(a) Verify Rule 7.4.5(c).   (b) Differentiate $y =$ tan 3x.

Solution:

(a) $\dfrac{d}{dt}$ (sec t) $= \dfrac{d}{dt}\left(\dfrac{1}{\cos t}\right) = \dfrac{(\cos t)(0) - (-\sin t)(1)}{\cos^2 t}$

$\qquad = \dfrac{(\sin t)(1)}{(\cos t)(\cos t)} = $ (tan t)(sec t)

(b) $y' = 3$ sec^2 3x

7.4.7   Example in Biology

A homing pigeon is released from a boat at point B on a lake, and it flies to its coop, which is situated at point C on the shore. The bird avoids flying directly, perhaps because it requires twice as much energy to fly over water as it does over land, owing to the presence of downdrafts. Consequently, its path is from B to a point P on the shore, and from there to the coop at C, as in Figure 7.4–4. What angle at P will <u>minimize the</u> <u>energy expended?</u>

Solution:

In this kind of problem, it is often best to choose an angle as our variable; we shall use the size of angle APB as the variable t, although the problem could be solved if we chose instead ABP.

Referring to Figure 7.4–4, we find that $\overline{BP} = d$ csc t and $\overline{PC} = L - \overline{AP} = L - d$ cot t. Hence, the total energy required is a multiple of $2\overline{BP} + \overline{PC}$ or $E(t) = (2d$ csc $t) + (L - d$ cot $t) = L + d(2$ csc $t -$ cot $t)$. Setting

$\qquad E'(t) = d[-2(\text{csc } t)(\text{cot } t) - (-\text{csc}^2 \, t)]$
$\qquad\qquad = -d$ csc $t(2$ cot $t -$ csc $t) = 0$

since csc $t \neq 0$, we must have 2 cot $t -$ csc $t = 0$. Hence, 2 cot $t =$ csc t; or 2 cos $t/\sin t = 1/\sin t$; or 2 cos $t = 1$; or cos $t = \frac{1}{2}$.

Figure 7.4–4

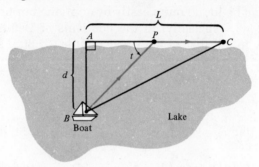

Hence, $t = \pi/3$, and the reader can check that this yields a relative and absolute minimum of $E(t)$.

Rule 7.4.5 gives us certain antiderivatives immediately. Since we have already found antiderivatives for tan t and cot t (in Rule 7.3.4 and problem 2 of Exercises 7.3.C), we can summarize these results in the following rule.

7.4.8   Rule

(a) $\displaystyle\int \sec^2 t \; dt = \tan t + k$ (d) $\displaystyle\int (\csc t)(\cot t) \; dt = -\csc t + k$

(b) $\displaystyle\int \csc^2 t = -\cot t + k$ (e) $\displaystyle\int \tan t \; dt = -\ln|\cos t| + k$

(c) $\displaystyle\int (\sec t)(\tan t) \; dt = \sec t + k$ (f) $\displaystyle\int \cot t \; dt = \ln|\sin t| + k$

where k is an arbitrary constant.

7.4.9   Example

Evaluate $\displaystyle\int \sec^2 3t \; dt$.

Solution:

$$\int \sec^2 3t \; dt = \tfrac{1}{3} \int \sec^2 3t(3 \; dt) = \tfrac{1}{3} \tan 3t + k$$

SUMMARY

We have discussed the derivatives, integrals, and graphs of the functions tan t, cot t, sec t, and csc t, the four remaining circular functions. In the following exercises we will ask you to practice with these derivatives and integrals, as well as with certain relationships existing between these functions. Also, Exercises 7.4.C discuss the inverse circular functions.

EXERCISES 7.4. A

1. Evaluate
    (a) $\sec (\pi/4)$ (b) $\tan (\pi/4)$ (c) $\cot (-\pi/4)$
    (d) $\tan \pi$ (e) $\tan (\pi/3)$
2. Evaluate
    (a) $\sec (7\pi/4)$ (b) $\tan (7\pi/4)$
3. If a right triangle ABC has a right angle at C, side $AC = 4$, and side $BC = 3$, find the measures of angles A and B. (Use Table VII.)
4. If the right triangle ABC has a right angle at C, the measure of angle A is $57°$, and $AC = 10$, find BC.
5. Find the derivatives:
    (a) $\tan t + \sec t$ (b) $3 \csc t + 2 \cot t$
6. Find the critical points of the function $f(t) = \sec t$ and determine whether they give relative maxima, minima, or neither.
7. Evaluate

    (a) $\int \sec^2 (t/2) \, dt$ (b) $\int (\sec 2t)(\tan 2t) \, dt$

    (Hint: See Rule 7.4.8.)
8. Derive the following identities.
    (a) $\cot (t + \pi) = \cot t$ (b) $\tan t = \cot (\pi/2 - t)$
    (c) $(\sin t)(\csc t) = 1$ (d) $\cot (\pi - t) = -\cot t$
    (e) $\cot^2 t + 1 = \csc^2 t$
    (Hint: See Definition 7.4.2 and Example 7.4.3.)
9. Differentiate the composite functions
    (a) $\tan (t^2)$ (b) $\sqrt{\sec t}$
10. If $F(x) = \cot x$, and $\dfrac{dF}{dt} = 2$ when $x = \pi/4$, use implicit differentiation to find $\dfrac{dx}{dt}$.

11. If a spermatozoon of Example 7.4.1 has a lateral displacement wave given by $y = 3 \sin 2\pi (x + t)$, find the angle of inclination r of the tangent line to this wave at the point whose lateral position is $x = \frac{1}{8}$ when the time is $t = 0$.

EXERCISES 7.4. B

1. Evaluate
    (a) $\csc (\pi/4)$ (b) $\cot (\pi/4)$ (c) $\cot \pi$
    (d) $\sec (\pi/6)$ (e) $\tan (-\pi/4)$
2. Evaluate
    (a) $\sec (-7\pi/4)$ (b) $\tan (-7\pi/4)$
3. If the right triangle ABC has a right angle at C, side $AC = 5$, and side $BC = 12$, find the measures of angles A and B.

4. If a right triangle RST has a right angle at T, the measure of the angle R is $28°$, and $RT = 100$, find ST.

5. Find the derivatives:
    (a) $3 \sec t - 2 \tan t$ (b) $\csc t + \cot t$

6. Find the critical points of the function $f(t) = \csc t$ and determine whether they give relative maxima, minima, or neither.

7. Evaluate

    (a) $\displaystyle\int \csc^2 (t + 1)\, dt$ (b) $\displaystyle\int (\csc 2t)(\cot 2t)\, dt$

    (Hint: See Rule 7.4.8.)

8. Derive the following identities.
    (a) $\tan t = 1/\cot t$ (b) $(\sec t)(\cos t) = 1$
    (c) $\tan (t + \pi) = \tan t$ (d) $\sec (\pi/2 - t) = \csc t$
    (e) $\tan^2 t + 1 = \sec^2 t$

    (Hint: See Definition 7.4.2 and Example 7.4.3.)

9. Differentiate the composite functions
    (a) $\cot (-t^3 + 3)$ (b) $\csc^2 (-t)$

10. If $F(x) = \tan x$ and $\dfrac{dF}{dt} = 1$ when $x = \pi/6$, find $\dfrac{dx}{dt}$ by implicit differentiation.

11. If a spermatozoon of Example 7.4.1 has a lateral displacement wave given by $y = 2 \sin \pi(x + t)$, find the angle of inclination r of the tangent line to this wave at the point whose lateral position is $x = \frac{1}{2}$ when the time is $t = 0$.

EXERCISES 7.4. C

1. Refer to the graph of the tangent function given in Figure 7.4–3.
    (a) Prove that $y = \tan x$ is increasing wherever the derivative exists.
    (b) How can you tell that the function gets very large in absolute value near the points $\pi/2 + n\pi$ for all integers n?
    HC (c) Notice that the tangent function was proved to have period π in Example 7.4.3. Use your hand calculator to plot the points on the graph of $y = \tan t$ for $t = -1.5, -1.3, -1.0, -0.8, -0.5, -0.3, 0, 0.3, 0.5, 0.8, 1.0, 1.3,$ and 1.5 radians.

HC 2. (a)

	t (radians)				
t	1	0.5	0.2	0.08	0.006
$(\tan t)/t$					

(b) What does this indicate about $\displaystyle\lim_{t \to 0} \frac{\tan t}{t}$?

3. Since the functions $\sin t$, $\cos t$, and $\tan t$ repeat themselves infinitely often as t varies throughout the real numbers, they cannot, strictly speaking, have inverses. However, the sine function, for example, achieves each of its possible values exactly once in the interval $-\pi/2 \leq t \leq \pi/2$. Hence, in this interval it has an inverse. Similarly, the functions $\cos t$ and $\tan t$ achieve their values exactly once in the intervals $0 \leq t \leq \pi$ and $-\pi/2 < t < \pi/2$, respectively. (See Figure 7.4–3.) Thus, we can define the functions.

$$\text{Arcsin } t = y \ (-1 \le t \le 1) \quad \text{if and only if} \quad \sin y = t \left(-\frac{\pi}{2} \le y \le \frac{\pi}{2}\right)$$

$$\text{Arccos } t = y \ (-1 \le t \le 1) \quad \text{if and only if} \quad \cos y = t \ (0 \le y \le \pi)$$

$$\text{Arctan } t = y \ (-\infty < t < \infty) \quad \text{if and only if} \quad \tan y = t \left(-\frac{\pi}{2} < y < \frac{\pi}{2}\right)$$

(a) Sketch the graphs of the inverse functions given above.
Evaluate
(b) Arcsin $(\frac{1}{2})$ (c) Arccos $(-\frac{1}{2})$
(d) Arctan (1) (e) Arctan $(\sin \frac{3}{5})$

4.   Assuming the validity of $\frac{d}{dt}$ (Arcsin t) $= 1/\sqrt{1-t^2}$:

(a) Differentiate Arcsin $(t^2 + 1)$.

(b) Evaluate $\displaystyle\int \frac{dt}{\sqrt{1 - 4t^2}}$ and $\displaystyle\int \frac{t \, dt}{\sqrt{1 - t^4}}$.

5.   If $\frac{d}{dt}$ (Arctan t) $= \frac{1}{1+t^2}$:

(a) Differentiate Arctan $(t^2 + 1)$.

(b) Evaluate $\displaystyle\int \frac{dt}{9t^2 + 1}$ and $\displaystyle\int \frac{t^2 \, dt}{1 + t^6}$.

ANSWERS   7.4. A

1. (a) $\sqrt{2} \approx 1.414$ (b) 1   (c) -1 (d) 0   (e) $\sqrt{3} \approx 1.732$
2. (a) $\sqrt{2} \approx 1.414$ (b) -1
3. $A \approx 37°, B \approx 53°$
4. $BC \approx 15.399$
5. (a) $\sec^2 t + (\sec t)(\tan t)$ (b) $-3(\csc t)(\cot t) - 2 \csc^2 t$
6. Critical points at $n\pi$ for all integers n. If n is even, there is a local minimum; if n is odd, there is a local maximum.
7. (a) $2 \tan (t/2) + k$ (b) $\frac{1}{2} \sec 2t + k$

8. (a) $\cot (t + \pi) = \dfrac{\cos (t + \pi)}{\sin (t + \pi)} = \dfrac{-\cos t}{-\sin t} = \cot t$

(b) $\tan t = \dfrac{\sin t}{\cos t} = \dfrac{\cos (\pi/2 - t)}{\sin (\pi/2 - t)} = \cot (\pi/2 - t)$

(c) $(\sin t)(\csc t) = \dfrac{\sin t}{\sin t} = 1$

(d) $\cot (\pi - t) = \dfrac{\cos (\pi - t)}{\sin (\pi - t)} = \dfrac{-\cos t}{\sin t} = -\cot t$

(e) Divide $\cos^2 t + \sin^2 t = 1$ through by $\sin^2 t$.
9. (a) $2t \sec^2 (t^2)$ (b) $\frac{1}{2}(\tan t)\sqrt{\sec t}$
10. -1
11. About $84°$

ANSWERS 7.4. C

1. (a) If the derivative exists, a function is increasing whenever the derivative is positive; since $y' = \sec^2 x > 0$, the function is everywhere increasing, provided that it is defined.

    (b) $\tan t = \dfrac{\sin t}{\cos t}$ and $\cos t$ takes the value 0 at $\pi/2 + n\pi$ but $\sin t = \pm 1$.

    (c) About $-14.1, -3.6, -1.6, -1, -0.5, -0.3, 0, 0.3, 0.5, 1, 1.6, 3.6,$ and 14.1

2. (a)

	t (radians)				
	1	0.5	0.2	0.08	0.006
$(\tan t)/t$	1.557408	1.092605	1.013550	1.002139	1.000012

    (b) $\lim\limits_{t \to 0} \dfrac{\tan t}{t} = 1$

3. (a)

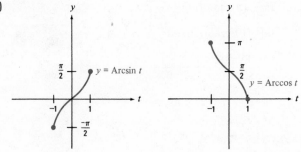

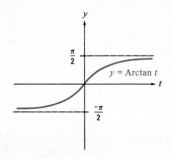

    (b) $\pi/6$ (c) $2\pi/3$ (d) $\pi/4$ (e) $\frac{3}{4}$

4. (a) $\dfrac{2}{\sqrt{-2 - t^2}}$ (b) $\frac{1}{2}$ Arcsin $2t + k$ (c) $\frac{1}{2}$ Arcsin $t^2 + k$

5. (a) $\dfrac{2t}{t^4 + 2t^2 + 2}$ (b) $\frac{1}{3}$ Arctan $3t + k$ (c) $\frac{1}{3}$ Arctan $t^3 + k$

SAMPLE QUIZ   **Sections 7.3 and 7.4**

1. (20 pts) Differentiate
    (a) $y = \tan x + \csc x$ (b) $y = \tan 3x$
    (c) $y = \sec (x^2)$ (d) $y = \cot^2 x$

2. (25 pts) Find the antiderivatives indicated:

(a) $\int \cos 3x \, dx$ (b) $\int \sin(-4x) \, dx$

(c) $\int \sec^2(x+2) \, dx$ (d) $\int \dfrac{\sin 2x}{\cos 2x} \, dx$

3. (15 pts) Graph $y = \tan x$.
4. (20 pts) Find the area between one lobe of the curve $y = \sin 3x$ and the x-axis.
5. (20 pts) Suppose that the birth rate of a population is about $B(t) = 10 \cos 2x$ and the death rate is approximately $D(t) = 6 \sin 2x$. If the rate of change in population is given by the difference between the birth rate and the death rate and $P(0) = 100$, find a formula for the population function $P(t)$.

The answers are at the back of the book.

SAMPLE TEST   ## Chapter 7

Each question is worth 10 points.

1. If $\sin 45° = 0.7071$, find
   (a) $\sin 135°$ (b) $\cos 45°$ (c) $\sin 315°$
   (d) $\sin 405°$ (e) $\cos(-45°)$
2. Sketch the graph of $y = \sin 2t°$ from $t = 0$ to $t = 360°$.
3. (a) Convert 30° to radians. (b) Convert 3 radians to degrees.
4. Differentiate the following:
   (a) $\sin t$ (b) $\cos t$ (c) $3 \sin t - \cos t$
   (d) $\tan t$ (e) $\sin^2 3t$
5. Find all antiderivatives of the following:
   (a) $\sec^2 t$ (b) $\cos 3t$

   (c) $t \sin(t^2)$ (d) $\dfrac{\sin t}{\cos^2 t}$

6. What are the periods of the following functions in radians:
   (a) $\sin t$ (b) $\tan t$ (c) $\sin t + \cos t$
   (d) $\sin 4t$ (e) $\sec t$
7. Find the absolute maximum of the function $f(t) = 2 + \sin 3t$.
8. Find the area between the graph of $y = 2 + \sin 3t$, the t-axis, $t = 0$, and $t = \pi/3$.
9. What is the general solution of the differential equation $\sin y \, dy = \cos x \, dx$?
10. Sketch the graph of $y = 3 \sin^2 t + 3 \cos^2 t$.

The answers are at the back of the book.

REVIEW OF ALGEBRA

R.1 Real Numbers, Signed Numbers, and Absolute Value

The first numbers that a child learns are the <u>counting numbers</u> — one, two, three, etc. And the first arithmetic process that we study is addition. It soon becomes apparent that when any two counting numbers are added together, the answer is another counting number.

But when subtraction is studied, the answers are no longer necessarily counting numbers. What, for example, is 3 minus 5? If there are to be answers to all subtraction problems, the set of counting numbers must be extended to include negative integers and zero. Thus, through subtraction, we obtain the <u>integers</u>.

When two integers are multiplied, the answer is always an integer. But this is not true for division; if there are to be answers to all division problems, the set of integers must be extended to a still larger set called the rational numbers. Every <u>rational number</u> is the quotient of two integers. Notice that, in particular, every integer is the quotient of two integers (for example, $2 = \frac{6}{3}$) and therefore a rational number.

In elementary algebra you learned to plot numbers on a line with zero on the "middle" of the line, the negative numbers to the left of zero, and the positive numbers to the right of zero. Every rational number corresponds to a particular point on this "number line." Does every point correspond to a rational number? No! This answer is not obvious, but more than two thousand years ago the Greeks had a proof that $\sqrt{2}$ is not the quotient of two integers; since $\sqrt{2}$ corresponds to a point on the line lying between 1.4 and 1.5, we conclude that not all such points correspond to rational numbers. All numbers that correspond to some point on the number line are called <u>real numbers</u>.

Set of numbers	Examples
Counting numbers	1, 2, 3, 4
Integers	$\ldots, -2, -1, 0, 1, 2, \ldots$
Rational numbers	$\frac{1}{2}, -\frac{1}{2}, \frac{13}{17}, \frac{8}{4}$
Real numbers	$\sqrt{2}, \sqrt{3}, \pi, \frac{7}{2}$
Complex numbers	$\sqrt{-1}, 2 + 3\sqrt{-1}$

Each set includes the preceding sets.

This course uses the set of real numbers. By this we mean that every number mentioned in the text is real, and when we ask what solutions an equation has, we are asking what *real* numbers substituted into the equation make it true.

You might wonder if there are numbers other than real numbers. In some situations there are. For example, in this book we say that the equation $x^2 + 1 = 0$ does not have any solutions. But in some books the solution to this equation is assumed to exist and is written $\sqrt{-1}$ or i or j. Such numbers are useful (indispensable, actually) in electrical engineering but are beyond the scope of this text. The number $\sqrt{-1}$ is an example of a complex number.

Every real number is complex, but not every complex number is real $(\sqrt{-1})$. Every rational number is real, but not every real number is rational $(\sqrt{2})$. Every integer is a rational number, but not every rational number is an integer $(\frac{1}{2})$. Every counting number is an integer, but not every integer is a counting number.

We repeat. This text uses the set of real numbers, those numbers that correspond to points of the real line.

Signed Numbers

Numbers preceded by a "+" or "−" are often called signed numbers. When adding signed numbers, it is sometimes helpful to think of arrows along the number line. To add any two numbers of the same sign, add the magnitudes and keep the same sign. To add two numbers of the opposite sign, subtract the magnitudes and use the sign of the larger.

Figure R.1–1

To subtract, merely change the sign of the subtrahend (the second number) and add:

$$5 - (-7) = 5 + 7 = 12 \qquad 5 - (+7) = 5 - 7 = -2 \qquad -5 - (-7) = 2$$

Multiplication by a positive number keeps the sign of the other number being multiplied.

Multiplication by a negative number changes the sign of the other number being multiplied.

$$(-3)4 = -12 \qquad (-3)(-4) = 12$$

Figure R.1–2

$$(3)(4) = 12 \qquad\qquad (3)(-4) = -12$$

The product of two numbers of the same sign is positive. The product of two numbers of opposite sign is negative. Division follows the same rules as multiplication, since to divide by some number (for example, 2) is the same as to multiply by its reciprocal (for example, $\frac{1}{2}$).

Absolute Value

The expression $|x|$ is read "the absolute value of x" and it denotes the distance of x from zero on the real line. For example, both 3 and −3 have absolute value 3; they both lie 3 units from zero. Similarly, $|4| = |-4| = 4$. The absolute value of any positive number is the number itself. To find the absolute value of a negative number, you merely drop the minus sign. As you might guess, $|0| = 0$.

Figure R.1–3

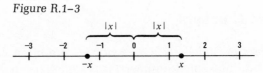

Just as $|x| = |x - 0|$ denotes the distance between the numbers x and 0 on the number line, the expression $|x - y|$ is *the distance between the numbers x and y*. Thus the distance between the numbers −2 and 3 is $|-2 - 3| = 5$ or $|3 - (-2)| = 5$. Both equations show that there are 5 units between −2 and 3. Similarly, the distance between the numbers 3 and 1 can be written either $|3 - 1| = 2$ or $|1 - 3| = 2$. It works both ways.

Figure R.1–4

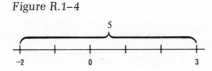

Example: Let us find all the numbers x such that $|x - 3| = 4$. We must realize that such an x is 4 units from 3. Hence, there are exactly two numbers, 7 and −1, that satisfy $|x - 3| = 4$.

Figure R.1–5

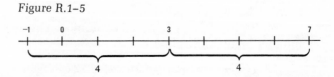

EXERCISES FOR SECTION R.1

For each of the following numbers, tell if it is (a) a counting number, (b) an integer, (c) a rational number, (d) a real number, or (e) a complex number. (More than one answer may be correct.)

1. 0

2. $\frac{3}{7}$

3. -1

4. $\sqrt{2}$

5. $\frac{4}{2}$

6. $\frac{2}{4}$

7. $\sqrt{4}$

8. $\sqrt{-1}$

9 $\sqrt{(-2)/(-2)}$

10. $\sqrt{0}$

Perform the following computations.

11. $2 - (-6) =$

12. $-2 - 6 =$

13. $2(-7) + 8 =$

14. $2(-5) - (3) =$

15. $4 + (-6)(-2) =$

16. $(-2)(-3) - 4 =$

17. $(-2)(-4) - (-9) =$

18. $8 - 9 =$

19. $(-8)(2) + 7 =$

20. $(-2)(-4) - 5 =$

21. $-7 - 2 =$

22. $(-4)(-3) - (-2)(-4) =$

23. $-9 + 8(-2) =$

24. $-5 + (-4) =$

25. $(-5)(-3) + (-4)(2) =$

26. $(-7)(-4) + (-5)(-5) =$

27. $-2 - 3 =$

28. $(-6)(3) - (-4)(-2) =$

29. $(-2)(-3)(-4) + 4 =$

30. $-6 - (-8) =$

Evaluate the following.

31. $|5| =$ 32. $|-5| =$ 33. $|\frac{1}{2}| =$ 34. $|-\frac{4}{5}| =$ 35. $|-\pi| =$

Find all the numbers x that satisfy each of the following.

36. $|x| = 3$

37. $|x| = -1$

38. $|x| = 0$

39. $|x - 2| = 0$

40. $|x - 3| = 2$

41. $|x + 2| = 5$

ANSWERS

1. (b)-(e)	12. -8	23. -25	34. $\frac{4}{5}$
2. (c)-(e)	13. -6	24. -9	35. π
3. (b)-(e)	14. -13	25. 7	36. $3, -3$
4. (d) and (e)	15. 16	26. 53	37. none
5. (a)-(e)	16. 2	27. -5	38. 0
6. (c)-(e)	17. 17	28. -26	39. 2
7. (a)-(e)	18. -1	29. -20	40. 1, 5
8. (e)	19. -9	30. 2	41. $-7, 3$
9. (a)-(e)	20. 3	31. 5	
10. (b)-(e)	21. -9	32. 5	
11. 8	22. 4	33. $\frac{1}{2}$	

R.2 Cartesian Coordinates and the Distance Formula

In the sixteenth century, René Descartes put two real lines (called <u>axes</u>) at right angles to each other and suggested how to label points on the resulting Cartesian plane. The point where the lines cross is called the origin. The horizontal line going through the origin is often called the x-axis. The vertical line going through the origin is often called the y-axis. Sometimes it is convenient to change these letters. For example, the vertical axis will sometimes designate cost; then we might call it the C-axis.

Each ordered pair of numbers [for example, (2, 3)] corresponds to exactly one point on the plane. The first number of an ordered pair (often called the x-coordinate) tells how far left or right the given point is from the vertical axis. The second number (often called the y-coordinate) tells how far the point is above or below the horizontal axis. (See Figure R.2–1.)

Figure R.2–1

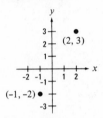

If the x-coordinate is negative, the point will lie to the left of the vertical axis. If the y-coordinate is negative, the point lies below the horizontal axis.

The Distance Formula

When we know the coordinates of a pair of points, we can calculate the distance between them. The distance between the points with coordinates (x_1, y_1) and (x_2, y_2) is given by the distance formula:

$$d = \sqrt{(x_1 - x_2)^2 + (y_1 - y_2)^2}$$

This formidable-looking formula is really quite easy to remember when you realize that it is only a disguised form of the Pythagorean theorem: The sum of the squares of the sides of a right triangle is equal to the square of the hypotenuse. (See Figure R.2–2.)

Figure R.2–2

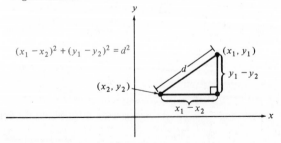

R.2.1   Example

Let us find the distance between the points having coordinates (1, 9) and (4, 5). Thus, we either write

$$d = \sqrt{(1-4)^2 + (9-5)^2} = \sqrt{(-3)^2 + (+4)^2} = 5$$

or

$$d = \sqrt{(4-1)^2 + (5-9)^2} = \sqrt{(+3)^2 + (-4)^2} = 5$$

depending on whether we chose the point (1, 9) or the point (4, 5) to be (x_1, y_1).

R.2.2   Example

Let us find all the points (x, y) which are at a distance of 3 units from the origin (0, 0). We can use the distance formula to write the condition defining the points (x, y) by

$$\sqrt{(x-0)^2 + (y-0)^2} = \sqrt{x^2 + y^2} = 3$$

Upon squaring, we get

$$x^2 + y^2 = 9$$

The last equation is an equation of the circle with its center at the origin and having a radius of 3. The required points (x, y) form this circle.

EXERCISES FOR SECTION   R.2

1. Plot the following points on a Cartesian plane.
   (1, 2) (1, 3) (1, 4) (2, −3) (−1, 3) (−4, −2) (2, −π)

Find the distance between the following pairs of points.

2. (1, 3) and (2, 3)
3. (1, 3) and (2, 0)
4. (1, 3) and (2, −1)
5. (−1, 3) and (2, −1)
6. (−1, −3) and (2, −1)
7. (−1, −3) and (−2, −1)
8. Show that the three points (1, 0), (3, 0), and (2, 5) are the vertices of an isosceles triangle. (*Hint:* Show that two of the sides are equal in length.)
9. Show that the points (0, 0), (1, 1), (2, 1), and (1, 0) form the vertices of a parallelogram. (*Hint:* Show that two pairs of sides are equal in length.)
10. Show that the points (1, 9), (4, 5), and (1, 5) form the vertices of a right triangle. (*Hint:* Show that the lengths of the sides satisfy the Pythagorean theorem.)
11. Write an equation satisfied by all the points (x, y) that are 5 units from the origin (and, of course, in the Cartesian plane.)
12. Find an equation of the circle with center at (0, 0) and having a radius of 2 units.

ANSWERS

1.

2. 1
3. $\sqrt{10}$
4. $\sqrt{17}$
5. 5
6. $\sqrt{13}$
7. $\sqrt{5}$
8. The lengths of sides are 2, $\sqrt{26}$, and $\sqrt{26}$.
9. The lengths of sides are 1, 1, $\sqrt{2}$, and $\sqrt{2}$.
10. $3^2 + 4^2 = 5^2$
11. $x^2 + y^2 = 25$
12. $x^2 + y^2 = 4$

R.3 Linear Equations

The equation $3x - 4 = x + 6$ is a linear equation in one unknown, x. A solution to such an equation is a number that when substituted for x makes the equation true. To solve such equations, you first add and subtract terms on both sides of the equation until all the terms containing x are on one side and all the terms not containing x are on the other side. Then you divide both sides of the equation by the coefficient of x to isolate the x.

$$3x - 4 = x + 6$$

Adding 4:
$$\frac{+4 \qquad +4}{3x = x + 10}$$

Subtracting x:
$$\frac{-x = -x}{2x = 10}$$

Dividing by 2:
$$x = 5$$

This procedure often reveals that a given linear equation in one unknown has precisely one solution. But not always: Every number is a solution of $2x = 2x$, and no number is a solution of $2x + 1 = 2x$.

Linear Equations in Two Unknowns

The equation $2x + 3y = 7$ is a linear equation in two unknowns, x and y. Two linear equations in two unknowns often have exactly one solution in common. But this time a single solution is a pair of numbers — one value for x and one value for y — that when substituted into the equations make both equations true.

Two linear equations in two unknowns can be solved by any one of three methods: the addition–subtraction method, the substitution method, or the graphing method.

R.3.1   Example Using the Addition–Subtraction Method

$$2x + 3y = 7$$
$$2x + y = 3$$

Subtracting the second equation from the first: $2y = 4$

Solving for y: $y = 2$

Substituting into first equation: $2x + 3(2) = 7$

Solving for x: $2x = 7 - 6 = 1$

$$x = \tfrac{1}{2}$$

R.3.2   Harder Example Using the Addition–Subtraction Method

$$2x + 3y = 8$$
$$5x + 2y = 9$$

To obtain the same coefficient for x, multiply the first equation by 5 and the second by 2:

$$10x + 15y = 40$$
$$10x + 4y = 18$$

Now subtract the second from the first: $11y = 22$

Divide by 11 to solve for y: $y = 2$

Substitute $y = 2$ into the first equation: $2x + 3(2) = 8$

Subtract 6 from both sides: $2x = 8 - 6 = 2$

Divide by 2 to solve for x: $x = 1$

R.3.3   Example Using the Substitution Method

$$2x + 3y = 7$$
$$2x + y = 3$$

Solving for y in the second equation: $y = 3 - 2x$

Substituting this expression for y back into first equation: $2x + 3(3 - 2x) = 7$

Solving for x: $2x + 9 - 6x = 7$

$$2 = 4x$$
$$\tfrac{1}{2} = x$$

Substituting into the third equation to find y: $y = 3 - 2(\tfrac{1}{2})$

$$= 3 - 1$$
$$= 2$$

R.3.4   **Example Using the Graphing Method**

Methods for graphing equations are described in detail in Chapter 1. Briefly, it is possible to graph a linear equation by plotting two points that make the equation true (possibly by setting $x = 0$ and solving for y and then setting $y = 0$ and then solving for x) and drawing a straight line through these two points.

To use the graphing method for solving simultaneous equations, graph each equation and locate the point of intersection.

Figure R.3–1

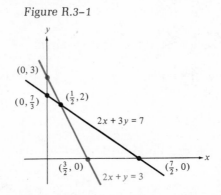

Like the situation of one linear equation in one unknown, a system of two linear equations in two unknowns does not necessarily have a unique solution. There are two ways that things can go wrong. One is that we may have a contradiction or inconsistency:

$$x + y = 1$$
$$x + y = 2$$

If the sum of two numbers is 1, then obviously it cannot be 2, so this system of equations does not have a solution. In less obvious cases, if we try to solve the system by either the addition–subtraction or the substitution method, we shall eventually arrive at a contradiction and so discover that the system cannot be solved. If we graph two such equations, we shall find that the lines are parallel. (See Figure R.3–2.)

The other situation where there is not a unique solution to a system of two linear equations in two unknowns is when the two equations are redundant:

$$x + \ y = 1$$
$$2x + 2y = 2$$

If the sum of any two numbers is 1, then the sum of their doubles is always 2; the second equation in this pair puts no more restrictions on the numbers than was already imposed by the first. If we try to graph these two equations, we will discover that they describe the same line; the same pairs of numbers (x, y) are solutions to both.

Figure R.3–2

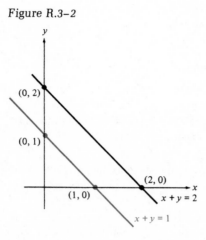

R.3.5   Example

Suppose that the revenue from selling x items is $R = 2x + 5$ and the cost of producing x items is $C = x + 10$. Then we can graph these two expressions on the same graph if we call the y-axis a $-axis. We discover that the cost is greater than the revenue for $x = 0$. But the lines come closer together as x grows until they intersect; the point where the revenue equals the cost is called the <u>break-even point</u>. We find this point by setting $R = C$ — that is, by setting $2x + 5 = x + 10$. In this case, we find that for $R = C$, we have $x = 5$ and $R = C = 15$. The point $(5, 15)$ is, therefore, the break-even point.

Figure R.3–3

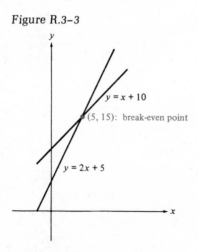

EXERCISES FOR SECTION   R.3

Solve for x:

1. $2x + 4 = x + 1$
2. $x + 4 = 3x - 6$
3. $4 - 3x = 5x - 7$
4. $5x + 5 = 8x - 3$
5. $-x + 4 = 2x + 10$

6. Solve the following systems of equations using the addition–subtraction method.

    (a) $2x + 7y = 20$ (b) $4x + 3y = 10$ (c) $5x + 2y = 23$

        $3x + 6y = 21$ $3x + 5y = 13$ $4x - 3y = 0$

7. Solve the following systems of equations using the substitution method.

    (a) $x + 4y = -2$ (b) $3x + 5y = -9$ (c) $3x - y = 2$

        $2x - 5y = 9$ $x + 7y = -3$ $4x + 3y = 20$

8. Solve the following systems of equations using the addition–subtraction method.

    (a) $2x + 5y = 13$ (b) $3x + 5y = 31$ (c) $5x - 2y = 4$

        $3x + 2y = 14$ $6x + 3y = 27$ $3x + 4y = 18$

9. Solve the following systems of equations using the substitution method.

    (a) $x + 3y = 0$ (b) $3x - 4y = 8$ (c) $5x + y = 22$

        $4x + 3y = 9$ $x + 2y = -4$ $7x - 2y = 7$

For each pair of equations, find the unique solution if it exists; otherwise, tell whether the pair is a contradiction or a redundancy.

10. $x + 3y = 10$
     $2x - y = 6$

11. $x + 2y = 11$
     $3x - 2y = 9$

12. $2x + 3y = 7$
     $4x + 6y = 11$

13. $5x - 2y = 11$
     $2x + 4y = -10$

14. $2x - y = 3$
     $6x - 3y = 9$

15. $4x + 7y = 10$
     $2x + 3y = 4$

16. If the cost of producing x hats is given by $C = 2x + 3$ and the revenue obtained from the sale of these hats is found using $R = 3x$, find the break-even point.

17. If the revenue obtained from the sale of x sweaters is given by $R = 2x$ and the cost of producing these sweaters is found using $C = x + 2$, find the break-even point.

ANSWERS

1. -3 3. $\frac{11}{8}$ 5. -2

2. 5 4. $\frac{8}{3}$

6. (a) $x = 3, y = 2$ (b) $x = 1, y = 2$ (c) $x = 3, y = 4$

7. (a) $x = 2, y = -1$ (b) $x = -3, y = 0$ (c) $x = 2, y = 4$

8. (a) $x = 4, y = 1$ (b) $x = 2, y = 5$ (c) $x = 2, y = 3$

9. (a) $x = 3, y = -1$ (b) $x = 0, y = -2$ (c) $x = 3, y = 7$

10. $x = 4, y = 2$ 14. redundant

11. $x = 5, y = 3$ 15. $x = -1, y = 2$

12. contradiction   16. (3, 9)

13. $x = 1, y = -3$ 17. (2, 4)

R.4   Exponents, Algebraic Operations, and Rational Fractions

Exponents

We define $x^2 = x \cdot x$, $x^3 = x \cdot x \cdot x$, $x^4 = x \cdot x \cdot x \cdot x$, and so on. The "4" in the expression "x^4" is called an <u>exponent</u>. The following examples illustrate a number of well-known rules for exponents:

Examples	Rules
$x^2 \cdot x^3 = (x \cdot x)(x \cdot x \cdot x) = x^5$	$x^n \cdot x^m = x^{n+m}$
$\dfrac{x^5}{x^2} = \dfrac{x \cdot x \cdot x \cdot x \cdot x}{x \cdot x} = x^3$	$x^n / x^m = x^{n-m}$
$x^0 = x^{2-2} = x^2/x^2 = 1$	$x^0 = 1$
$x^1 = x^{3-2} = x^3/x^2 = x$	$x^1 = x$
$x^{-2} = x^{0-2} = x^0/x^2 = 1/x^2$	$x^{-n} = 1/x^n$
$(x^2)^3 = (x \cdot x)(x \cdot x)(x \cdot x) = x^6$	$(x^m)^n = x^{mn}$
$(\sqrt{x})^2 = x = x^1$, so $\sqrt{x} = x^{1/2}$	$x^{1/n} = \sqrt[n]{x}$

By setting $x^{m/n} = (x^{1/n})^m = (x^m)^{1/n}$ we can define $x^{m/n}$ for every rational number m/n. In this book, to avoid the use of complex numbers, we never allow x to be negative if n is even.

R.4.1   Example

Evaluate
(a) $16^{3/2}$
(b) $16^{3/4}$
(c) $16^{-3/4}$

Solution:
(a) $16^{3/2} = (\sqrt{16})^3 = 4^3 = 64$
(b) $16^{3/4} = (\sqrt[4]{16})^3 = 2^3 = 8$
(c) $16^{-3/4} = 1/(16)^{3/4} = \frac{1}{8}$

We can use the values of $x^{m/n}$ to approximate x^a for any positive x and any real number a.

R.4.2   Example

$3^{14/10} < 3^{\sqrt{2}} < 3^{15/10}$ because $14/10 < \sqrt{2} < 15/10$.

Multiplying Algebraic Expressions

When a variable with an exponent is multiplied by a number, such as in the expression $3x^4$, we say that the expression is a <u>monomial</u>. When several monomials are added together, such as in $x^2 + 3x + 2$, we call the resulting expression a <u>polynomial</u>. It is often important to be able to multiply polynomials together easily; we now present some patterns for doing this.

The <u>distributive law</u> (sometimes called multiplying by a common <u>monomial factor</u>) is used to multiply a monomial by another polynomial. It says that $a(b + c + d) = ab + ac + ad$.

R.4.3   Examples

$$3(4x^2 - 2x + 7) = 12x^2 - 6x + 21$$
$$5x(2x^2 + 4x - 3) = 10x^3 + 20x^2 - 15x$$

To multiply together two <u>binomials</u> (that is, two polynomials, each of which has two terms), we apply the distributive law twice: $(a + b)(c + d) = (a + b)c + (a + b)d = ac + bc + ad + bd$. A device for remembering this pattern is the <u>FOIL</u> rule:

$$
\begin{array}{ccccccc}
 & \text{First} & + & \text{Outside} & + & \text{Inside} & + & \text{Last} \\
 & \text{terms} & & \text{terms} & & \text{terms} & & \text{terms} \\
(a + b)(c + d) = & ac & + & ad & + & bc & + & bd
\end{array}
$$

R.4.4   Example

$$(3x^2 + 2)(4x + 7) = 12x^3 + 21x^2 + 8x + 14$$

If the two binomials we are multiplying together are the same except for one sign, then "outside terms" + "inside terms" = 0, and we have

$$(a + b)(a - b) = a^2 - ab + ab - b^2 = a^2 - b^2$$

R.4.5   Example

$$(x + 3)(x - 3) = x^2 - 9$$
$$(2x - 5)(2x + 5) = 4x^2 - 25$$

If the two binomials are the same, then two of the terms in the expansion are the same:

$$(a + b)^2 = a^2 + ab + ab + b^2 = a^2 + 2ab + b^2$$

R.4.6   Examples

$$(x + 3)^2 = x^2 + 6x + 9$$
$$(2x + 3)^2 = 4x^2 + 12x + 9$$

Rational Expressions

Quotients of polynomials are called <u>rational</u> expressions. Thus, $\frac{x^2 + 3x - 1}{x^3 - 2x + 3}$ is an example of a rational expression. The arithmetic of rational expressions (addition, subtraction, multiplication, and division) follows the same rules as the arithmetic of ordinary fractions of integers.

To multiply fractions we simply multiply their numerators to form the new numerator and multiply their denominators to form the new denominator:

$$\frac{a}{b} \cdot \frac{c}{d} = \frac{a \cdot c}{b \cdot d}$$

To divide fractions we invert the divisor and multiply:

$$\frac{a}{b} \div \frac{c}{d} = \frac{a}{b} \cdot \frac{d}{c} \qquad \text{provided that } c \neq 0$$

R.4.7 **Examples**

$$\frac{x + 2}{x^2} \cdot \frac{x}{x + 3} = \frac{(x + 2) \cdot x}{x^2 \cdot (x + 3)} = \frac{x + 2}{x \cdot (x + 3)}$$

$$\frac{x}{3} \div \frac{x}{x + 6} = \frac{x}{3} \cdot \frac{x + 6}{x} = \frac{x \cdot (x + 6)}{3x} = \frac{x + 6}{3}$$

To add fractions we must find a common denominator. This can always be done by multiplying the denominators together, although if we see something simpler we may use that instead.

Thus

$$\frac{a}{b} + \frac{c}{d} = \frac{ad}{bd} + \frac{bc}{bd} = \frac{ad + bc}{bd}$$

R.4.8 **Examples**

$$\frac{x}{3} + \frac{4}{x} = \frac{x^2}{3x} + \frac{12}{3x} = \frac{x^2 + 12}{3x}$$

$$\frac{x + 1}{3x} + \frac{4}{x^2} = \frac{x(x + 1)}{3x^2} + \frac{12}{3x^2} = \frac{x(x + 1) + 12}{3x^2}$$

$$\frac{x}{3} - \frac{x + 1}{2x} = \frac{2x^2}{6x} - \frac{3(x + 1)}{6x} = \frac{2x^2 - 3(x + 1)}{6x}$$

EXERCISES FOR SECTION **R.4**

1. $x^3 \cdot x^5 =$
2. $(x^3)^5 =$

3. $x^7/x^4 =$
4. $2^0 =$

5. $2^1 =$

6. $2^{-3} =$

7. $3^{-2} =$

8. $4^{1/2} =$

9. $4^{3/2} =$

10. $9^{1/2} =$

11. $9^{3/2} =$

12. $8^{1/3} =$

13. $8^{2/3} =$

14. $8^{-1/3} =$

15. $8^{-2/3} =$

16. $16^{-1/2} =$

17. $3(x - 2) =$

18. $5(x^2 + 4x) =$

19. $2x(x^2 - \pi) =$

20. $x^2(5x + 7) =$

21. $(2x + 4)(3x + 5) =$

22. $(x - 1)(2x + 5) =$

23. $(x^2 - 8)(2x - 3) =$

24. $(2x^2 - 3x)(x + 4) =$

25. $(x - 4)(x + 4) =$

26. $(x^2 - 2)(x^2 + 2) =$

27. $(3x + 1)(3x - 1) =$

28. $(x - 4)^2 =$

29. $(2x + 5)^2 =$

30. $(3x - 4)^2 =$

31. $\dfrac{5}{x} + \dfrac{x}{2} =$

32. $7/x^2 - x/4 =$

33. $3/x^3 + x^2/(x + 1) =$

34. $\dfrac{x + 1}{3x^2} + \dfrac{2}{x^3} =$

35. $\dfrac{x + 1}{x + 2} \cdot \dfrac{x^2 - 4}{x + 1} =$

36. $\dfrac{x^3 + x + 3}{x - 4} \cdot \dfrac{x^2 - 16}{x^2} =$

37. $\dfrac{x}{4} \div \dfrac{x}{8} =$

38. $\dfrac{x + 2}{6} \div \dfrac{x^2 + 4x + 4}{3x} =$

39. $\dfrac{x}{5} - \dfrac{x - 1}{x} =$

40. $\dfrac{x^2 - x}{x + 1} - \dfrac{x + 3}{2x} =$

ANSWERS

1. x^8

2. x^{15}

3. x^3

4. 1

5. 2

6. $\frac{1}{8}$

7. $\frac{1}{9}$

8. 2

9. 8

10. 3

11. 27

12. 2

13. 4

14. $\frac{1}{2}$

15. $\frac{1}{4}$

16. $\frac{1}{4}$

17. $3x - 6$

18. $5x^2 + 20x$

19. $2x^3 - 2\pi x$

20. $5x^3 + 7x^2$

21. $6x^2 + 22x + 20$

22. $2x^2 + 3x - 5$

23. $2x^3 - 3x^2 - 16x + 24$

24. $2x^3 + 5x^2 - 12x$

25. $x^2 - 16$

26. $x^4 - 4$

27. $9x^2 - 1$

28. $x^2 - 8x + 16$

29. $4x^2 + 20x + 25$

30. $9x^2 - 24x + 16$

31. $\dfrac{10 + x^2}{2x}$

32. $\dfrac{28 - x^3}{4x^2}$

33. $\dfrac{3(x + 1) + x^5}{x^3(x + 1)}$

34. $\dfrac{x(x + 1) + 6x}{3x^3}$

35. $\dfrac{(x + 1)(x^2 - 4)}{(x + 2)(x + 1)} = x - 2$

36. $\dfrac{(x^3 + x + 3)(x^2 - 16)}{(x - 4)(x^2)} = \dfrac{(x^3 + x + 3)(x + 4)}{x^2}$

37. $\dfrac{8x}{4x} = 2$

39. $\dfrac{x^2 - 5(x - 1)}{5x}$

38. $\dfrac{(x + 2)(3x)}{6(x^2 + 4x + 4)} = \dfrac{x}{2(x + 2)}$

40. $\dfrac{2x(x^2 - x) - (x + 1)(x + 3)}{2x(x + 1)}$

R.5 Factoring, Quadratic Equations, and Simplifications

Factoring is the opposite of multiplying. We try to envision an algebraic expression as a product by discovering what two expressions multiplied together equal it. This is usually a matter of observation and practice, but there are three special cases that one should learn to recognize at once.

The difference of two squares: $a^2 - b^2 = (a + b)(a - b)$.

R.5.1   Example

$$4x^2 - 9 = (2x + 3)(2x - 3)$$

Perfect squares:

$$a^2 + 2ab + b^2 = (a + b)(a + b) = (a + b)^2$$
$$a^2 - 2ab + b^2 = (a - b)(a - b) = (a - b)^2$$

R.5.2   Example

$$x^2 - 8x + 16 = x^2 - 2(4)x + (4)^2 = (x - 4)^2$$

The difference (or sum) of two cubes:

$$a^3 - b^3 = (a - b)(a^2 + ab + b^2)$$
$$a^3 + b^3 = (a + b)(a^2 - ab + b^2)$$

R.5.3   Example

$$x^3 + 8 = (x + 2)(x^2 - 2x + 4)$$

Another useful rule is that if there is a common factor in a series of terms, we can extract it from each of them.

R.5.4   Example

$$2x^3 + 5x^2 + x = x(2x^2 + 5x + 1)$$

Educated guessing is, however, the most common way to factor. This involves applying the FOIL rule in reverse and, in particular, looking for the integer factors of the constant term.

R.5.5   Example

To factor $x^2 + 5x + 6$, we notice that

$$6 = (2)(3) = (-2)(-3) = (6)(1) = (-6)(-1)$$

Consider the first two factors. If you multiply out $(x + 2)(x + 3)$, you will get $x^2 + 5x + 6$, as required. Hence, $x^2 + 5x + 6 = (x + 2)(x + 3)$.

If the coefficient of the x^2-term is not 1, then factoring is much more difficult because we must guess about how to factor it as well as about how to factor the constant term.

R.5.6   Example

Factor $2x^2 + 7x + 6$. The coefficient of x^2 can be factored in only one way $(2 \cdot 1)$, but the constant 6 can be factored in two ways $(6 \cdot 1$ or $2 \cdot 3)$, and each of these ways can be paired with the factorization of $(2 \cdot 1)$ in two orders. Thus we can guess

$$(2\ 6)(1\ 1)\quad \text{or}\quad (2\ 1)(1\ 6)\quad \text{or}\quad (2\ 2)(1\ 3)\quad \text{or}\quad (2\ 3)(1\ 2)$$

Trying each possibility, we see that the last choice gives a $7x$ in the product. Thus we have $2x^2 + 7x + 6 = (2x + 3)(x + 2)$.

Quadratic Equations

Quadratic equations, those of the form $ax^2 + bx + c = 0$, can sometimes be solved by factoring. (Also see Section 1.4.) The procedure is to factor the quadratic expression $ax^2 + bx + c = a(x - r_1)(x - r_2)$, and then set the factored expression equal to zero: $a(x - r_1)(x - r_2) = 0$. If a product equals zero, at least one of its factors must be zero.

$$x - r_1 = 0 \quad \text{or} \quad x - r_2 = 0$$

Hence

$$x = r_1 \quad \text{and} \quad x = r_2$$

are the solutions.

R.5.7   Example

To solve $x^2 + 5x + 6 = 0$, we write $(x + 2)(x + 3) = 0$. Hence we must have $x + 2 = 0$ or $x + 3 = 0$, and the solutions are $x = -2$ and $x = -3$.

Simplifications

Sometimes we can use factoring to simplify rational expressions because common factors appear in both the numerator and denominator and they can be "canceled."

R.5.8   Examples

$$\frac{2x^2 + 5x}{x} = \frac{x(2x + 5)}{x} = 2x + 5$$

$$\frac{2x^3 + 6x^2}{x} = \frac{2x^2(x + 3)}{x} = 2x(x + 3)$$

$$\frac{4x^2 - 25}{2x - 5} = \frac{(2x - 5)(2x + 5)}{2x - 5} = 2x + 5$$

EXERCISES FOR SECTION **R.5**

Factor each of 1–20 as completely as possible.

1. $x^2 - 4 =$
2. $x^2 - 49 =$
3. $x^2 - 25 =$
4. $x^2 - 1 =$
5. $x^3 - x^2 =$
6. $x^3 + x^2 =$
7. $x^2 - x =$
8. $x^2 - 2x + 1 =$
9. $3x^2 - 12x + 9 =$
10. $8x^3 - 8x =$

11. $x^2 + 4x + 4 =$
12. $x^2 + 6x + 9 =$
13. $x^2 - 6x + 9 =$
14. $x^4 - 6x^2 + 9 =$
15. $x^3 - 27 =$
16. $x^3 + 8 =$
17. $x^3 - 8 =$
18. $2x^3 - 8x =$
19. $x^3 + 6x^2 + 9x =$
20. $2x^3 + 10x^2 + 12x =$

Solve each of the following equations by factoring:

21. $x^2 + 2x + 1 = 0$
22. $x^2 - 4x + 4 = 0$
23. $x^2 - 16 = 0$
24. $x^2 + 7x + 12 = 0$
25. $x^3 + 7x^2 + 12x = 0$ (*Hint:* Factor completely.)

Simplify the following expressions by factoring and cancelling:

26. $\dfrac{x^2 + x}{x}$

27. $\dfrac{x^3 - 3x}{x}$

28. $\dfrac{4x^2 - 2x}{2x}$

29. $\dfrac{4x^3 - 2x^2}{2x}$

30. $\dfrac{3x^3 - 9x^2}{3x}$

31. $\dfrac{x^2 - 1}{x + 1}$

32. $\dfrac{x^2 - 25}{x + 5}$

33. $\dfrac{x^2 - 49}{x - 7}$

ANSWERS

1. $(x - 2)(x + 2)$
2. $(x - 7)(x + 7)$
3. $(x - 5)(x + 5)$
4. $(x - 1)(x + 1)$
5. $x^2(x - 1)$
6. $x^2(x + 1)$
7. $x(x - 1)$

8. $(x - 1)^2$
9. $3(x - 1)(x - 3)$
10. $8x(x - 1)(x + 1)$
11. $(x + 2)(x + 2)$
12. $(x + 3)(x + 3)$
13. $(x - 3)(x - 3)$
14. $(x^2 - 3)(x^2 - 3)$

15. $(x - 3)(x^2 + 3x + 9)$
16. $(x + 2)(x^2 - 2x + 4)$
17. $(x - 2)(x^2 + 2x + 4)$
18. $2x(x + 2)(x - 2)$
19. $x(x + 3)(x + 3)$
20. $2x(x + 2)(x + 3)$
21. $x = -1$
22. $x = 2$
23. $x = 4$ and $x = -4$
24. $x = -3$ and $x = -4$

25. $x = 0$, $x = -3$, and $x = -4$
26. $x + 1$
27. $x^2 - 3$
28. $2x - 1$
29. $x(2x - 1)$
30. $x(x - 3)$
31. $x - 1$
32. $x - 5$
33. $x + 7$

R.6 Inequalities

We say that a number x is <u>less than</u> another number z if x lies to the left of z on the number line. Equivalently, we say that x is less than z if $z - x$ is positive. We write this $x < z$. If x is less than z, then z is <u>greater than</u> x; we write this $z > x$.

R.6.1 Examples

$$2 < 3; -3 < 2; -3 < -2.$$

Positive numbers follow ordinary usage. But negative numbers live in "through-the-looking-glass-land" and we must be careful not to get things reversed. For example, $-2 > -3$, even though $|-2| < |-3|$. We see this by examining the line. The number -2 is to the right of -3, so -2 is greater than -3. Equivalently, we can calculate $-2 - (-3) = -2 + 3 = 1 > 0$.

Figure R.6-1

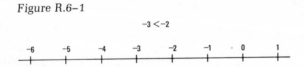

A statement of the form $x < 3$ is read "x is less than 3." This statement is true for any number, such as 2, 0, or -10, which lies to the left of three on the number line. *Warning!* It is not true for the number 3 itself. (See Figure R.6-2.)

Figure R.6-2

An inequality that contains a variable can often be "solved" in a sense similar to that of solving equations. The <u>solution</u> of an inequality is that set of numbers, each of which when substituted into the inequality <u>makes the inequality true</u>. But the solution set of an inequality is an interval (finite or infinite), in contrast to the solution set of an equation, which is a finite number of points.

To solve an inequality, we can add or subtract the same number to both sides of the inequality without changing the sense of the inequality. We can also multiply or divide an inequality by a *positive* number. (But if we multiply or divide by a negative number, we must reverse the inequality; we do not use this fact in this text.)

R.6.2   Example

$$5x - 2 < 3x + 4$$

Subtracting 3x and adding 2: $\qquad 2x < 6$

Dividing by 2: $\qquad x < 3$

Thus the solution to the inequality $5x - 2 < 3x + 4$ is the set of numbers less than 3.

Quadratic Inequalities

To solve an inequality such as $x^2 + 3x + 2 > 0$, it is first necessary to factor the quadratic expression. (See Section R.5.) Since, $x^2 + 3x + 2 = (x + 2)(x + 1)$, we can rewrite our inequality as

$$(x + 2)(x + 1) > 0$$

We must now realize that for a product of two factors to be positive, both of the factors must be positive, or both of the factors must be negative. Examining the factors separately, we have

$$x + 2 > 0 \quad \text{if and only if} \quad x > -2$$
$$x + 1 > 0 \quad \text{if and only if} \quad x > -1$$

Thus, if $x > -1$, both $(x + 2) > 0$ and $(x + 1) > 0$, so $x^2 + 3x + 2 > 0$. Moreover,

$$x + 2 < 0 \quad \text{if and only if} \quad x < -2$$
$$x + 1 < 0 \quad \text{if and only if} \quad x < -1$$

Thus, if $x < -2$, both $(x + 2) < 0$ and $(x + 1) < 0$, so $x^2 + 3x + 2 > 0$. Between the points $x = -2$ and $x = -1$ (that is, when $-2 < x < -1$), the factor $(x + 2)$ is positive and the factor $(x + 1)$ is negative, so the product $(x^2 + 3x + 2)$ is negative.

The cases are summarized in the following table:

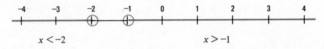

	−2		−1		x
x + 1	negative	negative	positive
x + 2	negative	positive	positive
(x + 1)(x + 2)	positive	negative	positive

The solution is seen to be x < −2 or x > −1.

Figure R.6–3

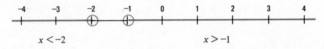

To solve an inequality such as (x − 5)(x − 3) < 0, we must realize that for a product of two factors to be negative, one of the factors, but not both, must be negative. Thus it is again helpful to examine each factor separately. We get

x − 3 < 0   if and only if   x < 3
x − 5 < 0   if and only if   x < 5

Thus, if x ⩾ 5, neither factor is negative, so the product is not negative. If x < 3, both factors are negative (or zero if x = 3), so the product again is not negative.

	3		5		x
x − 3	negative	positive	positive
x − 5	negative	negative	positive
(x − 5)(x − 3)	positive	negative	positive

But for 3 < x < 5, we have x − 3 > 0 and x − 5 < 0, so the product is negative. Thus the solution is

3 < x < 5

EXERCISES FOR SECTION R.6

For each of the following pairs of numbers, indicate which is greater.

1. −1 and −3
2. 4 and 8
3. −4 and 6

4. −5 and −2
5. −3 and −5
6. −1 and 0

Solve the following inequalities.

7. x − 7 > 0
8. x + 3 < 4
9. x − 1 > −4

10. 3x > 12
11. 2x − 4 < 6

Solve the following inequalities.

12. $(x - 1)(x - 3) < 0$

13. $(x + 1)(x - 2) < 0$

14. $(x + 3)(x + 2) < 0$

15. $(x - 1)(x - 3) > 0$

16. $(x + 1)(x - 2) > 0$

17. $x^2 - 4 < 0$

18. $x^2 - 1 > 0$

19. $x^2 - 1 < 0$

20. $x^2 + 2x + 1 > 0$

21. $x^2 + 4x + 4 < 0$

22. $x^2 + 4x + 3 > 0$

23. $x^2 - 5x + 6 < 0$

ANSWERS

1. $-3 < -1$

2. $4 < 8$

3. $-4 < 6$

4. $-5 < -2$

5. $-5 < -3$

6. $-1 < 0$

7. $x > 7$

8. $x < 1$

9. $x > -3$

10. $x > 4$

11. $x < 5$

12. $1 < x < 3$

13. $-1 < x < 2$

14. $-3 < x < -2$

15. $x < 1$ or $x > 3$

16. $x < -1$ or $x > 2$

17. $-2 < x < 2$

18. $x < -1$ or $x > 1$

19. $-1 < x < 1$

20. $x \neq -1$

21. no solution

22. $x < -3$ or $x > -1$

23. $2 < x < 3$

Appendix I

MORE ON LIMITS

The definitions of "limit" and "continuity" given in Section 1.5 are sufficient for the purposes of the material presented in this text. Indeed, they were all that were used by mathematicians during the first two hundred years of the history of calculus. But mathematicians and philosophers during that time were often bothered by the vagueness of the ideas—as you may, legitimately, be. During the last part of the nineteenth century, mathematicians such as Georg Cantor (1845–1918) evolved rigorous, more precise definitions that avoid logical paradoxes. The following definitions are now universally accepted by mathematicians and are presented here, along with some of their consequences.

Definition

A function $f(x)$ approaches a limit L as x approaches x_1 [denoted $\lim_{x \to x_1} f(x) = L$] if for each positive number ϵ, there exists a $\delta > 0$ such that whenever $0 < |x - x_1| < \delta$,

$$|f(x) - L| < \epsilon$$

In other words, if x is within δ units of x_1, then $f(x)$ will be within a pre-assigned ϵ units of the limit L. For different desired levels of "closeness," ϵ, between $f(x)$ and L, varying degrees of "closeness" δ between x and x_1 may be needed, as indicated in Figure A.I–1. We may not be able to "get" $f(x)$ exactly equal to L. But we can force it to be no more than any specified distance, ϵ, away from L by choosing x to be sufficiently near (at least as near as the corresponding δ) to x_1.

Figure A.I–1

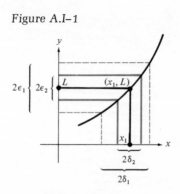

Example

Let $f(x) = 2x$.

(a) Show, using the definition of limit, that $\lim_{x \to 3} f(x) = 6$.

(b) If we want $f(x)$ to be within 0.1 of 6, how close to 3 should x be?

(c) If $\epsilon = 0.0001$, what should δ be?

Solution:

(a) We need to show that given any $\epsilon > 0$, there exists a $\delta > 0$ such that whenever $0 < |x - 3| < \delta$, then $|f(x) - 6| < \epsilon$.

Thus we first assume that $\epsilon > 0$ has been given. A little bit of foresight (and experience with this type of problem) leads us to suggest that $\delta = \epsilon/2$ is a good choice for δ. Now we verify this. Suppose that

$0 < \|x - 3\| < \delta$	Then, in particular,
$\|x - 3\| < \delta$	This means the same as
$-\delta < (x - 3) < \delta$	Multiplying by 2, which is positive, we have
$-2\delta < (2x - 6) < 2\delta$	Using the fact that $\delta = \epsilon/2$, we have
$-\epsilon < (2x - 6) < \epsilon$	This means the same as
$\|2x - 6\| < \epsilon$	which was what we wanted to prove.

(b) If we want $f(x)$ to be within 0.1 of 6, we take $\epsilon = 0.1$. The previous part of the problem showed that if $|x - 3| < \epsilon/2$, then $|f(x) - 6| < \epsilon$. Thus if we take x within $0.1/2 = 0.05$ of 3, $f(x)$ will be within 0.1 of 6.

(c) Using again the result of part (a), we take $\delta = 0.00005$.

The preceding example involved an unusually simple function and you might conclude some things from it that are not necessarily so. Since it is clear that $f(3) = 6$, you may think that we can always find a limit by merely plugging in the value of x_1 and obtain $f(x_1)$. Wrong! Actually the limit may exist when $f(x_1)$ is not even defined; Examples 1.5.4 and 1.5.5 in the text illustrate how this can happen.

If, in fact, the limit does exist and it equals $f(x_1)$, then the function is continuous at x_1.

Definition

A function $f(x)$ is <u>continuous at a point</u> $x = x_1$ if x_1 is in the domain of the function and for any $\epsilon > 0$, there exists $\delta > 0$ such that whenever $|x - x_1| < \delta$,

$$|f(x) - f(x_1)| < \epsilon$$

Another consequence of the precise definition is that we can prove a number of theorems whose truth we have merely assumed in this book.

Theorem 1

If $f(x) = k$ is a constant function, $\lim_{x \to x_1} f(x) = k$ for any x_1 in the domain.

Proof:

Suppose that $\epsilon > 0$ has been given. Then for any x, $|f(x) - k| = |k - k| = 0 < \epsilon$. Thus $|f(x) - k|$ is less than ϵ no matter what δ is chosen. We can, therefore, choose δ to be 1 or 2 or 200 or whatever we please.

Theorem 2

If $\lim_{x \to x_1} f(x) = L$ and $\lim_{x \to x_1} g(x) = M$, then $\lim_{x \to x_1} (f(x) \pm g(x)) = L \pm M$.

Proof:

Suppose that $\epsilon > 0$ is given. Then $\epsilon/2$ is also positive. Thus, by definition of limit, there is a δ_f such that

$$\text{whenever } 0 < |x - x_1| < \delta_f, \qquad |f(x) - L| < \frac{\epsilon}{2}$$

Also, by definition of limit, there is a δ_g such that

$$\text{whenever } 0 < |x - x_1| < \delta_g, \qquad |g(x) - M| < \frac{\epsilon}{2}$$

Then we choose δ to be the smaller of δ_f and δ_g; this means that whenever $0 < |x - x_1| < \delta$, we have both $0 < |x - x_1| < \delta_f$ and $0 < |x - x_1| < \delta_g$. It follows that

$$|(f(x) \pm g(x)) - (L \pm M)| = |(f(x) - L) \pm (g(x) - M)|$$
$$\leq |f(x) - L| + |g(x) - M| < \frac{\epsilon}{2} + \frac{\epsilon}{2} = \epsilon$$

which is what was to be proved.

Example: $\lim_{x \to 2} (3x + x^3) = \lim_{x \to 2} 3x + \lim_{x \to 2} x^3 = 6 + 8 = 14$

In the theorems that follow we shall omit the proofs and merely give examples, so you can see how they are used. The proofs are even more interesting than those above and can be found in many calculus textbooks. If this kind of theory fascinates you, you would certainly enjoy math courses designed for math majors.

Theorem 3

If $\lim_{x \to x_1} f(x) = L$ and $\lim_{x \to x_1} g(x) = M$, then $\lim_{x \to x_1} f(x)g(x) = L \cdot M$.

Example: $\lim_{x \to 2} (3x \cdot x^3) = (\lim_{x \to 2} 3x)(\lim_{x \to 2} x^3) = 6 \cdot 8 = 48$

Theorem 4

If $\lim_{x \to x_1} f(x) = L$ and $\lim_{x \to x_1} g(x) = M \neq 0$, then $\lim_{x \to x_1} f(x)/g(x) = L/M$.

Example: $\lim_{x \to 2} (3x/x^3) = \lim_{x \to 2} 3x / \lim_{x \to 2} x^3 = \frac{6}{8} = \frac{3}{4}$

Appendix II

PROOFS OF SOME DIFFERENTIATION FORMULAS

Theorem 1

(Rule 2.2.1, page 77) If $f(x) \equiv b$, where b is a constant, then $f'(x) = 0$.

Proof:

$$f'(x) = \lim_{\Delta x \to 0} \frac{f(x + \Delta x) - f(x)}{\Delta x} = \lim_{\Delta x \to 0} \frac{b - b}{\Delta x} = 0$$

Theorem 2

(Rule 2.2.2, page 78) If $f(x) = bx^n$, where b and n are any real numbers, then $f'(x) = bnx^{n-1}$.

We prove this here only for the case where n is a positive integer and $b = 1$. The theorem is true in the general case, but the proof is more complicated.

Proof:

In high school you probably learned the following facts, which taken together are called the *binomial theorem*.

$$(a + b)^2 = a^2 + 2ab + b^2$$
$$(a + b)^3 = a^3 + 3a^2b + 3ab^2 + b^3$$
$$(a + b)^4 = a^4 + 4a^3b + 6a^2b^2 + 4ab^3 + b^4$$
$$(a + b)^5 = a^5 + 5a^4b + 10a^3b^2 + 10a^2b^3 + 4ab^4 + b^5$$
$$(a + b)^n = a^n + na^{n-1}b + \frac{n(n - 1)}{2} a^{n-2}b^2$$
$$+ \frac{n(n - 1)(n - 2)}{3 \cdot 2} a^{n-3}b^3 + \cdots + b^n$$

For our purpose, it is sufficient to notice that

$$(a + b)^n = a^n + na^{n-1}b + b^2 \text{ (a sum of products of } a \text{ and } b)$$

Letting $a = x$ and $b = \Delta x$, this means in particular that

$$f(x + \Delta x) = (x + \Delta x)^n = x^n + nx^{n-1} \Delta x +$$
$$(\Delta x)^2 \text{ (a sum of products of } x \text{ and } \Delta x)$$

We use this fact to calculate the derivative of $f(x) = x^n$.

$$f'(x) = \lim_{\Delta x \to 0} \frac{f(x + \Delta x) - f(x)}{\Delta x}$$

$$= \lim_{\Delta x \to 0} \frac{x^n + nx^{n-1} \Delta x + (\Delta x)^2 \text{ (a sum of products)} - x^n}{\Delta x}$$

$$= \lim_{\Delta x \to 0} \frac{nx^{n-1} \Delta x + (\Delta x)^2 \text{ (a sum of products)}}{\Delta x}$$

$$= \lim_{\Delta x \to 0} [nx^{n-1} + \Delta x \text{ (a sum of products)}] = nx^{n-1}$$

Summarizing, we have that $f'(x) = nx^{n-1}$, which proves the theorem.

Theorem 3

(Rule 2.2.4, page 79) If $f(x) = g(x) + h(x)$, where both g and h have derivatives at x, then $f'(x) = g'(x) + h'(x)$.

Proof:

$$f'(x) = \lim_{\Delta x \to 0} \frac{f(x + \Delta x) - f(x)}{\Delta x}$$

$$= \lim_{\Delta x \to 0} \frac{[g(x + \Delta x) + h(x + \Delta x)] - [g(x) + h(x)]}{\Delta x}$$

$$= \lim_{\Delta x \to 0} \frac{g(x + \Delta x) - g(x) + h(x + \Delta x) - h(x)}{\Delta x}$$

$$= \lim_{\Delta x \to 0} \frac{g(x + \Delta x) - g(x)}{\Delta x} + \lim_{\Delta x \to 0} \frac{h(x + \Delta x) - h(x)}{\Delta x} = g'(x) + h'(x)$$

To prove the final two theorems in this appendix, we need a *lemma*, which is a preliminary theorem used to prove a later theorem. Theorem 4 can be considered to be a lemma.

Theorem 4

Any differentiable function is continuous. In other words, if $\lim_{\Delta x \to 0} [f(x + \Delta x) - f(x)]/\Delta x$ exists, then $\lim_{\Delta x \to 0} f(x + \Delta x) = f(x)$.

Proof:

By the hypothesis, $\lim_{\Delta x \to 0} [f(x + \Delta x) - f(x)]/\Delta x$ exists. But

$$f(x + \Delta x) = f(x + \Delta x) - f(x) + f(x) = \frac{f(x + \Delta x) - f(x)}{\Delta x} (\Delta x) + f(x)$$

Taking limits on both sides of this expression,

$$\lim_{\Delta x \to 0} f(x + \Delta x) = \lim_{\Delta x \to 0} \left[\frac{f(x + \Delta x) - f(x)}{\Delta x} (\Delta x) + f(x) \right]$$

$$= f'(x) \lim_{\Delta x \to 0} (\Delta x) + f(x) = f'(x) \cdot 0 + f(x)$$

$$= f(x)$$

Theorem 5

(Rule 2.3.2, page 82) If $f(x) = g(x)h(x)$ and g' and h' exist at x, then $f'(x) = g'(x)h(x) + g(x)h'(x)$.

Proof:

To prove this theorem, we use the device of adding and subtracting the term $g(x + \Delta x)h(x)$ to the numerator of the fraction.

$$f'(x) = \lim_{\Delta x \to 0} \frac{f(x + \Delta x) - f(x)}{\Delta x} = \lim_{\Delta x \to 0} \frac{g(x + \Delta x)h(x + \Delta x) - g(x)h(x)}{\Delta x}$$

$$= \lim_{\Delta x \to 0} \frac{g(x + \Delta x)h(x + \Delta x) - g(x + \Delta x)h(x) + g(x + \Delta x)h(x) - g(x)h(x)}{\Delta x}$$

$$= \lim_{\Delta x \to 0} \frac{g(x + \Delta x)h(x + \Delta x) - g(x + \Delta x)h(x)}{\Delta x}$$

$$+ \lim_{\Delta x \to 0} \frac{g(x + \Delta x)h(x) - g(x)h(x)}{\Delta x}$$

$$= \lim_{\Delta x \to 0} g(x + \Delta x) \lim_{\Delta x \to 0} \frac{h(x + \Delta x) - h(x)}{\Delta x} + \lim_{\Delta x \to 0} \frac{g(x + \Delta x) - g(x)}{\Delta x} h(x)$$

$$= g(x)h'(x) + g'(x)h(x)$$

Notice that to obtain the left term of this expression we used Theorem 4.

Theorem 6

(Rule 2.3.3, page 82) If $f(x) = g(x)/h(x)$, where both g' and h' exist at x and $h(x) \neq 0$, then $f'(x) = [h(x)g'(x) - g(x)h'(x)]/[h(x)]^2$.

Proof:

We use the same devices to prove this theorem as we did for the previous one, but the algebra is a bit more complicated.

$$f'(x) = \lim_{\Delta x \to 0} \frac{\dfrac{g(x + \Delta x)}{h(x + \Delta x)} - \dfrac{g(x)}{h(x)}}{\Delta x} = \lim_{\Delta x \to 0} \frac{\dfrac{g(x + \Delta x)h(x) - h(x + \Delta x)g(x)}{h(x + \Delta x)h(x)}}{\Delta x}$$

$$= \lim_{\Delta x \to 0} \frac{g(x + \Delta x)h(x) - g(x)h(x) + g(x)h(x) - h(x + \Delta x)g(x)}{\Delta x h(x + \Delta x)h(x)}$$

$$= \frac{\displaystyle\lim_{\Delta x \to 0} \frac{g(x + \Delta x) - g(x)}{\Delta x} h(x) - g(x) \lim_{\Delta x \to 0} \frac{h(x + \Delta x) - h(x)}{\Delta x}}{\displaystyle\lim_{\Delta x \to 0} h(x + \Delta x)h(x)}$$

$$= \frac{g'(x)h(x) - g(x)h'(x)}{[h(x)]^2}$$

Appendix III

THE MEAN VALUE THEOREM

In this section we shall discuss a result from differential calculus that sheds light on some other rules which are taken on faith in this text. The mean value theorem says that if the function $y = f(x)$ has a derivative, then for any two values $x = a$ and $x = b$, there is a point c between a and b such that the slope $f'(c)$ of the tangent line to the graph of $y = f(x)$ is the same as the slope of the straight line connecting $(a, f(a))$ and $(b, f(b))$. Figure A.III–1 should give you an idea of what the theorem says; it proclaims the existence of a c that makes the picture "work," that makes the two straight lines parallel. Below is a formal statement of the theorem.

Figure A.III–1

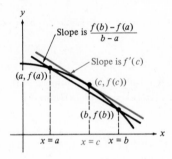

Theorem 1: The Mean Value Theorem

If $y = f(x)$ is differentiable between a and b and also continuous at points a and b, then there exists some $x = c$ between a and b such that

$$f'(c) = \frac{f(b) - f(a)}{b - a} \qquad a \neq b$$

The mean value theorem can be used, for example, to prove Rule 2.4.1. We prove here the first part of the rule. By mimicking this proof, you can probably prove the second and third parts yourself.

Theorem 2

If $f'(x)$ exists and is greater than zero between $x = a$ and $x = b$ and $f(x)$ is continuous at a and b, then $f(x)$ is increasing from a to b.

396

Proof:

To say that a function is "increasing" is in mathematical language to say that if $x_1 < x_2$, then $f(x_1) < f(x_2)$. We want to prove this.

Suppose, therefore, that $a \le x_1 < x_2 \le b$. By the mean value theorem there is a number c between x_1 and x_2 such that

$$\frac{f(x_2) - f(x_1)}{x_2 - x_1} = f'(c) > 0$$

Since $x_1 < x_2$, $(x_2 - x_1)$ is positive, and since the whole fraction is positive, we can conclude that $f(x_2) - f(x_1) > 0$. Thus $f(x_2) > f(x_1)$, which was what we wanted to prove.

The mean value theorem can also be used to prove some facts that we assumed to be "intuitively obvious" in Chapter 4. We prove two of these facts here.

Theorem 3

If $f'(x) = 0$ for all x such that $a \le x \le b$, then $f(x)$ is a constant function on that interval.

Proof:

Let x be any number such that $a < x < b$. By the mean value theorem there exists a number c such that $a < c < x$ and

$$\frac{f(x) - f(a)}{x - a} = f'(c)$$

But by the assumption of this theorem, $f'(c) = 0$. This implies that $f(x) = f(a)$. Since x was arbitrary in the interval, we have proved that $f(x)$ is constant on that interval [and, incidentally, constantly equal to $f(a)$].

Theorem 4

Any two antiderivatives of the same function, $f(x)$, must differ by a constant.

Proof:

We want to prove that if $F(x)$ and $G(x)$ are both such that their derivatives are $f(x)$, then $F(x) = G(x) + k$ for some constant k. In other words, we are assuming that

$$F'(x) = G'(x) = f(x)$$

It follows that if $H(x)$ is defined by $H(x) = F(x) - G(x)$,

$$H'(x) = F'(x) - G'(x) = 0$$

By the previous theorem,

$$H(x) \equiv k \qquad \text{where } k \text{ is some constant}$$

Thus, by definition of $H(x)$,

$$F(x) - G(x) \equiv k$$

or $F(x) = G(x) + k$, which is what was to be proved.

Appendix IV

DERIVATIVES OF THE LOGARITHM AND EXPONENTIAL FUNCTIONS

To find the derivative of the logarithm function we will use the following three properties of the logarithm which were introduced in Section 3.4 of the text and proved in your high school algebra course.

1. $\ln (a) - \ln (b) = \ln (a/b)$.
2. $b \ln (a) = \ln (a^b)$.
3. $\ln (e) = 1$.

Now let $f(x) = \ln (x)$, and begin calculating the derivative at an arbitrary point, x.

$$f'(x) = \lim_{\Delta x \to 0} \frac{\ln (x + \Delta x) - \ln (x)}{\Delta x} \qquad \text{(definition of derivative)}$$

$$= \lim_{\Delta x \to 0} \frac{1}{\Delta x} \ln \left(\frac{x + \Delta x}{x} \right) \qquad \text{(property 1 above)}$$

$$= \lim_{\Delta x \to 0} \frac{1}{x} \frac{x}{\Delta x} \ln \left(1 + \frac{\Delta x}{x} \right) \qquad \left[\text{multiplying } \frac{1}{\Delta x} \text{ by } \frac{x}{x}, \text{ and dividing} \right.$$
$$\left. (x + \Delta x) \text{ by } x \right]$$

$$= \lim_{\Delta x \to 0} \frac{1}{x} \ln \left(1 + \frac{1}{x/\Delta x} \right)^{x/\Delta x} \qquad \text{(property 2 above and algebra)}$$

$$= \frac{1}{x} \ln (e) \qquad \qquad \text{[definition of } e \text{ (given on page 50}$$
$$\text{of Section 1.5) and the observation}$$
$$\text{that as } \Delta x \to 0, \text{ we also have}$$

$$\frac{x}{\Delta x} \to \infty]$$

$$= \frac{1}{x} \qquad \qquad \text{(property 3 above)}$$

Thus we have shown that the derivative of $f(x) = \ln (x)$ is indeed $f'(x) = 1/x$.

To show that the derivative of $f(x) = e^x$ is $f'(x) = e^x$, we use the fact that the exponential function is the inverse of the logarithm function and the facts regarding derivatives of inverse functions that were presented in Section 3.2. Let

$$y = e^x$$
$$\ln (y) = x \qquad \text{(definition of "logarithm" as inverse of exponential}$$
$$\text{function)}$$

Now taking the derivative of this with respect to y, using the fact proved above, we discover

$$\frac{1}{y} = \frac{dx}{dy}$$

$$y = \frac{dy}{dx} \qquad \text{(by Section 3.2)}$$

$$e^x = \frac{dy}{dx} = f'(x) \qquad \text{(using the definition of "y")}$$

Thus we have shown that the derivative of $f(x) = e^x$ is indeed $f'(x) = e^x$.

Appendix V

L'HÔPITAL'S RULE

Appendix I gives us a little more depth on the theoretical aspects of finding limits. However, as early as 1696 the Marquis de L'Hôpital published the following *practical* rule for finding limits, which bears his name.

Theorem: L'Hôpital's Rule

(a) If the functions $f(x)$ and $g(x)$ are such that $\lim\limits_{x \to x_0} f(x) = 0$ and $\lim\limits_{x \to x_0} g(x) = 0$,

then $\lim\limits_{x \to x_0} \dfrac{f(x)}{g(x)} = \lim\limits_{x \to x_0} \dfrac{f'(x)}{g'(x)} = L$, provided that L and the derivatives exist. The

symbol x_0 can be any number or $\pm\infty$.

(b) If the functions $f(x)$ and $g(x)$ are such that $\lim\limits_{x \to x_0} f(x) = \infty$ and $\lim\limits_{x \to x_0} g(x) = \infty$,

then $\lim\limits_{x \to x_0} \dfrac{f(x)}{g(x)} = \lim\limits_{x \to x_0} \dfrac{f'(x)}{g'(x)} = L$, provided that L and the derivatives exist. The

symbol x_0 can be any number or $\pm\infty$.

Symbols of the form $0/0$ and ∞/∞ (which sometimes occur when computing limits) are called <u>indeterminate forms</u>. They are not numbers—no, not even the number 1! We encountered the first form quite frequently when we were discovering how to calculate derivatives.

Example

Evaluate the limit $\lim\limits_{x \to 1} \dfrac{x^2 - 1}{x - 1}$ using L'Hôpital's rule.

Solution:

Let $f(x) = x^2 - 1$ and $g(x) = x - 1$. Then $\lim\limits_{x \to 1} f(x) = \lim\limits_{x \to 1} g(x) = 0$, as required by part (a) of L'Hôpital's rule. Since $f'(x) = 2x$ and $g'(x) = 1$, we may write

$$\lim\limits_{x \to 1} \frac{x^2 - 1}{x - 1} = \lim\limits_{x \to 1} \frac{2x}{1} = 2(1) = 2.$$

$$\left[\text{We can check this result, since } \lim\limits_{x \to 1} \frac{x^2 - 1}{x - 1} = \lim\limits_{x \to 1} \frac{(x + 1)(x - 1)}{(x - 1)} = 2. \right]$$

Example

Evaluate the limit $\lim\limits_{x \to 0} \dfrac{e^x - 1}{x}$.

400

Solution:

Let $f(x) = e^x - 1$ and $g(x) = x$. Then $\lim\limits_{x \to 0} f(x) = \lim\limits_{x \to 0} g(x) = 0$, as required by part

(a) of L'Hôpital's rule. Since $f'(x) = e^x$ and $g'(x) = 1$, we may write

$$\lim_{x \to 0} \frac{e^x - 1}{x} = \lim_{x \to 0} \frac{e^x}{1} = e^0 = 1.$$

(Notice that the last example could not be done by an algebraic trick.)

Example

Evaluate the limit $\lim\limits_{x \to \infty} \dfrac{2 \ln (x) + 3}{x - 5}$ using L'Hôpital's rule.

Solution:

Let $f(x) = 2 \ln (x) + 3$ and $g(x) = x - 5$. Then $\lim\limits_{x \to \infty} f(x) = \infty$ and $\lim\limits_{x \to \infty} g(x) = \infty$, as

required by part (b) of L'Hôpital's rule. Since $f'(x) = 2/x$ and $g'(x) = 1$, we may
write

$$\lim_{x \to \infty} \frac{2 \ln (x) + 3}{x - 5} = \lim_{x \to \infty} \frac{2/x}{1} = 0$$

EXERCISES

Evaluate the following limits using L'Hôpital's rule:

1. $\lim\limits_{x \to 2} \dfrac{x^2 - 4}{x - 2} =$

6. $\lim\limits_{x \to 0} \dfrac{e^x - e^{-x}}{2x} =$

2. $\lim\limits_{x \to 2} \dfrac{x^3 - 8}{x - 2} =$

7. $\lim\limits_{x \to \infty} \dfrac{x + 1}{x - 1} =$

3. $\lim\limits_{x \to 0} \dfrac{x}{e^x - 1} =$

8. $\lim\limits_{x \to \infty} \dfrac{x + 1}{x^2} =$

4. $\lim\limits_{x \to 0} \dfrac{e^x + e^{-x} - 2}{e^x - e^{-x}} =$

9. $\lim\limits_{x \to \infty} \dfrac{\ln (x)}{x} =$

5. $\lim\limits_{x \to 0} \dfrac{x^2}{xe^x} =$

10. $\lim\limits_{x \to \infty} \dfrac{x^2}{e^x} =$

(*Hint*: Use L'Hôpital's rule twice.)

ANSWERS

1.	4	6.	1
2.	12	7.	1
3.	1	8.	0
4.	0	9.	0
5.	0	10.	0

Appendix VI

PROOFS OF SOME TRIGONOMETRIC FORMULAS

Our proof of the validity of the differentiation formula for the sine function in Section 7.2 depended upon two formulas from trigonometry. Here are the proofs of those formulas.

Theorem

For all real numbers a and b, we have

$$\cos (a - b) = \cos a \cos b + \sin a \sin b$$
$$\sin (a + b) = \sin a \cos b + \cos a \sin b$$

Proof:

In Figure A.VI–1, let $\overline{AB}$ be the distance between points A and B on the unit circle. Then the Pythagorean theorem gives us

$$(\overline{AB})^2 = (\cos b - \cos a)^2 + (\sin a - \sin b)^2$$
$$= \cos^2 a - 2 \cos a \cos b + \cos^2 b + \sin^2 a - 2 \sin a \sin b + \sin^2 b$$

Figure A.VI–1

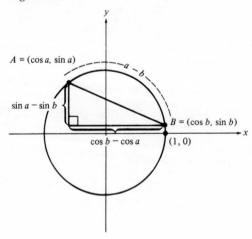

Since $\sin^2 a + \cos^2 a = \sin^2 b + \cos^2 b = 1$, this yields the formula

$$(\overline{AB})^2 = 2 - 2(\cos a \cos b + \sin a \sin b)$$

402

TABLE I 405

Table I (continued)

n	r = 0.03	r = 0.04	r = 0.05	r = 0.06	r = 0.07	r = 0.08
1	1.0300	1.0400	1.0500	1.0600	1.0700	1.0800
2	1.0609	1.0816	1.1025	1.1236	1.1449	1.1664
3	1.0927	1.1249	1.1576	1.1910	1.2250	1.2597
4	1.1255	1.1699	1.2155	1.2625	1.3108	1.3605
5	1.1593	1.2167	1.2763	1.3382	1.4026	1.4693
6	1.1940	1.2653	1.3401	1.4185	1.5007	1.5869
7	1.2299	1.3159	1.4071	1.5036	1.6058	1.7138
8	1.2668	1.3686	1.4775	1.5938	1.7182	1.8509
9	1.3048	1.4233	1.5513	1.6895	1.8385	1.9990
10	1.3439	1.4802	1.6289	1.7908	1.9672	2.1589
11	1.3842	1.5395	1.7103	1.8983	2.1049	2.3316
12	1.4258	1.6010	1.7959	2.0122	2.2522	2.5182
13	1.4685	1.6651	1.8856	2.1392	2.4048	2.7196
14	1.5126	1.7317	1.9799	2.2609	2.5785	2.9372
15	1.5580	1.8009	2.0790	2.3966	2.7590	3.1722
16	1.6047	1.8730	2.1829	2.5404	2.9522	3.4259
17	1.6528	1.9479	2.2920	2.6928	3.1588	3.7000
18	1.7024	2.0258	2.4066	2.8543	3.3799	3.9960
19	1.7535	2.1068	2.5270	3.0256	3.6165	4.3157
20	1.8061	2.1911	2.6533	3.2071	3.8697	4.6610
21	1.8603	2.2788	2.7860	3.3996	4.1406	5.0338
22	1.9161	2.3699	2.9253	3.6035	4.4304	5.5365
23	1.9736	2.4647	3.0715	3.8198	4.7405	5.8715
24	2.0328	2.5633	3.2251	4.0489	5.0724	6.3412
25	2.0938	2.6658	3.3864	4.2919	5.4274	6.8435
26	2.1566	2.7725	3.5557	4.5494	5.8074	7.3964
27	2.2213	2.8834	3.7335	4.8223	6.2139	7.9881
28	2.2879	2.9987	3.9201	5.1117	6.6488	8.6231
29	2.3566	3.1187	4.1161	5.4184	7.1143	9.3173
30	2.4273	3.2434	4.3219	5.7435	7.6123	10.0627

Table II

Values of $(1 + r)^{-n}$

n	r = 0.005	r = 0.0075	r = 0.01	r = 0.0125	r = 0.015	r = 0.02
1	0.9950	0.9926	0.9901	0.9877	0.9852	0.9804
2	0.9901	0.9852	0.9803	0.9755	0.9707	0.9612
3	0.9851	0.9778	0.9706	0.9634	0.9563	0.9432
4	0.9802	0.9706	0.9610	0.9515	0.9422	0.9238
5	0.9754	0.9633	0.9515	0.9398	0.9283	0.9057
6	0.9705	0.9562	0.9420	0.9282	0.9145	0.8880
7	0.9657	0.9490	0.9327	0.9167	0.9010	0.8706
8	0.9609	0.9420	0.9235	0.9054	0.8877	0.8535
9	0.9561	0.9350	0.9143	0.8942	0.8746	0.8368
10	0.9513	0.9280	0.9053	0.8832	0.8617	0.8203
11	0.9466	0.9211	0.8963	0.8723	0.8489	0.8043
12	0.9419	0.9142	0.8874	0.8615	0.8364	0.7885
13	0.9372	0.9074	0.8787	0.8509	0.8240	0.7730
14	0.9326	0.9007	0.8700	0.8404	0.8118	0.7579
15	0.9279	0.8940	0.8613	0.8300	0.7999	0.7430
16	0.9233	0.8873	0.8528	0.8197	0.7880	0.7284
17	0.9187	0.8807	0.8444	0.8096	0.7764	0.7142
18	0.9141	0.8742	0.8360	0.7996	0.7649	0.7002
19	0.9096	0.8676	0.8277	0.7898	0.7536	0.6864
20	0.9051	0.8612	0.8195	0.7800	0.7425	0.6730
21	0.9006	0.8548	0.8114	0.7704	0.7315	0.6598
22	0.8961	0.8484	0.8034	0.7609	0.7207	0.6468
23	0.8916	0.8421	0.7954	0.7515	0.7100	0.6342
24	0.8872	0.8358	0.7876	0.7422	0.6995	0.6217
25	0.8828	0.8296	0.7798	0.7330	0.6892	0.6095
26	0.8783	0.8234	0.7720	0.7240	0.6790	0.5976
27	0.8740	0.8173	0.7644	0.7150	0.6690	0.5859
28	0.8697	0.8112	0.7568	0.7062	0.6591	0.5744
29	0.8653	0.8052	0.7493	0.6975	0.6494	0.5631
30	0.8610	0.7992	0.7419	0.6889	0.6398	0.5521

TABLE II   407

Table II (continued)

n	r = 0.03	r = 0.04	r = 0.05	r = 0.06	r = 0.07	r = 0.08
1	0.9709	0.9615	0.9524	0.9434	0.9348	0.9259
2	0.9426	0.9246	0.9070	0.8900	0.8734	0.8573
3	0.9151	0.8890	0.8638	0.8396	0.8163	0.7938
4	0.8885	0.8548	0.8227	0.7921	0.7629	0.7350
5	0.8626	0.8219	0.7835	0.7473	0.7130	0.6806
6	0.8375	0.7903	0.7462	0.7050	0.6663	0.6302
7	0.8131	0.7599	0.7107	0.6651	0.6228	0.5835
8	0.7894	0.7307	0.6768	0.6274	0.5820	0.5403
9	0.7664	0.7026	0.6446	0.5919	0.5439	0.5002
10	0.7441	0.6756	0.6139	0.5584	0.5083	0.4632
11	0.7224	0.6496	0.5847	0.5268	0.4751	0.4289
12	0.7014	0.6246	0.5568	0.4970	0.4440	0.3971
13	0.6810	0.6006	0.5303	0.4688	0.4150	0.3677
14	0.6611	0.5775	0.5051	0.4423	0.3878	0.3405
15	0.6419	0.5553	0.4810	0.4173	0.3624	0.3152
16	0.6232	0.5340	0.4581	0.3936	0.3387	0.2919
17	0.6050	0.5134	0.4363	0.3714	0.3166	0.2703
18	0.5874	0.4936	0.4155	0.3503	0.2960	0.2510
19	0.5703	0.4746	0.3957	0.3305	0.2765	0.2317
20	0.5537	0.4564	0.3769	0.3118	0.2584	0.2145
21	0.5375	0.4388	0.3589	0.2942	0.2415	0.1987
22	0.5219	0.4220	0.3419	0.2775	0.2275	0.1839
23	0.5067	0.4057	0.3256	0.2618	0.2109	0.1703
24	0.4919	0.3901	0.3101	0.2470	0.1971	0.1577
25	0.4776	0.3751	0.2953	0.2330	0.1842	0.1460
26	0.4637	0.3607	0.2812	0.2198	0.1722	0.1352
27	0.4502	0.3468	0.2678	0.2074	0.1609	0.1252
28	0.4371	0.3335	0.2551	0.1956	0.1504	0.1159
29	0.4243	0.3207	0.2429	0.1846	0.1406	0.1073
30	0.4120	0.3083	0.2314	0.1741	0.1314	0.0994

Table III

Values of e^x and e^{-x}

x	e^x	e^{-x}	x	e^x	e^{-x}
0.01	1.0101	0.9900	0.41	1.5068	0.6637
0.02	1.0202	0.9802	0.42	1.5220	0.6570
0.03	1.0305	0.9704	0.43	1.5373	0.6505
0.04	1.0408	0.9608	0.44	1.5527	0.6440
0.05	1.0513	0.9512	0.45	1.5683	0.6376
0.06	1.0618	0.9418	0.46	1.5841	0.6313
0.07	1.0725	0.9324	0.47	1.6001	0.6250
0.08	1.0833	0.9231	0.48	1.6161	0.6188
0.09	1.0942	0.9139	0.49	1.6323	0.6126
0.10	1.1052	0.9048	0.50	1.6487	0.6065
0.11	1.1163	0.8958	0.51	1.6653	0.6005
0.12	1.1275	0.8869	0.52	1.6820	0.5945
0.13	1.1388	0.8781	0.53	1.6989	0.5886
0.14	1.1503	0.8694	0.54	1.7160	0.5827
0.15	1.1618	0.8607	0.55	1.7333	0.5769
0.16	1.1735	0.8521	0.56	1.7507	0.5712
0.17	1.1853	0.8437	0.57	1.7683	0.5655
0.18	1.1972	0.8353	0.58	1.7860	0.5599
0.19	1.2092	0.8270	0.59	1.8040	0.5543
0.20	1.2214	0.8187	0.60	1.8221	0.5488
0.21	1.2337	0.8106	0.61	1.8404	0.5434
0.22	1.2461	0.8025	0.62	1.8589	0.5379
0.23	1.2586	0.7945	0.63	1.8776	0.5326
0.24	1.2712	0.7866	0.64	1.8965	0.5273
0.25	1.2840	0.7788	0.65	1.9155	0.5220
0.26	1.2969	0.7711	0.66	1.9348	0.5169
0.27	1.3100	0.7634	0.67	1.9542	0.5117
0.28	1.3231	0.7558	0.68	1.9739	0.5066
0.29	1.3364	0.7483	0.69	1.9937	0.5016
0.30	1.3499	0.7408	0.70	2.0138	0.4966
0.31	1.3634	0.7334	0.71	2.0340	0.4916
0.32	1.3771	0.7261	0.72	2.0544	0.4868
0.33	1.3910	0.7189	0.73	2.0751	0.4819
0.34	1.4049	0.7118	0.74	2.0950	0.4771
0.35	1.4191	0.7047	0.75	2.1170	0.4724
0.36	1.4333	0.6977	0.76	2.1353	0.4677
0.37	1.4477	0.6907	0.77	2.1593	0.4630
0.38	1.4623	0.6839	0.78	2.1815	0.4584
0.39	1.4770	0.6771	0.79	2.2034	0.4538
0.40	1.4918	0.6703	0.80	2.2255	0.4493

TABLE III   409

Table III (*continued*)

x	e^x	e^{-x}	x	e^x	e^{-x}
0.81	2.2479	0.4449	1.1	3.0042	0.3329
0.82	2.2705	0.4404	1.2	3.3201	0.3012
0.83	2.2933	0.4360	1.3	3.6693	0.2725
0.84	2.3164	0.4317	1.4	4.0552	0.2466
0.85	2.3396	0.4274	1.5	4.4817	0.2231
0.86	2.3632	0.4232	1.6	4.9530	0.2019
0.87	2.3869	0.4190	1.7	5.4739	0.1827
0.88	2.4109	0.4148	1.8	6.0496	0.1653
0.89	2.4351	0.4107	1.9	6.6859	0.1496
0.90	2.4596	0.4066	2.0	7.3891	0.1353
0.91	2.4843	0.4025	2.1	8.1662	0.1225
0.92	2.5098	0.3985	2.2	9.0250	0.1108
0.93	2.5345	0.3946	2.3	9.9742	0.1003
0.94	2.5600	0.3906	2.4	11.0232	0.0907
0.95	2.5857	0.3867	2.5	12.1825	0.0821
0.96	2.6117	0.3829	2.6	13.4637	0.0743
0.97	2.6379	0.3791	2.7	14.8797	0.0672
0.98	2.6645	0.3753	2.8	16.4446	0.0608
0.99	2.6912	0.3716	2.9	18.1741	0.0550
1.0	2.7183	0.3679	3.0	20.0855	0.0498

Table IV

Natural Logarithms of N

N	0.0	0.1	0.2	0.3	0.4	0.5	0.6	0.7	0.8	0.9
0		−2.3026	−1.6094	−1.2040	−0.9163	−0.6931	−0.5108	−0.3567	−0.2231	−0.1054
1	0.0000	0.0953	0.1823	0.2624	0.3365	0.4055	0.4700	0.5306	0.5878	0.6419
2	0.6931	0.7419	0.7885	0.8329	0.8755	0.9163	0.9555	0.9933	1.0296	1.0647
3	1.0986	1.1314	1.1632	1.1939	1.2238	1.2528	1.2809	1.3083	1.3350	1.3610
4	1.3863	1.4110	1.4351	1.4586	1.4816	1.5041	1.5261	1.5476	1.5686	1.5892
5	1.6094	1.6292	1.6487	1.6677	1.6864	1.7047	1.7228	1.7405	1.7579	1.7750
6	1.7918	1.8083	1.8245	1.8405	1.8563	1.8718	1.8871	1.9021	1.9169	1.9315
7	1.9459	1.9601	1.9741	1.9879	2.0015	2.0149	2.0281	2,0412	2.0541	2.0669
8	2.0794	2.0919	2.1041	2.1163	2.1282	2.1401	2.1518	2.1633	2.1748	2.1861
9	2.1972	2.2083	2.2192	2.2300	2.2407	2.2513	2.2618	2.2721	2.2824	2.2925
10	2.3026	2.3125	2.3224	2.3321	2.3418	2.3514	2.3609	2.3702	2.3795	2.3888
11	2.3979	2.4069	2.4159	2.4243	2.4336	2.4423	2.4510	2.4596	2.4681	2.4765
12	2.4842	2.4912	2.5014	2.5096	2.5177	2.5257	2.5337	2.5416	2.5494	2.5572
13	2.5649	2.5726	2.5802	2.5878	2.5953	2.6027	2.6101	2.6174	2.6247	2.6319
14	2.6391	2.6462	2.6532	2.6603	2.6672	2.6741	2.6810	2.6873	2.6946	2.7014
15	2.7031	2.7147	2.7213	2.7279	2.7344	2.7408	2.7473	2.7537	2.7600	2.7663
16	2.7726	2.7788	2.7850	2.7912	2.7973	2.8034	2.8094	2.8154	2.8214	2.8273
17	2.8332	2.8391	2.8449	2.8507	2.8565	2.8622	2.8679	2.8736	2.8792	2.8848
18	2.8904	2.8959	2.9014	2.9069	2.9124	2.9178	2.9232	2.9285	2.9339	2.9392
19	2.9344	2.9497	2.9549	2.9601	2.9653	2.9704	2.9755	2.9806	2.9857	2.9907
20	2.9957	3.0007	3.0057	3.0106	3.0155	3.0204	3.0253	3.0301	3.0350	3.0397
21	3.0445	3.0493	3.0540	3.0587	3.0624	3.0681	3.0727	3.0773	3.0819	3.0865
22	3.0910	3.0956	3.1001	3.1046	3.1091	3.1135	3.1179	3.1224	3.1263	3.1311
23	3.1355	3.1398	3.1442	3.1485	3.1527	3.1570	3.1612	3.1655	3.1697	3.1739
24	3.1781	3.1822	3.1864	3.1905	3.1946	3.1987	3.2027	3.2068	3.2108	3.2149
25	3.2189	3.2229	3.2263	3.2308	3.2347	3.2387	3.2426	3.2465	3.2504	3.2542

Note: To find the logarithm of a number N not in the table above, write $N = 10^n M$, where $1 \leq M \leq 10$. Then compute $\ln (N) = \ln (M) + n \ln (10)$. For example, $\ln (110) = \ln (1.1) + 2 \ln (10) = 0.0953 + 4.6052 = 4.7005$.

TABLE V   411

Table V

The Integrals

1. $\displaystyle\int \frac{du}{a^2 - u^2} = \frac{1}{2a} \ln\left|\frac{a+u}{a-u}\right| + k$ 2. $\displaystyle\int \frac{du}{u^2 - a^2} = \frac{1}{2a} \ln\left|\frac{u-a}{u+a}\right| + k$

3. $\displaystyle\int \sqrt{u^2 \pm a^2}\, du = \tfrac{1}{2}(u\sqrt{u^2 \pm a^2}) \pm \frac{a^2}{2} \ln|u + \sqrt{u^2 \pm a^2}| + k$

4. $\displaystyle\int \frac{du}{\sqrt{u^2 \pm a^2}} = \ln|u + \sqrt{u^2 \pm a^2}| + k$

5. $\displaystyle\int ue^u\, du = ue^u - e^u + k$ 6. $\displaystyle\int u^2 e^u\, du = e^u(u^2 - 2u + 2) + k$

7. $\displaystyle\int u^n e^{au}\, du = \frac{u^n}{a} e^{au} - \frac{n}{a} \int u^{n-1} e^{au}\, du$

8. $\displaystyle\int \ln(u)\, du = u \ln(u) - u + k$

9. $\displaystyle\int \ln^n(u)\, du = u \ln^n(u) - n \int \ln^{n-1}(u)\, du$

10. $\displaystyle\int u^n \ln(u)\, du = \frac{u^{n+1}}{n+1}\left(\ln(u) - \frac{1}{n+1}\right) + k,\ n \neq -1$

11. $\displaystyle\int \frac{\ln^n(u)}{u}\, du = \frac{1}{n+1} \ln^{n+1}(u) + k,\ n \neq -1$

12. $\displaystyle\int \frac{du}{u^2 \sqrt{u^2 \pm a^2}} = \mp \frac{\sqrt{u^2 \pm a^2}}{a^2 u} + k$ 13. $\displaystyle\int \frac{du}{(u^2 \pm a^2)^{3/2}} = \frac{\pm u}{a^2 \sqrt{u^2 \pm a^2}} + k$

14. $\displaystyle\int u^n [\ln(u)]^m\, du = \frac{u^{n+1}}{n+1}[\ln(u)]^m - \frac{m}{n+1}\int u^n [\ln(u)]^{m-1}\, du$

Trigonometric Integrals (Chapter 7)

15. $\displaystyle\int \sin u\, du = -\cos u + k$ 16. $\displaystyle\int \cos u\, du = \sin u + k$

17. $\displaystyle\int \tan u\, du = -\ln|\cos u| + k$ 18. $\displaystyle\int \cot u\, du = \ln|\sin u| + k$

19. $\displaystyle\int \sec u\, du = \ln|\sec u + \tan u| + k$

20. $\displaystyle\int \csc u\, du = \ln|\csc u - \cot u| + k$

21. $\displaystyle\int \sec^2 u\, du = \tan u + k$ 22. $\displaystyle\int \csc^2 u\, du = -\cot u + k$

23. $\displaystyle\int \sec u \tan u\, du = \sec u + k$ 24. $\displaystyle\int \csc u \cot u\, du = -\csc u + k$

25. $\displaystyle\int \frac{du}{a^2 + u^2} = \frac{1}{a} \operatorname{Arctan} \frac{u}{a} + k$ 26. $\displaystyle\int \frac{du}{\sqrt{a^2 - u^2}} = \operatorname{Arcsin} \frac{u}{a} + k$

Table VI

Standard Normal Density Function: Integrals

$$A = \frac{1}{\sqrt{2\pi}} \int_{-\infty}^{b} e^{-z^2/2} \, dz$$

b	A	b	A
0.1	0.5398	1.6	0.9452
0.2	0.5793	1.7	0.9554
0.3	0.6179	1.8	0.9641
0.4	0.6554	1.9	0.9713
0.5	0.6915	2.0	0.9772
0.6	0.7257	2.1	0.9821
0.7	0.7580	2.2	0.9861
0.8	0.7881	2.3	0.9893
0.9	0.8159	2.4	0.9918
1.0	0.8413	2.5	0.9938
1.1	0.8643	2.6	0.9953
1.2	0.8849	2.7	0.9965
1.3	0.9032	2.8	0.9974
1.4	0.9192	2.9	0.9981
1.5	0.9332	3.0	0.9987

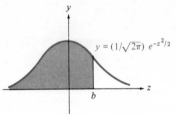

$y = (1/\sqrt{2\pi}) \, e^{-z^2/2}$

A is the shaded area corresponding to
the given b.

Table VII

Values of the Trigonometric Functions

Angle	Sin	Cos	Tan	Angle	Sin	Cos	Tan
1°	0.0175	0.9998	0.0175	46°	0.7193	0.6947	1.0355
2°	0.0349	0.9994	0.0349	47°	0.7314	0.6820	1.0724
3°	0.0523	0.9986	0.0524	48°	0.7431	0.6691	1.1106
4°	0.0698	0.9976	0.0699	49°	0.7547	0.6561	1.1504
5°	0.0872	0.9962	0.0875	50°	0.7660	0.6428	1.1918
6°	0.1045	0.9945	0.1051	51°	0.7771	0.6293	1.2349
7°	0.1219	0.9925	0.1228	52°	0.7880	0.6157	1.2799
8°	0.1392	0.9903	0.1405	53°	0.7986	0.6018	1.3270
9°	0.1564	0.9877	0.1584	54°	0.8090	0.5878	1.3764
10°	0.1736	0.9848	0.1763	55°	0.8192	0.5736	1.4281
11°	0.1908	0.9816	0.1944	56°	0.8290	0.5592	1.4826
12°	0.2079	0.9781	0.2126	57°	0.8387	0.5446	1.5399
13°	0.2250	0.9744	0.2309	58°	0.8480	0.5299	1.6003
14°	0.2419	0.9703	0.2493	59°	0.8572	0.5150	1.6643
15°	0.2588	0.9659	0.2679	60°	0.8660	0.5000	1.7321
16°	0.2756	0.9613	0.2867	61°	0.8746	0.4848	1.8040
17°	0.2924	0.9563	0.3057	62°	0.8829	0.4695	1.8807
18°	0.3090	0.9511	0.3249	63°	0.8910	0.4540	1.9626
19°	0.3256	0.9455	0.3443	64°	0.8988	0.4384	2.0503
20°	0.3420	0.9397	0.3640	65°	0.9063	0.4226	2.1445
21°	0.3584	0.9336	0.3839	66°	0.9135	0.4067	2.2460
22°	0.3746	0.9272	0.4040	67°	0.9205	0.3907	2.3559
23°	0.3907	0.9205	0.4245	68°	0.9272	0.3746	2.4751
24°	0.4067	0.9135	0.4452	69°	0.9336	0.3584	2.6051
25°	0.4226	0.9063	0.4663	70°	0.9397	0.3420	2.7475
26°	0.4384	0.8988	0.4877	71°	0.9455	0.3256	2.9042
27°	0.4540	0.8910	0.5095	72°	0.9511	0.3090	3.0777
28°	0.4695	0.8829	0.5317	73°	0.9563	0.2924	3.2709
29°	0.4848	0.8746	0.5543	74°	0.9613	0.2756	3.4874
30°	0.5000	0.8660	0.5774	75°	0.9659	0.2588	3.7321
31°	0.5150	0.8572	0.6009	76°	0.9703	0.2419	4.0108
32°	0.5299	0.8480	0.6249	77°	0.9744	0.2250	4.3315
33°	0.5446	0.8387	0.6494	78°	0.9781	0.2079	4.7046
34°	0.5592	0.8290	0.6745	79°	0.9816	0.1908	5.1446
35°	0.5736	0.8192	0.7002	80°	0.9848	0.1736	5.6718
36°	0.5878	0.8090	0.7265	81°	0.9877	0.1564	6.3138
37°	0.6018	0.7986	0.7536	82°	0.9903	0.1392	7.1154
38°	0.6157	0.7880	0.7813	83°	0.9925	0.1219	8.1443
39°	0.6293	0.7771	0.8098	84°	0.9945	0.1045	9.5144
40°	0.6428	0.7660	0.8391	85°	0.9962	0.0872	11.4301
41°	0.6561	0.7547	0.8693	86°	0.9976	0.0698	14.3007
42°	0.6691	0.7431	0.9004	87°	0.9986	0.0523	19.0811
43°	0.6820	0.7314	0.9325	88°	0.9994	0.0349	28.6363
44°	0.6947	0.7193	0.9657	89°	0.9998	0.0175	57.2900
45°	0.7071	0.7071	1.0000	90°	1.0000	0.0000	

1. The following relationships enable one to find the sine, cosine, and tangent of any angle between 0° and 360° as well as the values of the secant, cosecant, and cotangent of these angles.

	$\sin r° > 0$	$\sin t° > 0$		
$r = 180 - t$	$\cos r° < 0$	$\cos t° > 0$	$0 \leqq t < 90$	$\sec t° = \dfrac{1}{\cos t°}$
	$\tan r° < 0$	$\tan t° > 0$		
	$\sin r° < 0$	$\sin r° < 0$		$\csc t° = \dfrac{1}{\sin t°}$
$r = 180 + t$	$\cos r° < 0$	$\cos r° > 0$	$r = 360 - t$	
	$\tan r° > 0$	$\tan r° < 0$		$\cot t° = \dfrac{1}{\tan t°}$

2. Moreover, 2π radians of arc on the unit circle equals 360° of angle measure.

ANSWERS TO DIAGNOSTIC TEST

1. real
2. (a) -7 (b) -7 (c) $+1$ (d) -1
3. (a) 18 (b) -18 (c) -18
4. Yes
5. (a) 3 (b) 3

(If you are insecure about the facts above, you should read Section R.1, which begins on page 368, and do the accompanying exercises, on page 371.)

6.

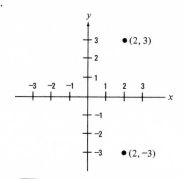

7. $\sqrt{34}$

(If these two problems confuse you, turn to page 371 and read Section R.2, which begins there.)

8. (a) $x = 5$ (b) $x = -2$

(For a further discussion of these ideas, turn to Section R.3 on page 374.)

9. (a) x^5 (b) x^3 (c) 1 (d) x (e) $1/x^2$ (f) $\frac{1}{2}$ (g) 3 (h) 2 (i) $\frac{1}{4}$ (j) $\frac{1}{81}$
10. (a) $12x^2 - 6x + 21$ (b) $ac + ad + bc + bd$ (c) $a^2 - b^2$ (d) $a^2 + 2ab + b^2$
    (e) $(6 + x^2)/2x$ (f) $8x^2 + 8x - 6$

(If you find these problems confusing, turn to page 379 and study Section R.4.)

11. (a) $(x + 3)(x - 3)$ (b) $(x - 6)(x - 1)$ (c) $(2x - 3)(x - 2)$ (d) $x(x + 1)(x - 1)$
    (e) $x(2 + x)(2 - x)$ (f) $x(2x + 1)(x + 1)$
12. (a) $x = 3$, $x = -3$ (b) $x = 6$, $x = 1$ (c) $x = 2$, $x = \frac{3}{2}$ (d) $x = 0$, $x = -1$,
    $x = 1$ (e) $x = 0$, $x = -2$, $x = 2$ (f) $x = 0$, $x = -\frac{1}{2}$, $x = -1$
13. (a) $x + 3$ (b) $x - 3$

(These topics are covered in Section R.5, which begins on page 383.)

14. (a) $x < 2$ (b) $x < 3$ (c) $x < 1$
15. (a) $0 < x < 1$ (b) $x < 0$ or $x > 1$ (c) $-2 < x < 2$

(Inequalities are reviewed in Section R.6, which begins on page 386; you will need to be able to solve these problems to understand Chapter 2.)

ANSWERS TO SAMPLE QUIZZES AND TESTS

SAMPLE QUIZ, SECTIONS 1.1, 1.2, 1.3 (page 33)

1. See page 3.
2. 23
3. marginal, fixed
4. right does, left does not
5.

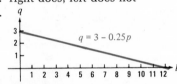

6. $p = -4q + 12$
7. positive, increasing

8.

9. $-\frac{1}{2}$
10. See page 32, answer 1.

SAMPLE QUIZ, SECTIONS 1.4, 1.5, 1.6 (page 62)

1. See page 34.
2. $C = 0.25x + 75, \$225$
3. (a) 10 (b) 0 (c) 3x

4. $(256 - 16)/(4 - 1) = 80$ ft/sec
5. $(76 - 29)/(20 - 10) = 4.7$

SAMPLE TEST, CHAPTER 1 (pages 62–63)

1. (a) -1 (b) 7 (c) 17
2. (a) yes (b) yes (c) no
3. (a) $450 (b) $5 (c) $200 (d) 5 and 200
4. $y = -2x + 2$
5. $y = 50x + 500; \$1,250$
6. (a) 0 (b) 4x (c) 6

7. (a) $\Delta x = 3$ (b) $\Delta y = 42$ (c) $\Delta y = 4x_1 \, \Delta x + 2 \, \Delta x^2$ (d) $\dfrac{\Delta y}{\Delta x} = 4x_1 + 2 \, \Delta x$

8. The vertical line has the equation $x = 0$. All the others are of the form $y = mx$, where m is the slope of the line.

9. $\frac{9}{2}$

416

SAMPLE QUIZ, SECTIONS 2.1, 2.2, 2.3, 2.4 (pages 98–99)

1. $f'(x) = \lim\limits_{\Delta x \to 0} \dfrac{f(x + \Delta x) - f(x)}{\Delta x}$

$= \lim\limits_{\Delta x \to 0} \dfrac{2(x + \Delta x)^2 - 4(x + \Delta x) + 7 - (2x^2 - 4x + 7)}{\Delta x}$

$= \lim\limits_{\Delta x \to 0} \dfrac{2x^2 + 4x\,\Delta x + 2\,\Delta x^2 - 4x - 4\,\Delta x + 7 - 2x^2 + 4x - 7}{\Delta x}$

$= \lim\limits_{\Delta x \to 0} (4x + 2\,\Delta x - 4) = 4x - 4$

2. $y' = 12x^2 + \tfrac{1}{2}x^{-1/2} - x^{-2}$, $y'' = 24x - \tfrac{1}{4}x^{-3/2} + 2x^{-3}$

3. $y' = (6x + 7)(4x^7 - 8) + (3x^2 + 7x)(28x^6)$

4. $f'(x) = \dfrac{(x^2 - x)3 - (3x + 2)(2x - 1)}{(x^2 - x)^2}$

5. (a) $x < 150$ (b) $x > 150$ (c) $x = 150$ (d) $P(150) = 4{,}490$

SAMPLE QUIZ, SECTIONS 2.5, 2.6, 2.7, 2.8 (page 138)

1. critical point $(1, 1)$ is a local maximum and an absolute maximum; concave downward everywhere; no inflection points

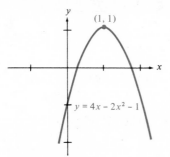

2. critical points are $x = -1$ and $x = 1$; $(-1, \tfrac{5}{3})$ is a local maximum; $(1, \tfrac{1}{3})$ is a local minimum; the only inflection point is $(0, 1)$; concave downward if $x < 0$; concave upward if $x > 0$

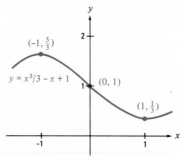

3. (a) $R = 20q - 2q^2$ (b) $R' = 20 - 4q$ (c) $E = (10 - q)/q$ (d) $q = 5$
    (e) $q < 5$

4. $q = 2$; $P(2) = 13$

5. (a) $x = 50$ (b) $C(50) = \$4{,}400$ (c) 20 times a years

SAMPLE TEST, CHAPTER 2 (page 139)

1. $y' = 2 - x^{-2} + 2x^{-1/2}$, $y'' = 2x^{-3} - x^{-3/2}$, $y''' = -6x^{-4} + \frac{3}{2}x^{-5/2}$
2. $y' = (2x)(4x^3 + 2x) + (x^2 + 3)(12x^2 + 2)$
3. (a) critical points: $x = 0$ and $x = 1$; inflection point at $(\frac{1}{2}, -\frac{1}{2})$
    (b)

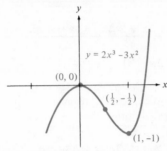

(c) $y = 12x + 7$
4. (a) $P'(x) = 6 - 0.06x$ (b) $x = 100$ (c) $P'' = -0.06 < 0$, so the second derivative test assures us that it is a maximum. (d) $P(100) = 100$
5. (a) 60   (b) $R(q) = 100q - 0.2q^2$ (c) $R'(q) = 100 - 0.4q$ (d) $q = 250$
    (e) $E = (100 - 0.2q)/0.2q$ (f) $E = \frac{3}{2} > 1$, so the demand is elastic.
    (g) $q = 250$
6. $y' = \lim\limits_{\Delta x \to 0} \dfrac{3(x + \Delta x)^2 + 5(x + \Delta x) - (3x^2 + 5x)}{\Delta x}$

    $= \lim\limits_{\Delta x \to 0} \dfrac{3x^2 + 6x\,\Delta x + 3\,\Delta x^2 + 5x + 5\,\Delta x - 3x^2 - 5x}{\Delta x}$

    $= \lim\limits_{\Delta x \to 0} \dfrac{6x\,\Delta x + 3\,\Delta x^2 + 5\,\Delta x}{\Delta x}$

    $= \lim\limits_{\Delta x \to 0} (6x + 3\,\Delta x + 5)$

    $= 6x + 5$

SAMPLE QUIZ, SECTIONS 3.1, 3.2, 3.3, 3.4, 3.5 (page 180)

1. (a) \$112.49 (b) \$112.68 (c) \$112.75
2. 7 percent
3. See page 172.
4. (a) $9e^{9x}$ (b) $(6x - 4)e^{3x^2 - 4x}$ (c) $\ln(x) + 1$ (d) $(3x^2 + 5)/(x^3 + 5x)$
5. $R' = 2x \ln x + x$
6. (a) $7(x^3 - 5x^2 + 6)^6(3x^2 - 10x)$ (b) $(4x^3 - 3)/\sqrt{2x^4 - 6x}$
7. $\dfrac{1}{7(2x^2 + 3x)^6(4x + 3)}$
8. $y = -x + 4$

SAMPLE TEST, CHAPTER 3 (page 198)

1. (a) $y' = 5(3x^6 - 9x^3)^4(18x^5 - 27x^2)$ (b) $y' = 5e^{5x}$
2. See page 172.
3. (a) \$74.73 (b) \$74.25 (c) \$74.08
4. 23 years

5. (a) $C' = (3x^2 + 6)/(x^3 + 6x)$ (b) $C' = 2x \ln (x) + x$
6. $y = 2x - 1$
7. (a) $\dfrac{dx}{dy} = \dfrac{\sqrt{2x^4 - 5x^2}}{4x^3 - 5x}$ (b) $\dfrac{dx}{dy} = \dfrac{e^{x^4 - 4x}}{4 - 4x^3}$
8. (a) $(6x + 5)/[2(3x^2 + 5x)]$ (b) $8x + 3$
9. 3 seconds
10. (a) $y' = 6(3^{6x}) \ln (3)$ (b) $y' = \dfrac{2}{x} \log_3 e$
11. $t = 8$
12. (a) $C(150) = 600$ (b) $dC = 27$ (c) $\frac{27}{600} = 0.045$, or 4.5 percent

SAMPLE QUIZ, SECTIONS 3.6, 3.7, 3.8, 4.1, 4.2 (pages 211–212)

1. $dy = (6x + 6e^{-3x})\, dx$ 3. $\frac{1}{7}e^{7x} + k$
2. $\frac{1}{8}x^8 + \frac{2}{3}x^{3/2} + \ln |x| + k$ 4. $\frac{1}{150}(5x^3 + 7)^{10} + k$
5. (a) $P(x) = 3x - 0.03x^2 - 10$ (b) $P(6) = 6.92$
6. (a) \$950 (b) $dC = 24$ (c) $\frac{24}{950} \approx 2.5$ percent
7. (a) 64 ft/sec (b) $t = 5$ sec (c) 400 ft
8. $t = 16$
9. 3.07

SAMPLE QUIZ, SECTIONS 4.4, 4.5, 4.6, 4.7 (pages 253–254)

1. (a) $-\frac{1}{3}y^{-3} = \frac{1}{4}x^4 + k;\ y = Pe^{0.05x}$ (b) $y = 200e^{0.05x};\ \$200e \approx \$544$
2. (a) 62.5 (b) $\frac{1}{2}(e^4 - e)$
3. (a) $\frac{65}{12} = 5\frac{5}{12}$ (b) $\frac{125}{6} = 20\frac{5}{6}$
4. (a) $(2, 5); 5\frac{1}{3}$ (b) $\frac{63}{32} = 1\frac{31}{32}$

SAMPLE TEST, CHAPTER 4 (page 254)

1. $R(x) = x^2 - \frac{2}{3}x + 9\frac{2}{3};\ R(4) = 23$
2. (a) $\frac{1}{5}x^5 + \frac{3}{4}x^{4/3} + 3 \ln |x| + k$ (b) $-\frac{1}{3}e^{-3x} + k$ (c) $\frac{1}{10}(x^2 + 7)^5 + k$
   (d) $-\frac{1}{6}xe^{-6x} - \frac{1}{36}e^{-6x} + k$
3. $V = 300e^{0.08t};\ \$300e \approx \$815$
4.

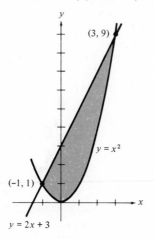

$$\int_{-1}^{3} (2x + 3 - x^2)\, dx = 10\frac{2}{3}$$

5. 18; $25\frac{1}{3}$

SAMPLE QUIZ, SECTIONS 5.1, 5.2 (page 269)

1. (a) $\frac{3}{2}$ (b) $\frac{1}{2}e^{-9}$ (c) does not converge
2. 1

3. (a)

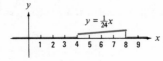

(b) $\frac{1}{4}$

4. (a)

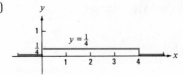

(b) $k = \frac{1}{24}$

5. (a) $e^{-0.3} \approx 0.74$

(b)

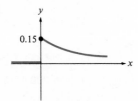

SAMPLE QUIZ, SECTIONS 5.3, 5.4 (pages 284–285)

1. $\mu = 5, \sigma = 4$
2. 0.1151
3. $m = \sqrt{40}; \mu = 6\frac{2}{9}$

4. $m = \dfrac{\ln(2)}{5} \approx 0.139$
5. $\mu = 6; \text{var}(x) = 1\frac{1}{3}$

SAMPLE TEST, CHAPTER 5 (page 285)

1. (a) $\frac{1}{5}$ (b) We let $u = -5x$, so $du = -5\,dx$ and $-du = 5\,dx$. $\displaystyle\int_0^{\infty} 5e^{-5x}\,dx =$

$$-\int_0^{\infty} e^u\,du = -e^u \Big|_0^{\infty} = -e^{-5x}\Big|_0^{\infty} = 0 - (-e^0) = 1$$

2. See pages 270 and 272.

3. (a)

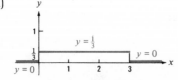

(b) $\mu = \displaystyle\int_0^3 \frac{1}{3}x\,dx = 1.5$

4. (a) $e^{-6} = \dfrac{1}{e^6}$

    (b)

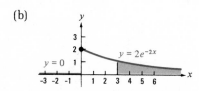

5. (a)

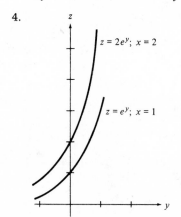

    (b) $k = \frac{1}{8}$ (c) $m = \sqrt{17}$
6. 0.6915
7. 0.1327

SAMPLE QUIZ, SECTIONS 6.1, 6.2, 6.3 (page 308)

1. 9.8
2. local minimum at $x = 6$, $y = -4$
3. $f_x = 2xe^y + \dfrac{1}{xy} + e^x$; $f_y = x^2e^y - \dfrac{\ln(x)}{y^2}$; $f_{xx} = 2e^y - \dfrac{1}{x^2y} + e^x$; $f_{xy} = 2xe^y -$

    $\dfrac{1}{xy^2} = f_{yx}$; $f_{yy} = x^2e^y + 2\,\dfrac{\ln(x)}{y^3}$

4.

$z = 2e^y;\ x = 2$

$z = e^y;\ x = 1$

SAMPLE QUIZ, SECTIONS 6.4, 6.5, 6.6 (page 326)

1. (a) $e - \frac{1}{4}$ (b) $e - \frac{1}{4}$ 3. $y = \frac{5}{2}x - \frac{8}{3}$
2. $x = 9$, $y = 1$ gives 101   4. $y = \frac{19}{14}x$

SAMPLE TEST, CHAPTER 6 (page 327)

1. (a) $f(1, 2, -1) = 11$ (b) $C(x, y) = 1.29x + 0.98y + 100$
   (c)

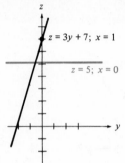

2. (a) $f_x(1, 2) = 4$; $f_y(1, 2) = 13$ (b) $f_{xx} = 2$; $\dfrac{\partial^2 f}{\partial y\,\partial x} = 1$ (c) $x = 1$ and $z - 10 = 13(y - 2)$

3. (a) There is a minimum of 30 at $(-3, -1)$ (b) $C_{xy}^2 - C_{xx}C_{yy} = 0 - 4 = -4 < 0$ and $C_{xx} = 2 > 0$ (c) The minimum is then taken at $(0, 0)$ and is 40.

4. (a) $y = \frac{3}{2}x - \frac{1}{6}$ (b) $y = \frac{51}{35}x$ (c) the first

5. $C(2, 2) = 4 + 4 = 8$ is a minimum.

6. 4

7. 4

SAMPLE QUIZ, SECTIONS 7.1, 7.2 (page 351)

1. (a) 2π (b) $\pi/4$

2. (a) $180°$ (b) $-30°$

3. $120°$

4. (a) 0.5   (b) -0.5 (c) 0.5   (d) 0.5   (e) -0.5 (f) $4/\sqrt{3}$ meters

5. $y = -x + \pi$

6.

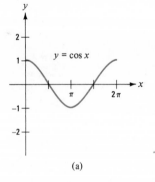

(a)

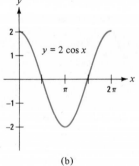

(b)

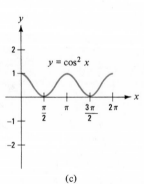

(c)

7. (a) $y' = -3\sin x + 2\cos x$ (b) $y' = -3\sin 3x$ (c) $y' = 2x\cos(x^2)$
   (d) $y' = -4\cos^3 x \sin x$

SAMPLE QUIZ, SECTIONS 7.3, 7.4 (pages 366–367)

1. (a) $y' = \sec^2 x - \cot x \csc x$ (b) $y' = 3 \sec^2 3x$
    (c) $y' = 2x \tan (x^2) \sec (x^2)$ (d) $y' = -2 \cot x \csc^2 x$

2. (a) $\frac{1}{3} \sin 3x + k$ (b) $\frac{1}{4} \cos (-4x) + k$ (c) $\tan (x + 2) + k$
    (d) $-\frac{1}{2} \ln |\cos 2x| + k$

3.

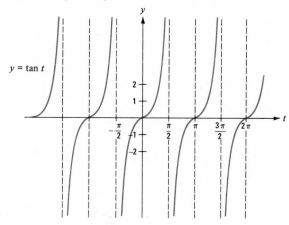

4. $\frac{2}{3}$

5. $P(t) = 5 \sin 2t + 3 \cos 2t + 97$

SAMPLE TEST, CHAPTER 7 (page 367)

1. (a) 0.7071 (b) 0.7071 (c) −0.7071 (d) 0.7071 (e) 0.7071

2.

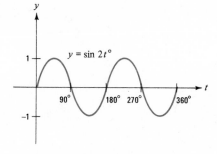

3. (a) $\pi/6$ (b) $(540/\pi)°$

4. (a) $\cos t$ (b) $-\sin t$ (c) $3 \cos t + \sin t$ (d) $\sec^2 t$ (e) $6 \sin 3t \cos 3t$

5. (a) $\tan t + k$ (b) $\frac{1}{3} \sin 3t + k$ (c) $-\frac{1}{2} \cos (t^2) + k$ (d) $(\cos t)^{-1} + k$

6. (a) 2π (b) π (c) 2π (d) $\pi/2$ (e) 2π

7. 3

8. $(2\pi/3) + (2/3)$

9. $-\cos y = \sin x + k$

10.

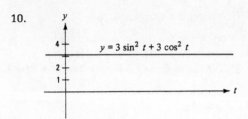

GLOSSARY

Antiderivative of $f(x)$: function $F(x)$ such that $F'(x) = f(x)$.

Average rate of change [of $f(x)$]: number given by $\dfrac{f(x + \Delta x) - f(x)}{\Delta x}$, where Δx denotes an increment in x.

Boundary condition (of a differential equation): equation of the form $f(x_1) = k_1$ (that is, one which says that the solution function at the value x_1 equals k_1).

Common logarithm: logarithm of a number to base 10.

Concave upward: graph shaped like a bowl holding water.

Constant function: function of the form $f(x) = c$, a constant (that is, a function yielding the same number for each x).

Constant of integration: indefinite constant which, when added to a particular antiderivative of a function, enables one to express *all* antiderivatives of the function.

Cosine function: trigonometric function that can be defined as the x-coordinate of an angle's intercept on the unit circle when one side of the angle is on the positive x-axis and its vertex is at the origin.

Definite integral [of $f(x)$ from a to b]: total change $F(b) - F(a)$ of any antiderivative $F(x)$ of $f(x)$ $\left[\text{denoted } \displaystyle\int_a^b f(x)\, dx\right]$.

Derivative [of $f(x)$]: instantaneous rate of change of $f(x)$ at a point x, or that number as given by a function $f'(x)$.

Exponential function: function of the form a^u, where a is a positive constant and u is a function [especially $f(x) = e^x$].

Function: correspondence between two sets of elements A and B such that for each element in A there corresponds exactly one element in B.

Growth rate [of $f(x)$]: ratio between $f'(x)$ and $f(x)$ at x.

Increasing function: function f such that $x_1 < x_2$ implies that $f(x_1) < f(x_2)$ over some interval (that is, a function that gets bigger as x gets bigger).

Increment: change in a variable from one number to another.

Indefinite integral [of $f(x)$]: symbol $\displaystyle\int f(x)\, dx$, which represents all antiderivatives of $f(x)$.

Inflection point (of a graph): point at which the concavity of the graph is changing.

Inverse [of a function $y = f(x)$]: function $g(y)$ such that $y = f(g(y))$.

Linear function: function of the form $y = mx + b$, where m and b are constants (that is, the graph is a straight line).

Local extreme point: point $(x, f(x))$ on a curve which is either a peak (local maximum) or a valley (local minimum).

Logarithm function: inverse of an exponential function.

Marginal cost (revenue, profit): sometimes the derivative of the total cost (revenue, profit) function; sometimes the average rate of change of these over a unit interval or when the interval is "small."

Natural logarithm: logarithm to the base $e \approx 2.71828$.

Periodic function: function for which there is a number p such that $f(t + p) = f(t)$ for all t.

Quadratic function: function of the form $y = ax^2 + bx + c$, where a, b, and c are constants, $a \neq 0$ (the graph is a parabola).

Sine function: trigonometric function that can be defined as the y-coordinate of an angle's intercept on the unit circle when one side of the angle is on the positive x-axis and its vertex is at the origin.

Slope (of a linear function): average rate of change for any two points on the line (the m in $y = mx + b$).

Tangent function: trigonometric function that can be defined by $\tan t = \sin t/\cos t$ for $\cos t \neq 0$.

Tangent line [to the graph of $y = f(x)$]: line with the equation $y - y_1 = f'(x_1)(x - x_1)$; that is, a straight line through a point (x_1, y_1) on the graph whose slope is $f'(x_1)$, the derivative at x_1.

Torr: unit of pressure equal to $\frac{1}{760}$ of an atmosphere.

Velocity (at a point): instantaneous rate of change in distance with respect to time of a "moving object."

INDEX